Advances in Experimental Medicine and Biology

Volume 1044

More information about this series at http://www.springer.com/series/5584

Yu Zhang
Editor

Chromosome Translocation

Springer

Editor
Yu Zhang
National Institute of Biological Sciences
Beijing, China

ISSN 0065-2598 ISSN 2214-8019 (electronic)
Advances in Experimental Medicine and Biology
ISBN 978-981-13-0592-4 ISBN 978-981-13-0593-1 (eBook)
https://doi.org/10.1007/978-981-13-0593-1

Library of Congress Control Number: 2018946723

Printed on acid-free paper

This Springer imprint is published by the registered company Springer Nature Singapore Pte Ltd.
The registered company address is: 152 Beach Road, #21-01/04 Gateway East, Singapore 189721, Singapore

Contents

Historical and Clinical Perspectives on Chromosomal Translocations

1

Ellen S. Wilch and Cynthia C. Morton

Abstract

Chromosomal translocations, rearrangements involving the exchange of segments between chromosomes, were documented in humans in 1959. The first accurately reported clinical phenotype resulting from a translocation was that of Down syndrome. In a small percentage of Down syndrome cases, an extra 21q is provided by a Robertsonian translocation chromosome, either occurring *de novo* or inherited from a phenotypically normal parent with the translocation chromosome and a balanced genome of 45 chromosomes. Balanced translocations, including both Robertsonian and reciprocal translocations, are typically benign, but meiosis in germ cells with balanced translocations may result in meiotic arrest and subsequent infertility, or in unbalanced gametes, with attendant risks of miscarriage and unbalanced progeny. Most reciprocal translocations are unique. A few to several percent of translocations disrupt haploinsufficient genes or their regulatory regions and result in clinical phenotypes. Balanced translocations from patients with clinical phenotypes have been valuable in mapping disease genes and in illuminating *cis*-regulatory regions. Mapping of discordant mate pairs from long-insert, low-pass genome sequencing now permits efficient and cost-effective discovery and nucleotide-level resolution of rearrangement breakpoints, information that is absolutely necessary for interpreting the etiology of clinical phenotypes in patients with rearrangements. Pathogenic translocations and other balanced chromosomal rearrangements constitute a

E. S. Wilch (✉)
The Developmental Genome Anatomy Project, Brigham and Women's Hospital, Boston, MA, USA

Department of Obstetrics, Gynecology and Reproductive Biology, Brigham and Women's Hospital, Boston, MA, USA
e-mail: cmorton@bwh.harvard.edu

C. C. Morton
The Developmental Genome Anatomy Project, Brigham and Women's Hospital, Boston, MA, USA

Department of Obstetrics, Gynecology and Reproductive Biology, Brigham and Women's Hospital, Boston, MA, USA

Harvard Medical School, Boston, MA, USA

Department of Pathology, Brigham and Women's Hospital, Boston, MA, USA

Program in Medical and Population Genetics, Broad Institute of Harvard and MIT, Cambridge, MA, USA

Manchester Academic Health Science Centre, University of Manchester, Manchester, UK

Y. Zhang (ed.), *Chromosome Translocation*, Advances in Experimental Medicine and Biology 1044,
https://doi.org/10.1007/978-981-13-0593-1_1

class of typically highly penetrant mutation that is cryptic to both clinical microarray and exome sequencing. A significant proportion of rearrangements include additional complexity that is not visible by conventional karyotype analysis. Some proportion of patients with negative findings on exome/genome sequencing and clinical microarray will be found to have etiologic balanced rearrangements only discoverable by genome sequencing with analysis pipelines optimized to recover rearrangement breakpoints.

Keywords
Cytogenetics · Karyotype · Balanced translocation · Breakpoint · Congenital anomaly · Gene mapping · Mate-pair sequencing

1.1 Introduction and Background

Chromosomal translocations encompass a diverse set of rearrangements involving the exchange of segments between chromosomes, and are common in humans. Balanced translocations, those without accompanying copy number variation, usually have no phenotypic consequence. Estimates vary, but about one in every 300–500 individuals has a balanced reciprocal translocation, and about one per 1000 has a balanced Robertsonian translocation (the joining of complete long arms of two acrocentric chromosomes in β-satellite sequences in the short arms) ([16, 24, 27, 28, 36, 51, 57, 60, 89]). From a multicenter study of 377,357 amniocenteses, the incidence of a *de novo* reciprocal translocation has been estimated at about 1 per 2000, and 1 per 9000 for a Robertsonian translocation [92]. Warburton also estimated the risk of congenital abnormality associated with a balanced reciprocal translocation to be about 6% (an approximate 2- to 3-fold increase over the general population risk), and that from balanced Robertsonian translocation to be negligible. Congenital anomaly or other clinical phenotype in an individual with a balanced translocation may result from any of a number of possible effects of a translocation, including direct gene disruption, creation of a fusion gene, dysregulation of a gene separated from its normal extragenic *cis*-regulatory elements, or dysregulation of a gene placed in an environnment of altered chromatin modification. Balanced translocations constitute an important class of mutation that are etiologic in hundreds of Mendelian diseases, and many cancers. When ascertained from individuals with clinical phenotypes, translocations have been invaluable biological tools in mapping disease loci and *cis*-regulatory regions of disease genes. Because missegregation of balanced translocations in meiosis may result in meiotic arrest or unbalanced gametes, individuals with balanced translocations have higher risks of subfertility and infertility, miscarriage, and genomic imbalance in their offspring.

1.2 Early Observations of Translocations

The field of human cytogenetics was very young in 1959 and 1960, when the first human translocations were reported. The correct diploid human chromosome number had only been established by Tjio and Levan in 1956 [86], shortly afterwards confirmed by Ford and Hamerton [17]. This foundational achievement depended upon a number of technical advances, including improved methods of tissue culture, use of colchicine for inducing mitotic arrest [18], and most critically, incubation of cells in hypotonic solution for better chromosome spreading [29, 30]. Still, only slow progress was made in the late 1950s in identifying and characterizing human chromosomal abnormalities. Until the advent of quinacrine mustard banding [7] and Giemsa banding [76, 81] in the 1970s, chromosomes could be reliably ordered by size and centromere position only into seven groups (when this system was used, the groups were denoted A-G). Distinguishing chromosomes within groups was difficult; many specific chromosome assignments published in those years were speculative, and some were almost certainly wrong. Before the development and widespread adoption of techniques for generating sufficient numbers of mitotic cells in cultures of peripheral leukocytes [32, 53, 59], bone marrow or testis, neither trivial to ask of patients or their family members, were the tissues generally sought for karyotyping. Thus, early karyotyping was usually limited to patients, so it was not possible to assess the segregation of a chromosomal rearrangement with a phenotype in a pedigree. In spite of these difficulties, trisomy 21 was convincingly established as the chromosomal basis of Down syndrome [OMIM #190685] in 1959, on the collective evidence of 30 cases (in order of publication: [43, 35, 19, 4]), and identification of other aneuploidy syndromes quickly followed.

Lejeune's group was also the first to report a human translocation. They found, in a patient with intellectual and speech disabilities and "polydysspondylie" (spondylocostal dysostosis, a skeletal dysplasia with six known loci, most demonstrating recessive inheritance [e.g. OMIM #277300]), a karyotype of 45 chromosomes with what is now known to be a Robertsonian translocation chromosome, which they interpreted as composed of 22q and either 14q or 15q ([88], described in [87]). Common and easily identifiable, particularly in a balanced karyotype of 45 chromosomes, it is not surprising that the first reported human translocation was a Robertsonian. However, the phenotype in this case was almost certainly not related to the translocation; we now know that most individuals with a Robertsonian translocation and a balanced karyotype are phenotypically normal, and no clinical phenotypes have yet to be associated convincingly with any balanced Robertsonian translocation, with the important exceptions of increased risk of miscarriage and reduced fertility.

The second reported human translocation was also a Robertsonian, interpreted as comprising chromosomes 21 and 14, identified in an unbalanced karyotype of a girl with Down syndrome and 46 chromosomes, with the translocation chromosome providing the extra copy of 21q etiologic for Down syndrome [67]. Because the maternal age effect in Down syndrome births [64] was by then well known, Polani et al. chose the children of young mothers to investigate, hoping to increase the chances of uncovering additional karyotypically-visible etiologies for Down syndrome. From the two cases they successfully karyotyped, they were lucky to have ascertained one case of translocation Down syndrome; a translocation chromosome is present in only about 3–4% of Down syndrome karyotypes, even among young mothers. Because the phenotypes of trisomy 21 Down syndrome and translocation Down syndrome are indistinguishable, Polani et al. were able to make a strong and correct case for the participation of chromosome 21 in this rearrangement.

Penrose et al. [65] were the first to demonstrate segregation of a balanced translocation in a family. A grandmother, mother, and daughter, all phenotypically normal, each had 45 chromosomes with a translocation chromosome interpreted as rob(15;21) [although it may have been the much more common rob(14;21)] (Fig. 1.1). The mother also had two children with translocation Down syndrome, and had reported two miscarriages. Carter et al. [6] also reported transmission of rob(15;21), in a three-generation pedigree where the mothers of two first cousins with Down syndrome had each inherited the translocation chromosome from their mother. Other Robertsonian translocations besides those associated with chromosome 21 were also identified around the same time. For example, Lejeune et al. [44] found, in a man with a 46, XXY karyotype and Klinefelter syndrome, a translocation between a D group chromosome (chr13–15) and chr22. Transmission of a rob(13;15) was described in a pedigree including 10 individuals with balanced karyotypes and the translocated chromosome; among those with the translocation were eight phenotypically normal individuals. Although the primary amenorrhea in the proband was unlikely to have been related to the translocation, azoospermia in Robertsonian translocation carriers is very common and likely accounted

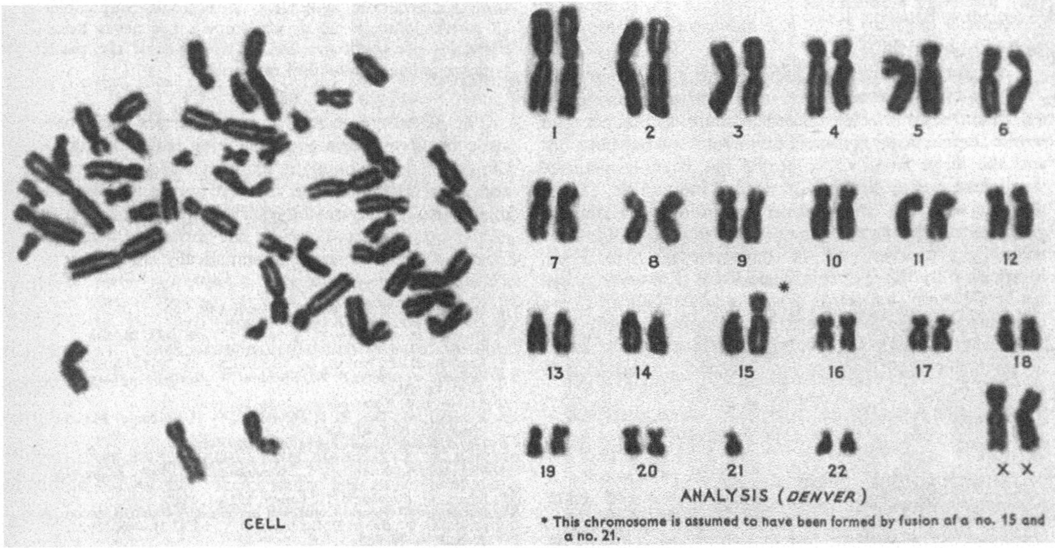

Fig. 1.1 The first chromosomal rearrangements to be identified in the early days of cytogenetics were Robertsonian translocations. These photomicrographs published in 1960 show a balanced karyotype of 45 chromosomes with a Robertsonian translocation involving chromosome 21, derived from skin cells of a phenotypically normal woman who had two children with translocation Down syndrome. This was the first publication to document inheritance of a translocation, in a three-generation family. (Reproduced from Ref. [65])

for that reported phenotype in one of the males in the pedigree [91].

It was recognized early on that translocations between chromosomes would only be detectable if the lengths of the exchanged chromosomal segments were sufficiently different, and large enough, to change the length of the translocated chromosome arms substantially, or to result in an appreciable change in the location of the centromere. In 1962, Edwards et al. [14] reported the first two such cases; the first, a translocation between chromosomes 4 and 9, and the second, between chromosomes 1 and 6. In each case, the balanced reciprocal translocation in a phenotypically normal parent was ascertained through investigation of genomically unbalanced offspring with mental retardation and congenital anomalies. Schmid [74] ascertained a balanced translocation carrier solely on the basis of a history of miscarriage. Among 10 couples who had experienced miscarriage and had one or more phenotypically normal children, one translocation carrier (a male) with an apparently balanced karyotype of 46 chromosomes was identified. The translocation involved one chromosome 21 or 22, and was non-Robertsonian, but the sizes of the exchanged segments were too small to reveal the identity of the larger translocation partner.

1.3 Clinical Relevance of Translocations

1.3.1 Balanced Translocation Carriers Have Higher Risks of Infertility, Miscarriage, and Unbalanced Progeny

Most reciprocal translocations arise on paternal chromosomes in spermatogenesis, and most Robertsonian translocations arise on maternal chromosomes in oogenesis [1, 63, 84]. Thomas et al. [84] detected a significant paternal age effect in the *de novo* occurrence of reciprocal translocations. Most individuals with balanced translocations are phenotypically normal, and balanced translocations may be transmitted in families through many generations. This was observed in the earliest pedigrees of translocation Down syndrome, and it is still the case that many balanced chromosomal rearrangements are ascertained in couples with an unbalanced fetus or child. In meiosis when a balanced translocation is present, the translocated chromosomes associate with their homologs as a trivalent (for Robertsonians) or tetravalent (for reciprocal translocations), and segregation results in normal and balanced, or unbalanced, gametes. Men with azoospermia or severe oligospermia (no viable sperm or low sperm count) have an incidence of balanced translocations that is many times higher than that in the general male population [8, 90, 96], evidence that checkpoints in male gametogenesis lead to apoptosis of unbalanced gametes. The proportion of normal, balanced, and unbalanced gametes produced by men with balanced translocations varies from practically none to nearly all unbalanced gametes [54, 66], with Robertsonian translocations typically yielding the lowest proportion of unbalanced gametes, and reciprocal translocations the highest. As most reciprocal translocations are unique, it is difficult to assess the reproductive risk of unbalanced progeny for couples in whom one partner is known to have a translocation [96], though Boué and Gallano [5], from assessing pregnancy outcomes in more than 1200 couples where one partner had a translocation, estimated the overall risk of unbalanced progeny to be 10% for a term pregnancy, and 11.5% for a fetus. A balanced translocation in one parent is significantly associated with miscarriage; parental balanced translocation was identified in ~3–4% of couples with recurrent miscarriage [62, 80]. In translocation Down syndrome, about two thirds of cases arise *de novo*, with about half due to rob(14;21) and half due to rea(21;21) [55], although most or all rea(21;21) are isochromosomes and not true Robertsonian translocations [77]. In inherited translocation Down syndrome, the translocation chromosome is more frequently inherited from the mother, indicating that checkpoints in oogenesis are not as stringent as those in spermatogenesis. Recurrence risk for Down syndrome from inherited Robertsonian translocations is believed to be 10–15% if the mother has the translocation, and less than 1% if the father has the translocation.

Only a few reciprocal translocations have been found to be recurrent. Emanuel syndrome [OMIM #609029] is characterized by multiple congenital anomalies and an unbalanced karyotype with a supernumerary der(22) inherited from an unaffected parent with a t(11;22)(q23;q11.2) [20, 94]. Breakpoints in palindromic AT-rich repeat (PATRR) sequences on chromosomes 11 and 22 suggest that palindrome-mediated formation of secondary structure promotes double-strand breakage and resulting translocation [37]. PATRRs have been recognized on chromosomes 1, 4, 8, 11, 17 and 22, and rare recurrent translocations between PATRRs on chromosomes 8 and 22 [that may result in progeny with supernumerary der(22)t(8;22) syndrome, OMIM #613700] have been reported [78], as well as rare translocations between PATRRs on chromosomes 17 and 22, in individuals with neurofibromatosis type I [NF1, OMIM #162200] [37].

1.3.2 Some Somatic Translocations Initiate Transformation of Cancer Cells

Although interest in cancer had driven much early progress in cytogenetic techniques, the complex, abundant, and variable chromosomal aberrations observed in many cancers were difficult to interpret, and also led investigators to believe that chromosomal rearrangements might all be secondary to the events that initiate tumorigenesis. Two early successes in deciphering the molecular biology of oncogenic translocations were groundbreaking in understanding the importance of chromosomal rearrangements in driving tumorigenesis.

Although not initially recognized as such, the first translocation associated to a specific cancer was the Philadelphia chromosome (Ph), uniquely associated with chronic myelogenous leukemia [CML, OMIM #608232] [59]. Its identity as a der(22) involved in a translocation with chr9 was only uncovered 13 years later, once banding techniques had been developed [73]. In 1982, the *ABL* oncogene [OMIM *189980], known to map to chr9, was shown to be translocated to the Ph chromosome 22 [10]. Fine-mapping of *ABL* [25] and the chr22 breakpoint cluster region [23] led to identification of an *ABL-BCR* fusion gene that results from translocation [79], and the fusion gene's oncogenic activity in CML was demonstrated [40]. More than 90% of CML cases have a t(9;22), and presence of the Ph chromosome aids in diagnosis; cryptic rearrangements are likely responsible for the remaining 10%.

Three translocations associated with Burkitt's lymphoma [OMIM #113970] and involving 8q24 were characterized in the late 1970s, with the causative gene rearrangement [83] identified prior to that of the Ph chromosome in CML. The common translocation, t(8;14)(q24;q32), places the intact coding exons of *MYC* [OMIM *190080], a gene encoding a cell growth and cell cycle transcription factor, close to the enhancer of the immunoglobulin heavy chain locus [*IgH* locus, see OMIM *147100], driving constitutive expression in B-lymphocytes and conferring oncogenicity. The t(2;8) and t(8;22) translocations place *MYC* in proximity to enhancers at the *IgK* or *IgL* loci, respectively, with similar effect. The pathogenicity of the CML and Burkitt's translocations is thus explained by two distinct, important, and generalizable models: in CML, the formation of a fusion gene results in pathogenic gain of function of a novel chimeric protein; in Burkitt's lymphoma, alteration of the regulatory environment of a gene drives its misexpression, resulting in pathogenic gain of function of the normal protein.

1.3.3 Balanced Reciprocal Translocations Are Etiologic for Many Clinical Phenotypes

The earliest pedigrees with translocations were those of translocation Down syndrome, where trisomy for 21q was clearly implicated in the clinical phenotype, and parents and other family members with the translocated chromosome and balanced karyotypes were phenotypically normal. Early pedigrees of reciprocal translocations likewise showed phenotypically normal parents with apparently balanced translocations whose children with clinical phenotypes had unbalanced karyotypes. These pedigrees, though perhaps surprising at first, established the paradigm that rearrangement without apparent loss or gain of chromosomal material had no consequence other than that of the contribution of unbalanced gametes to risks of infertility, miscarriage, and unbalanced offspring. Challenging that paradigm, Jacobs [34] presented epidemiological evidence suggesting that *de novo* balanced rearrangements (translocations and inversions) were overrepresented among mentally retarded individuals compared to consecutive or random newborns and individuals ascertained for unspecified reasons. Funderburk et al. [21] found balanced rearrangements to be significantly overrepresented among mentally retarded individuals when compared to children of normal intelligence with psychiatric indications. Other studies of outcomes where balanced rearrangements were ascertained from surveys of consecutive newborns [50, 56, 85] were underpowered, failing to ascertain large enough numbers of subjects with balanced rearrangements to interpret correctly an effect that we now know to apply to only a few to several percent of individuals with balanced, non-Robertsonian, rearrangements [92]; Warburton's robust estimate of the risk of congenital anomaly associated with such rearrangement required the ascertainment of outcomes of 377,357 amniocenteses from multiple clinical centers.

A high burden of proof is required for assigning etiology of phenotype to a balanced reciprocal translocation, given that most are both unique and without phenotypic consequence. Nonetheless, a diagnosis can be made when direct disruption of a known disease gene can be documented and correlated to a specific phenotype, as abundant case reports attest. Individuals with balanced translocations and clinical phenotypes have been useful in mapping a number of Mendelian disease loci, particularly those with severe phenotypes that usually occur *de novo*, precluding linkage analysis. For instance, the locus for neurofibromatosis type 1 (NF1, OMIM #162200) was identified on the basis of two balanced reciprocal translocations, both with breakpoints in

17q11.2 [42, 75]. Similarly, various rearrangements were reported in patients with Sotos syndrome [OMIM #117550], but two translocations with breakpoints in 5q35 [33, 52] directed investigation to that region; *NSD1* [OMIM *606681] was cloned, shown to be disrupted in the translocation of the tested patient, and point mutations and genomic deletions of *NSD1*, now known to explain about 90% of Sotos syndrome cases, were found in the majority of a cohort of patients [41].

Translocations have also been productive in identifying new candidate genes underlying common clinical phenotypes that may arise from dysfunction of any number of genes, as in autism spectrum disorder, cardiac defects, and orofacial clefting [26, 71]. Two cases described by Kim et al. [38] contributed to identifying a role for neurexin 1 [*NRXN1*, OMIM *600565] in autism spectrum disorder; *NRXN1* has since also been strongly implicated in other neurodevelopmental disorders, including schizophrenia and intellectual disability. Quintero-Rivera et al. [69] marshalled substantial evidence that matrin 3 [*MATR3*, OMIM *164015], previously associated with amyotrophic lateral sclerosis [ALS21; OMIM #606070], is etiologic for cardiac left ventricle outflow tract (LVOT) defects in a child with a t(1;5) disrupting *MATR3* on 5q. Interestingly, *AHDC1* [AT-hook DNA-binding motif-containing protein 1, OMIM *615790] on 1q was also disrupted by this translocation, and several aspects of the child's phenotype, including intellectual disability, facial dysmorphisms, and respiratory and sleep disturbances, were concordant with those reported for a newly-described syndrome [Xia-Gibbs syndrome, OMIM #615829] attributed to heterozygous mutation in *AHDC1* in only four cases [93].

1.3.4 Resolution of Translocation Breakpoints by Sequencing Provides New Information

Whole-genome sequencing now permits discovery and precise localization of rearrangement breakpoints [58, 71, 82] (Table 1.1). Particularly, analysis of discordant mate-pair mappings from low-coverage, long-insert whole-genome sequencing is a cost-effective means of doing so [82], and is likely to one day displace standard karyotyping. Sequence-level breakpoint mapping provides identities of genes directly disrupted in a rearrangement, and of nearby genes that may be dysregulated by altered positioning of *cis*-regulatory enhancers, other regulatory elements, or regions of chromatin modification. Sequencing discovers complexity that is cryptic to karyotyping and imbalances that are below the resolution of clinical microarrays. *De novo* balanced translocations detected on prenatal karyotype can be assessed in a timely manner by this approach; for example, in prenatal cases reported by Ordulu et al. [61], sequencing supported or confirmed a suspected genetic diagnosis in most of the cases referred for abnormal prenatal findings. Redin et al. [71] reported sequenced breakpoints in 248 of 273 subjects, the majority ascertained via the Developmental Genome Anatomy Project (DGAP), a long-running effort to identify genes important in development by investigating apparently balanced rearrangements in patients with a wide variety of phenotypes, including neurodevelopmental disorders and structural congenital anomalies [26]. Redin et al. were able to make high-confidence correlations of genes to phenotypes in about a quarter of the subjects, and identify, in another 20%, likely candidate genes based on gene disruption or predicted position effects. They also documented additional complexity, cryptic to karyotype, in more than 20% of the rearrangements that they analyzed. This complexity included genomic gains or losses, some proportion of which would be invisible to clinical microarray. Among 65 subjects (26% of the total) with complex rearrangements (three or more breakpoints), 13 had multiple breakpoints characteristic of the "shattering" phenomena of chromothripsis or chromoplexy; in one case, 57 breakpoints were mapped. Overall, about 80% of the 248 analyzed rearrangements were balanced or nearly so, with less than 10 kb of genomic imbalance, indicating that the majority of these pathogenic mutations fail to leave even a footprint on clinical microarray.

Table 1.1 The resolution of techniques to detect chromosomal structural variation has improved by several orders of magnitude since the 1950s

Technique	Introduced	Capable of detecting translocations?	Resolution
Karyotype - unbanded	1950s–1960s	Yes, but only Robertsonian translocations and unbalanced or balanced translocations resulting in significant change in length of the derivative chromosomes	~10–20 Mb, depending on chromosome length (although misidentification of chromosomes was common)
Karyotype - banded	1970s	Yes	~5–10 Mb
FISH	1980s–1990s	Yes, targeted to one or a few loci	<1 Mb to gene-level
Chromosomal microarray analysis (CMA)	2000s	No, except if unbalanced and suspected by a duplication-deficiency	~100 kb (depending on platform and laboratory cut-offs)
Next-generation sequencing (NGS)	Late 2000s	Yes, by discordant mate-pair mapping of whole-genome sequence	Base pair-level resolution if followed up with Sanger or amplicon sequencing; otherwise, resolution to level of insert (fragment) size

Some balanced translocations are also cryptic to karyotyping. In an unusual case [68], a child with multiple congenital anomalies and some features overlapping those of cri-du-chat syndrome [OMIM #123450] was found by FISH to have a 4.6 Mb deletion at the terminus of 5p, much smaller than the canonical cri-du-chat region. Two previous pregnancies of the mother had been terminated, one for complete lissencephaly [OMIM #607432], and one for intrauterine growth restriction. The karyotypes of both of the parents and the affected child were unremarkable. However, FISH of the mother's chromosomes, including a probe to 17p (a known lissencephaly locus), indicated a balanced translocation between 5p and 17p. The affected daughter had inherited the der(5) chromosome, resulting in gain of 17p as well as loss of 5p, with phenotypic features attributable to each. The authors hypothesized that the fetus with lissencephaly had inherited the der(17) chromosome, with presumed heterozygous loss of *PAFAH1B1* [OMIM *601545] responsible for the lissencephaly phenotype. A cautionary tale, this cryptic translocation was hypothesized and uncovered only through thoughtful assessment of a "peculiar" pedigree and distinctive phenotypes already strongly linked to known loci. Without exceptional circumstances such as these, cryptic balanced translocations will remain undetected until whole-genome sequencing designed to ascertain rearrangements becomes routine.

1.3.5 Mapping the Regulatory Genome with Translocations

Whole exome sequencing of patients with conditions of suspected single-gene etiology currently yields known or likely candidate pathogenic mutations in about half. Extragenic regulatory mutation is likely to constitute a significant proportion of this "missing" mutation. Mammalian genomes are now well-known to be looped, folded, and scaffolded into three-dimensional architectures that influence gene expression through the spatial control of interactions of gene promoters with extragenic enhancers and other regulatory elements [12, 47, 70]. Chromosomes are physically organized into topologically-associated domains (TADs), "neighborhoods" of typically a megabase or smaller, that demonstrate higher frequencies of chromatin-chromatin contacts within domains than across domains [48, 49]. Enhancer-promoter contact frequencies define TADs, delimited by boundary elements that discourage promiscuous interactions of enhancers with non-target neighboring promoters; critical gene-regulatory enhancers may be arrayed along a chromosome at distances of several 100 kilobases to a megabase or more from the genes they regulate, particularly in the case of genes important in developmental processes [11].

Translocations can displace important enhancers or other regulatory elements from the genes they regulate. This has been demonstrated for several genes where translocations were central in flagging locations of distant *cis*-regulatory elements. An enhancer controlling limb expression of *sonic hedgehog* (*Shh/SHH* [OMIM *600725]) was localized to an intron of a distant gene and implicated in *Shh/SHH* expression in both mouse and human by a combination of evidence, including a translocation in a patient with preaxial polydactyly type II [PPD2, OMIM #174500], where one breakpoint mapped very close to the enhancer, nearly 1 Mb away from *SHH* itself [45, 46]. A downstream regulatory region was identified by mapping translocation and inversion breakpoints in patients with aniridia [OMIM #106210] where the causative gene, *PAX6* [OMIM *607108], was found to be left intact by the rearrangements [15, 39]. Structural variants, including translocations, in patients with campomelic dysplasia or acampomelic campomelic dysplasia [CD and ACD, OMIM #114290] indicate that a complex *SOX9* [OMIM *608160] regulatory landscape exists as far as 2 Mb upstream and 500 kb downstream of the gene itself; severity of the phenotype in translocation patients is broadly correlated with distance of the translocation breakpoint from *SOX9* itself [22]. Other phenotypes are associated with copy number variation or rearrangements around *SOX9*, including Pierre-Robin sequence [PRS, OMIM %261800] and 46,XX and 46,XY disorders of sex development (DSDs) [OMIM #278850 and #613080], all of which can occur with or without accompanying CD or ACD. Translocations of patients with isolated PRS have been localized to two separate regions, about 1 Mb and 400 kb upstream of *SOX9* [2, 61]; likewise, deletions and duplications in DSD patients define critical regulatory regions for sex development between 500 and 640 kb upstream of *SOX9*.

Brief descriptions belie the complexity of *cis*-regulatory regions, the variations that may occur among phenotypes of patients with rearrangements variously disrupting a regulatory locus, and the many mechanisms proposed to explain the pathogenicity of those rearrangements [3]. Beyond simply removing an enhancer from its cognate promoter, translocations and other rearrangements change the physical conformation of the regulatory locus, altering or disrupting chromatin loops that may affect the function of enhancers and other elements that remain between a cognate promoter and a rearrangement breakpoint. Enhancers brought into a locus by a rearrangement may make spurious contacts with an existing promoter, altering its expression. New chromatin conformations may alter the maintenance of the epigenetic landscape, resulting in gene expression changes via classical position effect mechanisms. Thus, predicting the effects of translocations that disrupt extragenic sequences is not trivial. Zepeda-Mendoza et al. [95] collated genome-wide datasets of enhancer marks, DNAse-hypersensitivity sites, and TAD boundaries predicted by chromatin contacts [13], along with haploinsufficiency and triplosensitivity scores [31] and phenotype information from DECIPHER [9] and ClinGen [72] to identify and prioritize candidate genes that may be etiologic in patients harboring balanced rearrangements with intergenic breakpoints. The success of this approach relies on the existence of multiple, diverse datasets that inform our understanding of the complexities of gene regulation and function, recognizes that rearrangements may result in dysfunction of genes located at distances up to several Mb, and depends most critically on the availability of large collections of patient phenotype-genotype information.

1.4 Summary

Much Mendelian disease is rare. Rarer still are patients whose congenital anomalies or neurodevelopmental disorders are caused by balanced translocations that disrupt genes or their *cis*-regulatory regions. Approaches to diagnosing genetic etiologies that discount the contribution of balanced chromosomal rearrangements disenfranchise those patients, lengthening diagnostic odysseys and adding medical costs, and sometimes compromising patient care. Reciprocal translocations and other balanced rearrangements represent an important class of pathogenic variation that is cryptic to chromosomal microarray, cannot be ascertained from exome sequencing data, and may be incompletely described or even undetected by karyotyping. Nucleotide-level resolution of rearrangement breakpoints is essential for interpreting the etiology of phenotypes in patients with balanced rearrangements.

References

1. Bandyopadhyay R, Heller A, Knox-DuBois C, McCaskill C, Berend SA, Page SL, Shaffer LG (2002) Parental origin and timing of de novo Robertsonian translocation formation. Am J Hum Genet 71(6):1456–1462. https://doi.org/10.1086/344662
2. Benko S, Fantes JA, Amiel J, Kleinjan DJ, Thomas S, Ramsay J, Jamshidi N, Essafi A, Heaney S, Gordon CT, McBride D, Golzio C, Fisher M, Perry P, Abadie V, Ayuso C, Holder-Espinasse M, Kilpatrick N, Lees MM, Picard A, Temple IK, Thomas P, Vazquez MP, Vekemans M, Roest Crollius H, Hastie ND, Munnich A, Etchevers HC, Pelet A, Farlie PG, Fitzpatrick DR, Lyonnet S (2009) Highly conserved non-coding elements on either side of SOX9 associated with Pierre Robin sequence. Nat Genet 41(3):359–364. https://doi.org/10.1038/ng.329
3. Bhatia S, Kleinjan DA (2014) Disruption of long-range gene regulation in human genetic disease: a kaleidoscope of general principles, diverse mechanisms and unique phenotypic consequences. Hum Genet 133(7):815–845. https://doi.org/10.1007/s00439-014-1424-6
4. Book JA, Fraccaro M, Lindsten J (1959) Cytogenetical observations in mongolism. Acta Paediatr 48:453–468
5. Boué A, Gallano P (1984) A collaborative study of the segregation of inherited chromosome structural rearrangements in 1356 prenatal diagnoses. Prenat Diagn 4 Spec No:45–67
6. Carter CO, Hamerton JL, Polani PE, Gunalp A, Weller SD (1960) Chromosome translocation as a cause of familial mongolism. Lancet 2(7152):678–680
7. Caspersson T, Zech L, Johansson C, Modest EJ (1970) Identification of human chromosomes by DNA-binding fluorescent agents. Chromosoma 30(2):215–227
8. De Braekeleer M, Dao T-N (1990) Cytogenetic studies in couples experiencing repeated pregnancy losses. Hum Reprod 5(5):519–528
9. DECIPHER: Database of Chromosomal Imbalance and Phenotype in Humans using Ensembl Resources. Firth HV et al (2009) Am J Hum Genet 84 524–533. https://doi.org/10.1016/j.ajhg.2009.03.010
10. de Klein A, van Kessel AG, Grosveld G, Bartram CR, Hagemeijer A, Bootsma D, Spurr NK, Heisterkamp N, Groffen J, Stephenson JR (1982) A cellular oncogene is translocated to the Philadelphia chromosome in chronic myelocytic leukaemia. Nature 300(5894):765–767
11. de Laat W, Duboule D (2013) Topology of mammalian developmental enhancers and their regulatory landscapes. Nature 502(7472):499–506. https://doi.org/10.1038/nature12753
12. Dekker J, Rippe K, Dekker M, Kleckner N (2002) Capturing chromosome conformation. Science 295(5558):1306–1311. https://doi.org/10.1126/science.1067799
13. Dixon JR, Selvaraj S, Yue F, Kim A, Li Y, Shen Y, Hu M, Liu JS, Ren B (2012) Topological domains in mammalian genomes identified by analysis of chromatin interactions. Nature 485(7398):376–380. https://doi.org/10.1038/nature11082
14. Edwards J, Fraccaro M, Davies P, Young R (1962) Structural heterozygosis in man: analysis of two families. Annals Hum Genet Lond 26:163–178
15. Fantes J, Redeker B, Breen M, Boyle S, Brown J, Fletcher J, Jones S, Bickmore W, Fukushima Y, Mannens M (1995) Aniridia-associated cytogenetic rearrangements suggest that a position effect may cause the mutant phenotype. Hum Mol Genet 4(3):415–422
16. Forabosco A, Percesepe A, Santucci S (2009) Incidence of non-age-dependent chromosomal abnormalities: a population-based study on 88965 amniocenteses. Eur J Hum Genet 17(7):897–903. https://doi.org/10.1038/ejhg.2008.265
17. Ford CE, Hamerton JL (1956a) The chromosomes of man. Nature 178(4541):1020–1023
18. Ford CE, Hamerton JL (1956b) A colchicine, hypotonic citrate, squash sequence for mammalian chromosomes. Stain Technol 31(6):247–251
19. Ford CE, Jones KW, Miller OJ, Mittwoch U, Penrose LS, Ridler M, Shapiro A (1959) The chromosomes in a patient showing both mongolism and the Klinefelter syndrome. Lancet 1(7075):709–710
20. Fraccaro M, Lindsten J, Ford CE, Iselius L (1980) The 11q;22q translocation: a European collaborative analysis of 43 cases. Hum Genet 56(1):21–51
21. Funderburk SJ, Spence MA, Sparkes RS (1977) Mental retardation associated with "balanced" chromosome rearrangements. Am J Hum Genet 29(2):136–141
22. Gordon CT, Attanasio C, Bhatia S, Benko S, Ansari M, Tan TY, Munnich A, Pennacchio LA, Abadie V, Temple IK, Goldenberg A, van Heyningen V, Amiel J, FitzPatrick D, Kleinjan DA, Visel A, Lyonnet S (2014) Identification of novel craniofacial regulatory domains located far upstream of SOX9 and disrupted in Pierre Robin sequence. Hum Mutat 35(8):1011–1020. https://doi.org/10.1002/humu.22606

23. Groffen J, Stephenson JR, Heisterkamp N, de Klein A, Bartram CR, Grosveld G (1984) Philadelphia chromosomal breakpoints are clustered within a limited region, bcr, on chromosome 22. Cell 36(1):93–99
24. Hamerton JL, Canning N, Ray M, Smith S (1975) A cytogenetic survey of 14,069 newborn infants I. Incidence of chromosome abnormalities. Clin Genet 8(4):223–243
25. Heisterkamp N, Stephenson JR, Groffen J, Hansen PF, de Klein A, Bartram CR, Grosveld G (1983) Localization of the c-abl oncogene adjacent to a translocation break point in chronic myelocytic leukaemia. Nature 306(5940):239–242
26. Higgins AW, Alkuraya FS, Bosco AF, Brown KK, Bruns GA, Donovan DJ, Eisenman R, Fan Y, Farra CG, Ferguson HL, Gusella JF, Harris DJ, Herrick SR, Kelly C, Kim HG, Kishikawa S, Korf BR, Kulkarni S, Lally E, Leach NT, Lemyre E, Lewis J, Ligon AH, Lu W, Maas RL, MacDonald ME, Moore SD, Peters RE, Quade BJ, Quintero-Rivera F, Saadi I, Shen Y, Shendure J, Williamson RE, Morton CC (2008) Characterization of apparently balanced chromosomal rearrangements from the developmental genome anatomy project. Am J Hum Genet 82(3):712–722. https://doi.org/10.1016/j.ajhg.2008.01.011
27. Hook EB, Hamerton JL (1978) Analyses of data on rates of cytogenetic disorders in live births. Am J Hum Genet 30(3):330–331
28. Hook EB, Schreinemachers DM, Willey AM, Cross PK (1983) Rates of mutant structural chromosome rearrangements in human fetuses: data from prenatal cytogenetic studies and associations with maternal age and parental mutagen exposure. Am J Hum Genet 35(1):96–109
29. Hsu TC (1952) Mammalian chromosomes in vitro I. The karyotype of man. J Hered 44:167–172
30. Hsu TC, Pomerat CM (1953) Mammalian chromosomes in vitro II. A method for spreading the chromosomes of cells in tissue culture. J Heredity 44:23–29
31. Huang N, Lee I, Marcotte EM, Hurles ME (2010) Characterising and predicting haploinsufficiency in the human genome. PLoS Genet 6(10):e1001154. https://doi.org/10.1371/journal.pgen.1001154
32. Hungerford DA, Donnelly AJ, Nowell PC, Beck S (1959) The chromosome constitution of a human phenotypic intersex. Am J Hum Genet 11:215–236
33. Imaizumi K, Kimura J, Matsuo M, Kurosawa K, Masuno M, Niikawa N, Kuroki Y (2002) Sotos syndrome associated with a de novo balanced reciprocal translocation t(5;8)(q35;q24.1). Am J Med Genet 107(1):58–60
34. Jacobs PA (1974) Correlation between euploid structural chromosome rearrangements and mental subnormality in humans. Nature 249(453):164–165
35. Jacobs PA, Baikie AG, Court Brown WM, Strong JA (1959) The somatic chromosomes in mongolism. Lancet 1(7075):710
36. Jacobs PA, Browne C, Gregson N, Joyce C, White H (1992) Estimates of the frequency of chromosome abnormalities detectable in unselected newborns using moderate levels of banding. J Med Genet 29(2):103–108
37. Kato T, Kurahashi H, Emanuel BS (2012) Chromosomal translocations and palindromic AT-rich repeats. Curr Opin Genet Dev 22(3):221–228. https://doi.org/10.1016/j.gde.2012.02.004
38. Kim HG, Kishikawa S, Higgins AW, Seong IS, Donovan DJ, Shen Y, Lally E, Weiss LA, Najm J, Kutsche K, Descartes M, Holt L, Braddock S, Troxell R, Kaplan L, Volkmar F, Klin A, Tsatsanis K, Harris DJ, Noens I, Pauls DL, Daly MJ, MacDonald ME, Morton CC, Quade BJ, Gusella JF (2008) Disruption of neurexin 1 associated with autism spectrum disorder. Am J Hum Genet 82(1):199–207. https://doi.org/10.1016/j.ajhg.2007.09.011
39. Kleinjan DA, van Heyningen V (2005) Long-range control of gene expression: emerging mechanisms and disruption in disease. Am J Hum Genet 76(1):8–32. https://doi.org/10.1086/426833
40. Konopka JB, Watanabe SM, Witte ON (1984) An alteration of the human c-abl protein in K562 leukemia cells unmasks associated tyrosine kinase activity. Cell 37(3):1035–1042
41. Kurotaki N, Imaizumi K, Harada N, Masuno M, Kondoh T, Nagai T, Ohashi H, Naritomi K, Tsukahara M, Makita Y, Sugimoto T, Sonoda T, Hasegawa T, Chinen Y, Tomita Ha HA, Kinoshita A, Mizuguchi T, Yoshiura Ki K, Ohta T, Kishino T, Fukushima Y, Niikawa N, Matsumoto N (2002) Haploinsufficiency of NSD1 causes Sotos syndrome. Nat Genet 30(4):365–366. https://doi.org/10.1038/ng863
42. Ledbetter DH, Rich DC, O'Connell P, Leppert M, Carey JC (1989) Precise localization of NF1 to 17q11.2 by balanced translocation. Am J Hum Genet 44(1):20–24
43. Lejeune J, Gautier M, Turpin R (1959) Study of somatic chromosomes from 9 mongoloid children. C R Hebd Seances Acad Sci 248(11):1721–1722
44. Lejeune J, Turpin R, Decourt J (1960) Chromosome aberrations and human diseases XXY Klinefelter's syndrome from 46 chromosomes caused by T-T centromeric fusion. C R Hebd Seances Acad Sci 250:2468–2470
45. Lettice LA, Heaney SJ, Purdie LA, Li L, de Beer P, Oostra BA, Goode D, Elgar G, Hill RE, de Graaff E (2003) A long-range Shh enhancer regulates expression in the developing limb and fin and is associated with preaxial polydactyly. Hum Mol Genet 12(14):1725–1735
46. Lettice LA, Horikoshi T, Heaney SJ, van Baren MJ, van der Linde HC, Breedveld GJ, Joosse M, Akarsu N, Oostra BA, Endo N, Shibata M, Suzuki M, Takahashi E, Shinka T, Nakahori Y, Ayusawa D, Nakabayashi K, Scherer SW, Heutink P, Hill RE, Noji S (2002) Disruption of a long-range cis-acting regulator for Shh causes preaxial polydactyly. Proc Natl Acad Sci U S A 99(11):7548–7553. https://doi.org/10.1073/pnas.112212199
47. Liang S, Tippens ND, Zhou Y, Mort M, Stenson PD, Cooper DN, Yu H (2017) iRegNet3D: three-

dimensional integrated regulatory network for the genomic analysis of coding and non-coding disease mutations. Genome Biol 18(1):10. https://doi.org/10.1186/s13059-016-1138-2
48. Lupiáñez DG, Kraft K, Heinrich V, Krawitz P, Brancati F, Klopocki E, Horn D, Kayserili H, Opitz JM, Laxova R, Santos-Simarro F, Gilbert-Dussardier B, Wittler L, Borschiwer M, Haas SA, Osterwalder M, Franke M, Timmermann B, Hecht J, Spielmann M, Visel A, Mundlos S (2015) Disruptions of topological chromatin domains cause pathogenic rewiring of gene-enhancer interactions. Cell 161(5):1012–1025. https://doi.org/10.1016/j.cell.2015.04.004
49. Lupiáñez DG, Spielmann M, Mundlos S (2016) Breaking TADs: how alterations of chromatin domains result in disease. Trends Genet 32(4):225–237. https://doi.org/10.1016/j.tig.2016.01.003
50. MacGregor DJ, Imrie S, Tolmie JL (1989) Outcome of de novo balanced translocations ascertained prenatally. J Med Genet 26(9):590–591
51. Mackie Ogilvie C, Scriven PN (2002) Meiotic outcomes in reciprocal translocation carriers ascertained in 3-day human embryos. Eur J Hum Genet 10(12):801–806. https://doi.org/10.1038/sj.ejhg.5200895
52. Maroun C, Schmerler S, Hutcheon RG (1994) Child with Sotos phenotype and a 5:15 translocation. Am J Med Genet 50(3):291–293. https://doi.org/10.1002/ajmg.1320500313
53. Moorhead PS, Nowell PC, Mellman WJ, Battips DM, Hungerford DA (1960) Chromosome preparations of leukocytes cultured from human peripheral blood. Exp Cell Res 20:613–616
54. Morel F, Douet-Guilbert N, Le Bris MJ, Herry A, Amice V, Amice J, De Braekeleer M (2004) Meiotic segregation of translocations during male gametogenesis. Int J Androl 27(4):200–212. https://doi.org/10.1111/j.1365-2605.2004.00490.x
55. Mutton D, Alberman E, Hook EB (1996) Cytogenetic and epidemiological findings in down syndrome, England and Wales 1989 to 1993. National down Syndrome Cytogenetic Register and the Association of Clinical Cytogeneticists. J Med Genet 33(5):387–394
56. Nielsen J, Krag-Olsen B (1981) Follow-up of 32 children with autosomal translocations found among 11,148 consecutively newborn children from 1969 to 1974. Clin Genet 20(1):48–54
57. Nielsen J, Wohlert M (1991) Chromosome abnormalities found among 34,910 newborn children: results from a 13-year incidence study in Arhus, Denmark. Hum Genet 87(1):81–83
58. Nilsson D, Pettersson M, Gustavsson P, Förster A, Hofmeister W, Wincent J, Zachariadis V, Anderlid BM, Nordgren A, Mäkitie O, Wirta V, Käller M, Vezzi F, Lupski JR, Nordenskjöld M, Lundberg ES, Carvalho CMB, Lindstrand A (2017) Whole-genome sequencing of cytogenetically balanced chromosome translocations identifies potentially pathological gene disruptions and highlights the importance of microhomology in the mechanism of formation. Hum Mutat 38(2):180–192. https://doi.org/10.1002/humu.23146
59. Nowell PC, Hungerford DA (1960) Chromosome studies on normal and leukemic human leukocytes. J Natl Cancer Inst 25:85–109
60. Oliver-Bonet M, Navarro J, Carrera M, Egozcue J, Benet J (2002) Aneuploid and unbalanced sperm in two translocation carriers: evaluation of the genetic risk. Mol Hum Reprod 8(10):958–963
61. Ordulu Z, Kammin T, Brand H, Pillalamarri V, Redin CE, Collins RL, Blumenthal I, Hanscom C, Pereira S, Bradley I, Crandall BF, Gerrol P, Hayden MA, Hussain N, Kanengisser-Pines B, Kantarci S, Levy B, Macera MJ, Quintero-Rivera F, Spiegel E, Stevens B, Ulm JE, Warburton D, Wilkins-Haug LE, Yachelevich N, Gusella JF, Talkowski ME, Morton CC (2016) Structural chromosomal rearrangements require nucleotide-level resolution: lessons from next-generation sequencing in prenatal diagnosis. Am J Hum Genet 99(5):1015–1033. https://doi.org/10.1016/j.ajhg.2016.08.022
62. Ozawa N, Maruyama T, Nagashima T, Ono M, Arase T, Ishimoto H, Yoshimura Y (2008) Pregnancy outcomes of reciprocal translocation carriers who have a history of repeated pregnancy loss. Fertil Steril 90(4):1301–1304. https://doi.org/10.1016/j.fertnstert.2007.09.051
63. Page SL, Shaffer LG (1997) Nonhomologous Robertsonian translocations form predominantly during female meiosis. Nat Genet 15(3):231–232. https://doi.org/10.1038/ng0397-231
64. Penrose LS (1933) The relative effects of paternal and maternal age in mongolism. Reproduced in 2009. J Genet 88(1):9–14
65. Penrose LS, Ellis JR, Delhanty JD (1960) Chromosomal translocations in mongolism and in normal relatives. Lancet 2(7147):409–410
66. Perrin A, Morel F, Douet-Guilbert N, Le Bris MJ, Amice J, Amice V, De Braekeleer M (2010) A study of meiotic segregation of chromosomes in spermatozoa of translocation carriers using fluorescent in situ hybridisation. Andrologia 42(1):27–34. https://doi.org/10.1111/j.1439-0272.2009.00951.x
67. Polani PE, Briggs JH, Ford CE, Clarke CM, Berg JM (1960) A Mongol girl with 46 chromosomes. Lancet 1(7127):721–724
68. Primerano A, Colao E, Villella C, Nocera MD, Ciambrone A, Luciano E, D'Antona L, Vismara MFM, Loddo S, Novelli A, Perrotti N, Malatesta P (2015) A cryptic balanced translocation (5;17), a puzzle revealed through a critical evaluation of the pedigree and a FISH focused on candidate loci suggested by the phenotype. Mol Cytogenet 8:70. https://doi.org/10.1186/s13039-015-0172-1
69. Quintero-Rivera F, Xi QJ, Keppler-Noreuil KM, Lee JH, Higgins AW, Anchan RM, Roberts AE, Seong IS, Fan X, Lage K, Lu LY, Tao J, Hu X, Berezney R, Gelb BD, Kamp A, Moskowitz IP, Lacro RV, Lu

W, Morton CC, Gusella JF, Maas RL (2015) MATR3 disruption in human and mouse associated with bicuspid aortic valve, aortic coarctation and patent ductus arteriosus. Hum Mol Genet 24(8):2375–2389. https://doi.org/10.1093/hmg/ddv004

70. Rao SS, Huntley MH, Durand NC, Stamenova EK, Bochkov ID, Robinson JT, Sanborn AL, Machol I, Omer AD, Lander ES, Aiden EL (2014) A 3D map of the human genome at kilobase resolution reveals principles of chromatin looping. Cell 159(7):1665–1680. https://doi.org/10.1016/j.cell.2014.11.021
71. Redin C, Brand H, Collins RL, Kammin T, Mitchell E, Hodge JC, Hanscom C, Pillalamarri V, Seabra CM, Abbott MA, Abdul-Rahman OA, Aberg E, Adley R, Alcaraz-Estrada SL, Alkuraya FS, An Y, Anderson MA, Antolik C, Anyane-Yeboa K, Atkin JF, Bartell T, Bernstein JA, Beyer E, Blumenthal I, Bongers EM, Brilstra EH, Brown CW, Brüggenwirth HT, Callewaert B, Chiang C, Corning K, Cox H, Cuppen E, Currall BB, Cushing T, David D, Deardorff MA, Dheedene A, D'Hooghe M, de Vries BB, Earl DL, Ferguson HL, Fisher H, FitzPatrick DR, Gerrol P, Giachino D, Glessner JT, Gliem T, Grady M, Graham BH, Griffis C, Gripp KW, Gropman AL, Hanson-Kahn A, Harris DJ, Hayden MA, Hill R, Hochstenbach R, Hoffman JD, Hopkin RJ, Hubshman MW, Innes AM, Irons M, Irving M, Jacobsen JC, Janssens S, Jewett T, Johnson JP, Jongmans MC, Kahler SG, Koolen DA, Korzelius J, Kroisel PM, Lacassie Y, Lawless W, Lemyre E, Leppig K, Levin AV, Li H, Liao EC, Lim C, Lose EJ, Lucente D, Macera MJ, Manavalan P, Mandrile G, Marcelis CL, Margolin L, Mason T, Masser-Frye D, McClellan MW, Mendoza CJ, Menten B, Middelkamp S, Mikami LR, Moe E, Mohammed S, Mononen T, Mortenson ME, Moya G, Nieuwint AW, Ordulu Z, Parkash S, Pauker SP, Pereira S, Perrin D, Phelan K, Aguilar RE, Poddighe PJ, Pregno G, Raskin S, Reis L, Rhead W, Rita D, Renkens I, Roelens F, Ruliera J, Rump P, Schilit SL, Shaheen R, Sparkes R, Spiegel E, Stevens B, Stone MR, Tagoe J, Thakuria JV, van Bon BW, van de Kamp J, van Der Burgt I, van Essen T, van Ravenswaaij-Arts CM, van Roosmalen MJ, Vergult S, Volker-Touw CM, Warburton DP, Waterman MJ, Wiley S, Wilson A, Yerena-de Vega MC, Zori RT, Levy B, Brunner HG, de Leeuw N, Kloosterman WP, Thorland EC, Morton CC, Gusella JF, Talkowski ME (2017) The genomic landscape of balanced cytogenetic abnormalities associated with human congenital anomalies. Nat Genet 49(1):36–45. https://doi.org/10.1038/ng.3720
72. Rehm HL, Berg JS, Brooks LD, Bustamante CD, Evans JP, Landrum MJ, Ledbetter DH, Maglott DR, Martin CL, Nussbaum RL, Plon SE, Ramos EM, Sherry ST, Watson MS, ClinGen (2015) ClinGen--the clinical genome resource. N Engl J Med 372(23):2235–2242. https://doi.org/10.1056/NEJMsr1406261
73. Rowley JD (1973) Letter: a new consistent chromosomal abnormality in chronic myelogenous leukaemia identified by quinacrine fluorescence and Giemsa staining. Nature 243(5405):290–293
74. Schmid W (1962) A familial chromosome abnormality associated with repeated abortions. Cytogenetics 1:199–209
75. Schmidt MA, Michels VV, Dewald GW (1987) Cases of neurofibromatosis with rearrangements of chromosome 17 involving band 17q11.2. Am J Med Genet 28:771–777
76. Seabright M (1971) A rapid banding technique for human chromosomes. Lancet 2(7731):971–972
77. Shaffer LG, Jackson-Cook CK, Meyer JM, Brown JA, Spence JE (1991) A molecular genetic approach to the identification of isochromosomes of chromosome 21. Hum Genet 86(4):375–382
78. Sheridan MB, Kato T, Haldeman-Englert C, Jalali GR, Milunsky JM, Zou Y, Klaes R, Gimelli G, Gimelli S, Gemmill RM, Drabkin HA, Hacker AM, Brown J, Tomkins D, Shaikh TH, Kurahashi H, Zackai EH, Emanuel BS (2010) A palindrome-mediated recurrent translocation with 3:1 meiotic nondisjunction: the t(8;22)(q24.13;q11.21). Am J Hum Genet 87(2):209–218. https://doi.org/10.1016/j.ajhg.2010.07.002
79. Shtivelman E, Lifshitz B, Gale RP, Canaani E (1985) Fused transcript of abl and bcr genes in chronic myelogenous leukaemia. Nature 315(6020):550–554
80. Sugiura-Ogasawara M, Aoki K, Fujii T, Fujita T, Kawaguchi R, Maruyama T, Ozawa N, Sugi T, Takeshita T, Saito S (2008) Subsequent pregnancy outcomes in recurrent miscarriage patients with a paternal or maternal carrier of a structural chromosome rearrangement. J Hum Genet 53(7):622–628. https://doi.org/10.1007/s10038-008-0290-2
81. Sumner AT, Evans HJ, Buckland RA (1971) New technique for distinguishing between human chromosomes. Nat New Biol 232(27):31–32
82. Talkowski ME, Ernst C, Heilbut A, Chiang C, Hanscom C, Lindgren A, Kirby A, Liu S, Muddukrishna B, Ohsumi TK, Shen Y, Borowsky M, Daly MJ, Morton CC, Gusella JF (2011) Next-generation sequencing strategies enable routine detection of balanced chromosome rearrangements for clinical diagnostics and genetic research. Am J Hum Genet 88(4):469–481. https://doi.org/10.1016/j.ajhg.2011.03.013
83. Taub R, Kirsch I, Morton C, Lenoir G, Swan D, Tronick S, Aaronson S, Leder P (1982) Translocation of the c-myc gene into the immunoglobulin heavy chain locus in human Burkitt lymphoma and murine plasmacytoma cells. Proc Natl Acad Sci U S A 79(24):7837–7841
84. Thomas NS, Morris JK, Baptista J, Ng BL, Crolla JA, Jacobs PA (2010) De novo apparently balanced translocations in man are predominantly paternal in origin and associated with a significant increase in paternal age. J Med Genet 47(2):112–115. https://doi.org/10.1136/jmg.2009.069716
85. Tierney I, Axworthy D, Smith L, Ratcliffe SG (1984) Balanced rearrangements of the autosomes: results of a longitudinal study of a newborn survey population. J Med Genet 21(1):45–51
86. Tjio JH, Levan A (1956) The chromosome number of man. Hereditas 42:1–6

87. Turpin R, Lejeune J (1969) Human Afflications and Chromosomal Aberrations. Translation based on Les Chromosomes Humains (caryotype normal et variations pathologiques), Gauthier-Villars, 1965. International series of monographs in pure and applied biology. Modern trends in physiological sciences, vol 32. Pergamon Press, Oxford
88. Turpin R, Lejeune J, Lafourcade J, Gautier M (1959) Chromosome aberrations & human diseases; multiple spinal abnormalities with 45 chromosomes. C R Hebd Seances Acad Sci 248(25):3636–3638
89. Van Dyke DL, Weiss L, Roberson JR, Babu VR (1983) The frequency and mutation rate of balanced autosomal rearrangements in man estimated from prenatal genetic studies for advanced maternal age. Am J Hum Genet 35 (2):301–308.
90. Vincent MC, Daudin M, De MP, Massat G, Mieusset R, Pontonnier F, Calvas P, Bujan L, Bourrouillout G (2002) Cytogenetic investigations of infertile men with low sperm counts: a 25-year experience. J Androl 23(1):18–22. discussion 44–15
91. Walker S, Harris R (1962) Investigation of family showing transmission of a 13-15 chromosomal translocation (Denver classification). Br Med J 2(5296):25–26
92. Warburton D (1991) De novo balanced chromosome rearrangements and extra marker chromosomes identified at prenatal diagnosis: clinical significance and distribution of breakpoints. Am J Hum Genet 49(5):995–1013
93. Xia F, Bainbridge MN, Tan TY, Wangler MF, Scheuerle AE, Zackai EH, Harr MH, Sutton VR, Nalam RL, Zhu W, Nash M, Ryan MM, Yaplito-Lee J, Hunter JV, Deardorff MA, Penney SJ, Beaudet AL, Plon SE, Boerwinkle EA, Lupski JR, Eng CM, Muzny DM, Yang Y, Gibbs RA (2014) De novo truncating mutations in AHDC1 in individuals with syndromic expressive language delay, hypotonia, and sleep apnea. Am J Hum Genet 94(5):784–789. https://doi.org/10.1016/j.ajhg.2014.04.006
94. Zackai EH, Emanuel BS (1980) Site-specific reciprocal translocation, t(11;22) (q23;q11), in several unrelated families with 3:1 meiotic disjunction. Am J Med Genet 7(4):507–521. https://doi.org/10.1002/ajmg.1320070412
95. Zepeda-Mendoza CJ, Ibn-Salem J, Kammin T, Harris DJ, Rita D, Gripp KW, MacKenzie JJ, Gropman A, Graham B, Shaheen R, Alkuraya FS, Brasington CK, Spence EJ, Masser-Frye D, Bird LM, Spiegel E, Sparkes RL, Ordulu Z, Talkowski ME, Andrade-Navarro MA, Robinson PN, Morton CC (2017) Computational prediction of position effects of apparently balanced human chromosomal rearrangements. Am J Hum Genet. https://doi.org/10.1016/j.ajhg.2017.06.011
96. Zhang HG, Wang RX, Li LL, Sun WT, Zhang HY, Liu RZ (2015) Male carriers of balanced reciprocal translocations in Northeast China: sperm count, reproductive performance, and genetic counseling. Genet Mol Res 14(4):18792–18798. https://doi.org/10.4238/2015.December.28.28

2 Induction of Chromosomal Translocations with CRISPR-Cas9 and Other Nucleases: Understanding the Repair Mechanisms That Give Rise to Translocations

Erika Brunet and Maria Jasin

Abstract

Chromosomal translocations are associated with several tumor types, including hematopoietic malignancies, sarcomas, and solid tumors of epithelial origin, due to their activation of a proto-oncogene or generation of a novel fusion protein with oncogenic potential. In many cases, the availability of suitable human models has been lacking because of the difficulty in recapitulating precise expression of the fusion protein or other reasons. Further, understanding how translocations form mechanistically has been a goal, as it may suggest ways to prevent their occurrence. Chromosomal translocations arise when DNA ends from double-strand breaks (DSBs) on two heterologous chromosomes are improperly joined. This review provides a summary of DSB repair mechanisms and their contribution to translocation formation, the various programmable nuclease platforms that have been used to generate translocations, and the successes that have been achieved in this area.

Keywords

Double-strand break · CRISPR-Cas9 · Chromosomal translocation · NHEJ

2.1 Introduction

Chromosomal translocations join DNA segments derived from two heterologous chromosomes. Translocations influence the evolution of species, but they are mainly considered in the context of disease. In particular, they are prominent features of several types of cancers, from hematopoietic to solid tumors, leading to the expression of a new fusion oncogene or to the mis-regulation of a proto-oncogene. Models for translocation-associated cancers typically rely on ectopic expression of fusion genes in cell lines or on endogenous expression in transformed tumor cells. While valuable to the scientific community for providing insights into mechanisms of oncogenesis, these models may fail to fully recapitulate the human disease. For instance, mouse models overexpressing the NPM1-ALK fusion (implicated in Anaplasic Large Cell Lymphoma,

E. Brunet (✉)
Genome Dynamics in the Immune System Laboratory, Institut Imagine, INSERM UMR 1163, Université Paris Descartes, Sorbonne Paris Cité, Paris, France
e-mail: erika.brunet@inserm.fr

M. Jasin (✉)
Developmental Biology Program, Memorial Sloan Kettering Cancer Center, New York, NY, USA
e-mail: m-jasin@ski.mskcc.org

Y. Zhang (ed.), *Chromosome Translocation*, Advances in Experimental Medicine and Biology 1044,
https://doi.org/10.1007/978-981-13-0593-1_2

or ALCL) mostly induce B cell lymphomas rather than T cell lymphomas associated with the human disease, therefore failing to provide a robust preclinical model [56]. Moreover, patient-derived tumors cells will invariably have a number of tumor-acquired mutations.

Chromosomal translocations appear to arise from improper repair of DNA double-strand breaks (DSBs), which are highly toxic lesions. The "guardians" of genome integrity mostly ensure reliable repair of DSBs; also, unrepaired DSBs can lead to apoptosis or senescence. However, imprecise repair of DSBs has the potential to be highly deleterious, as it can lead to genome instability, including the formation of chromosomal rearrangements. In particular, chromosomal translocations can arise when DNA ends from DSBs on two heterologous chromosomes are improperly joined [45]. Given this, researchers have been taking advantage of various nucleases, especially the recently developed programmable nucleases, to deliberately induce DSBs at loci of interest to generate translocations. The goal is to ultimately generate faithful tumor models and also to understand the DSB "misrepair" mechanisms that lead to translocations.

2.2 Multiple DSB Repair Pathways: Repairing a Dangerous Lesion

Given that DSBs can compromise the integrity of the genome, it is perhaps not surprising that multiple pathways exist to repair DSBs [8, 21, 22] (Fig. 2.1). The two major DSB repair pathways in mammalian cells are nonhomologous end-joining (NHEJ) and homologous recombination (HR), also termed homology-directed repair. The relationship of these pathways is complex. In some cases, DSB repair is limited to one pathway, as in programmed DSBs in the immune system (NHEJ) or during meiosis (HR), but the pathways can also compete with each other for the repair of a single DSB and surprisingly even collaborate [26], such that DSB repair initiates by HR but is completed by NHEJ [12, 25, 42]

In cycling cells, an early determinant of DSB repair pathway choice is whether DNA ends undergo resection to generate 3′ single-stranded overhangs, which is promoted during the S/G2 phases of the cell cycle but suppressed during G1 [53]. The resected DNA is then coated with the RAD51 protein to form a nucleoprotein filament, which performs strand invasion of a homologous template to prime repair synthesis [36]. If templated by the sister chromatid, the preferred HR partner, repair synthesis leads to restoration of the original DNA sequence prior to breakage. By contrast, during canonical NHEJ (c-NHEJ), DNA ends are protected from resection and can be precisely joined, if ends do not require modification, or imprecisely joined after processing to make ends ligatable, giving rise to a variety of junctions.

End resection also provides single-stranded DNA intermediates for two other pathways (Fig. 2.1). Alternative NHEJ (alt-NHEJ), a major pathway of which is microhomology-mediated NHEJ (MMEJ), involves annealing at short sequence identities present near the DNA ends and thus only requires limited end resection (<100 bp) [46]. As with c-NHEJ, this pathway gives rise to a variety of junctions, although deletions may be longer with MMEJ due to resection [48]. Because any particular breakpoint junction can likely form by either NHEJ pathway, identifying which pathway is responsible requires statistical analysis to determine if microhomology is over represented or the use of pathway mutants, for example, in ligase IV (LIG4) for c-NHEJ and ligase III (LIG3) for alt-NHEJ [47, 48]. Single-strand annealing (SSA) also involves annealing at repeats flanking the DSB, but the repeats are much longer, and thus requires more extensive end resection than alt-NHEJ to uncover complementary single-strands. The physiological role of SSA in cells is unclear, but it has been used to distinguish whether HR mutants are defective in the early end-resection step of HR (defective in both HR and SSA) from those defective only at the strand invasion step (defective in HR but elevated SSA) [52].

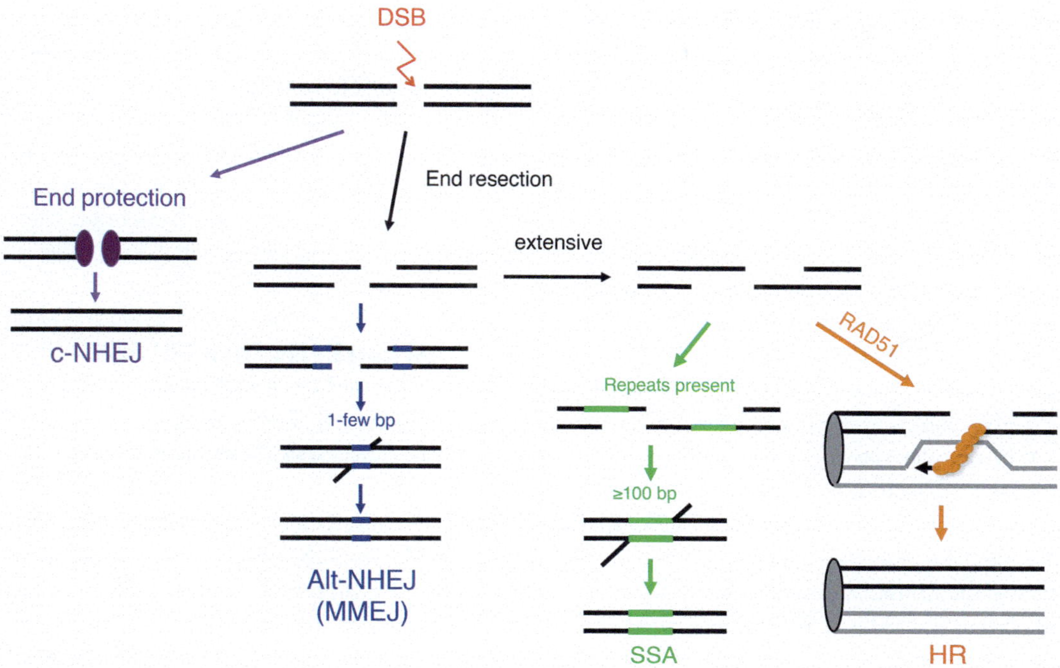

Fig. 2.1 DSB repair pathways. The two major DSB repair pathways in mammalian cells are nonhomologous end-joining (NHEJ) and homologous recombination (HR). In addition, end resection provides single-stranded DNA intermediates for two other pathways: Alternative NHEJ (Alt-NHEJ) using microhomology and Single Strand Annealing (SSA)

2.3 Elucidation of Translocation Mechanisms in Mouse Cells Using a Rare-Cutting Endonuclease

As the complexity of DSB repair pathways in mammalian cells was uncovered, investigators sought to determine how each pathway participates in translocation formation. In the era before programmable endonucleases, DSBs were introduced into the genome using the yeast homing endonuclease I-SceI [11], which has an ~18 bp recognition site and thus is suitable for studies in complex mammalian genomes. Several reporters were developed in mammalian cells to determine which DSB repair pathway(s) gives rise to translocations upon DSB formation [62]. I-SceI sites were introduced at specified chromosomal locations by gene targeting in mouse embryonic stem cells, which are diploid. Translocations were selected by reconstruction of a drug resistance marker and confirmed by fluorescence in situ hybridization.

Initial studies focused on HR between repeats on different chromosomes, given that translocations will form by HR in budding yeast [20] and that mammalian genomes are replete with sequence repeats. Introducing a DSB into a repeat on one chromosome, however, did not give rise to a translocation; rather the DSB was repaired by a simple gene conversion event with the other chromosome without exchange of flanking markers [44]. This study, as well as subsequent ones (e.g., [30, 51]), indicated that HR in mammalian cells is rarely associated with crossing over. A follow up study attempted to drive translocation formation by HR by truncating the repeats, such that restoration of the selectable marker would seem to require HR. However, in this case too, HR did not lead to translocations; rather, HR was coupled to NHEJ, involving a break-induced replication type of HR that was completed by NHEJ [42]. Presumably, the BLM helicase plays a major role in suppressing crossing over that would drive translocation formation, as it does between homologous chromosomes [30]. Thus, these studies indicated that while a DSB on one chromosome was sufficient to induce HR with another chromosome, it was not sufficient to drive translocation formation. Although HR between repeated sequences has been reported in several contexts, it is notable that many or most of these events likely occur during meiosis [27], when HR may be under different constraints.

Subsequent studies focused on introducing two DSBs, one on each chromosome. Using this approach, both NHEJ and SSA were found to give rise to reciprocal translocations; in fact, both derivative chromosome could form by NHEJ or SSA or one derivative chromosome could form by NHEJ and the other SSA [15, 43, 60]. SSA was highly proficient for translocation formation with identical sequence repeats and presumably because the DSBs occurred close by the repeats and on opposite sides [15]. However, when divergent repeats were used, specifically two different Alu elements from the MLL gene, the frequency of SSA-mediated translocations dropped substantially, resulting in more NHEJ-mediated events. Mismatch repair components likely suppress SSA between diverged, but homologous sequences, as it does in other contexts [16]. These studies highlight the constraints on SSA for mediating translocation formation, in particular, the degree of sequence identity between the repeats and the positions of the DSBs relative to the repeats.

Most translocation breakpoint junctions observed in tumors from patients join at sequences that do not share significant lengths of homology, indicating that they arose by NHEJ [32]. The role of c-NHEJ was investigated using the NHEJ-based translocation reporters in mouse embryonic stem cells. Translocations were found to be suppressed by c-NHEJ components LIG4 and Ku70 [48, 60], consistent with results in lymphoid systems where oncogenic translocations increased in the absence of these proteins [17]. Translocation junctions were biased towards the presence of microhomologies, with or without these c-NHEJ components, suggesting that alt-NHEJ (MMEJ) gave rise to translocations, even in wild-type cells [48]. Consistent with this, translocations were found to be largely dependent on the alt-NHEJ components LIG3 [47], the end resection factor CtIP [65], and polymerase theta [33], as determined using programmable nucleases. Taken together, these results indicate that alt-NHEJ is the major mechanism for translocation formation in mouse cells.

2.4 When ZFNs Then TALENs Arrived on the Scene: Tailored Nucleases for Tailored Translocations

The rare cutting I-SceI endonuclease proved to be a valuable tool to induce chromosomal translocations. The limitation of I-SceI is the necessity to target genomic loci with the recognition site, yet genome modification prior to 2005 was much more laborious in human cells than in mouse cells. The advent of programmable nucleases tailored to cleave any possible locus within genomes has opened tremendous possibilities to create de novo translocations to generate cancer models.

The development of tailored (programmable) endonucleases originated in 1996 with the report of a fusion of a zinc finger DNA binding domain with the cleavage domain of the FokI restriction enzyme to create a zinc finger nuclease (ZFN) [28]. Almost a decade later, a ZFN developed by Sangamo Biosciences was shown to cleave an endogenous locus in human cells to lead to its modification [57]. Building on the initial results obtained with I-SceI, our group harnessed ZFNs to induce chromosomal translocations at two endogenous loci in human cells [6, 61]. Breakpoint junctions were identified by PCR and clones carrying translocations could be recovered from tumor cell lines.

This study provided the first proof of concept of modeling translocations using custom-designed nucleases. Remarkably, translocations were also obtained in multipotential stem cells, both human embryonic stem cells and mesenchymal cells derived from them [6]. Following this work, a cancer-relevant translocation was recapitulated, t(11;22)(q24;q12), the most common rearrangement found in Ewing sarcoma, by designing ZFNs to target the most common breakpoints found in patients. Reciprocal translocations were readily recovered in mesenchymal precursor cells, leading to EWSR1-FLI1 fusion gene expression from the endogenous EWSR1 promoter. Notably, the joining characteristics – deletions, insertions, mutations – found in translocations resulting from ZFN cleavage fully recapitulated those from Ewing patient cells and demonstrated that the junctions arose by an NHEJ pathway. Of note, the FokI cleavage domain in a ZFN works as a dimer, with each monomer fused to a different assembly of zinc fingers for DNA recognition. This leaves open the possibility that incorrect ZFNs could form to cleave newly formed translocation junctions or off-target sites. Modified FokI domains that heterodimerize have been developed that strongly promote the use of the correct partner [14]; these are particularly valuable for the simultaneous use of pairs of ZFNs as required for translocation formation.

The technical complexity of designing and assembling highly specific and active zinc fingers, however, prohibited the wide spread use of ZFNs ("democratization"; [21]) by academic researchers. In 2010 the development of TALENs (Transcription Activator-like Effector Nucleases) extended the repertoire of tailored nucleases to one with a much more elementary code of base recognition [35]. TALENs use the same homo- or heterodimeric FokI cleavage domains as ZFNs, but assembling modules for DNA sequence recognition became much more trivial. As with ZFNs, the use of two TALENs enabled the formation of translocations [38]. Modeling t(2;5)(p23;q35), found in cases of ALCL, our group showed expression of oncogenic NPM1-ALK kinase activation in human cell lines. Conversely, the NPM1-ALK translocation in a patient cell line could be reverted with the same pair of TALENs, restoring the integrity of the two participating chromosomes and potentially permitting the analysis of phenotypic consequences of fusion protein loss once cells are transformed.

2.5 The CRISPR-Cas9 Revolution: When Easy Is Made Easier

Soon after TALEN development, CRISPR-Cas9 appeared upon the scene as a highly simplified tailored nuclease, using a guide RNA (gRNA or sgRNA) to recognize the complementary DNA sequence in the genome [24]. This new nuclease was quickly used to induce chromosomal translocations, including the previously described models of NPM1-ALK [18] and Ewing sarcoma [54] [41] and new models of lung cancer translocations [10] and acute myelogenous leukemia [41, 54].

CRISPR-Cas9 was also used by other teams to create chromosomal translocations in mouse embryonic stem cells [23] and in mouse myoblasts, the latter modeling the human alveolar rhabdyomyosarcoma Pax3-Foxo1 [29], but also in other organisms, namely *C. elegans* [9] and *Leishmania* [64].

Thus, it is now possible to faithfully model the full outcome of these chromosome rearrangements, including the formation of the reciprocal translocation, loss of one intact copy of each participating gene, recapitulating potential haploinsufficiency, and fusion gene expression from the endogenous promoter: basically the holy grail for those in quest of relevant translocation cancer models.

2.6 Isolating Translocation Clones

Despite their success, these studies also showed that isolation of translocation clones induced de novo remains tedious irrespective of the type of nuclease used and particularly in primary cells for which long sib-selection cycles are mostly unworkable. Whether it is just a matter of efficiency – translocation formation being much less efficient than simple intra-chromosomal repair – or whether expression of the fusion gene directly affects proliferation of the cells remains to be elucidated. Attempts to increase the translocation frequency have involved short single-strand oligonucleotides matching the DSB ends formed by

CRISPR-Cas9 to "guide" joining of the two chromosome ends for translocation formation [55].

A strategy for selecting translocation clones has also recently been developed, with EWSR1-WT1 found in desmoplastic round cell tumors as the model [58]. The approach uses CRISPR-Cas9 to induce integration of a homologous donor fragment containing a selectable marker at DSBs on the translocating chromosomes. The selectable marker is promoterless and contains an upstream splice acceptor to strongly enrich for HR events at the EWSR1 locus. A further refinement is that the selectable marker is flanked by LoxP sites, such that fusion protein expression is conditional and dependent on removal of the selectable marker cassette by expression of Cre recombinase. This strategy has also proved to be effective in a tumor cell line [50].

2.7 Modeling Oncogenesis Using Programmable Nucleases: First Steps Towards Full Transformation In Vivo

The first example of oncogenesis obtained *in vivo* from a nuclease-induced rearrangement was published in 2014 and involved formation of the EML4-ALK fusion through an 11 Mb inversion [31]. Although not a translocation *per se* involving two different chromosomes, this report provided direct evidence that the formation of the EML4-ALK fusion induced by CRISPR-Cas9 activity in the lungs of mice leads to lung adenocarcinomas. In parallel, another group obtained similar results [4]. While the high efficiency of transformation obtained in these studies is likely related to a high rate of intra-chromosomal rearrangements, this study suggests that translocations between two chromosomes could be tested in a similar manner.

While TALENs have been used to generate an MLL-AF9 fusion by knock-in [7], faithful *de novo* translocations giving rise to MLL-AF4 and MLL-AF9 have been produced by TALENs in CD34+ cells [5]. Although transformation to leukemia did not occur *in vitro*, a range of phenotypes was observed, from frequent loss of cells to the persistent proliferative advantage of a few cells, as well as some clones showing a transient proliferative advantage. More recently, another group reached similar conclusions with the MLL-ENL translocation in CD34+ cells *in vitro*, with cells forming normal hematopoietic colonies but eventually ceasing to proliferate, even if some clones shown extended plating capacity [39]. However, while the first round recipient mice initiated a "monocytic leukemia-like phenotype" but not immature AML, the second round recipients developed AML but with incomplete penetrance.

Despite the limitations uncovered in these studies, nuclease induced-translocation models are expected to reveal new aspects of tumorigenesis and will undoubtedly in the near future provide insights about the timing between the translocation occurrence and the appearance of the disease, more relevant to progression development found in patients by bypassing limitations of models expressing ectopically fusion genes. The role of the *in vivo* environment and the undeniably pivotal role of accumulation of secondary mutations certainly remains to be uncovered.

2.8 Elucidation of Translocation Mechanisms in Human Cells Using Programmable Nucleases

While translocations induced in mouse cells primarily arise by alt-NHEJ and are suppressed by c-NHEJ components, translocations induced by programmable nucleases (ZFNs, TALENs, CRISPR-Cas9) in human cells have breakpoint junctions characterized by little or no end processing, suggestive of c-NHEJ [6, 18]. In one study, half of the breakpoints in c-NHEJ-proficient cells demonstrated almost perfect joining of the ends (≤ 1 bp deletion) with few microhomologies [18]. Confirming the involvement of c-NHEJ, loss of LIG4 (or its partner XRCC4) reduced translocations in multiple cell lines. More recently, PARP3, which cooperates in the recruitment of c-NHEJ factors, has also been shown to participate to translocation formation [13]. Conversely, translocation frequency in human cells is not affected by loss of alt-NHEJ components LIG3 or CtIP [18]. In the absence of active c-NHEJ, the repertoire of breakpoint junctions is substantially modified with the appearance of numerous long deletions and the presence of longer microhomologies, consistent with a switch from c-NHEJ to alt-NHEJ (MMEJ). This conclusion is supported by the drastic reduction in translocation frequency when both c-NHEJ and alt-NHEJ components are lost.

Paired Cas9 nickases (nCas9) have also been used to induce translocations [18]. In this case, two gRNAs directed to opposite strands of each chromosome are used to generate two nicks which can be converted to DSBs with 5′ overhangs of ~40 bp that join to form translocations. At breakpoint junctions, deletions are substantially longer than found with unmodified Cas9, involving loss of sequences from the overhangs. The portions of the overhangs that are preserved are filled by DNA synthesis, which leads to duplications of sequences at breakpoint junctions. Microhomology is also increased, consistent with annealing of bases within the overhangs. Paired nickase-generated translocations are largely dependent on LIG4, demonstrating that c-NHEJ can give rise to junctions with microhomology.

Other groups have implicated PARP1, a reported alt-NHEJ component [1], in translocation formation in human cells. PARP1 knockdown and PARP1 inhibition (olaparib) have both been shown to reduce translocations [49, 63], while another report has shown that PARP1 overexpression can increase translocations in some, but not all, cell lines [55]. Translocations induced by irradiation have also been studied. Interestingly, CtIP has been shown to affect translocation formation in G1-irradiated cells at late time points after irradiation, but not in G2-irradiated cells [2, 3]. The authors suggest that these events arise from a subtype of c-NHEJ they called resection-dependent c-NHEJ, which is dependent on Artemis, DNA-PK and the exonuclease activity of MRE11. As IR induces complex DSBs which probably need a step of maturation before joining, we can wonder what the contribution of resection-dependent c-NHEJ is in the formation of nuclease-induced translocations where "cleaner" DSBs are induced. In conclusion, c-NHEJ is directly implicated in translocation formation in human cells, although a small contribution of alt-NHEJ cannot be excluded (Fig. 2.2). However, when c-NHEJ is impaired, alt-NHEJ becomes critical. It should be

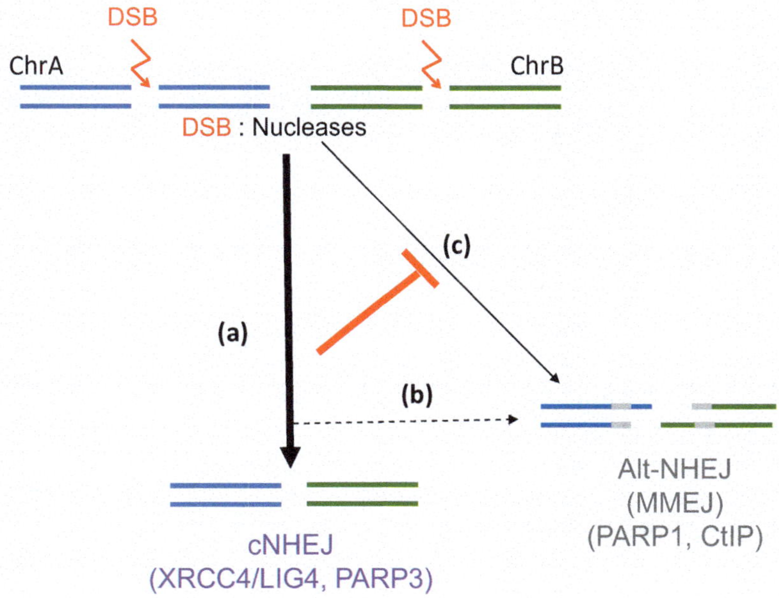

Fig. 2.2 Translocation mechanisms in human with programmable nucleases. Chromosomal translocations in human cells are principally formed by c-NHEJ (**a**) although a small contribution of alt-NHEJ cannot be excluded (**b**). In absence of active c-NHEJ, the contribution of alt-NHEJ become critical (**c**)

noted that the absence of NHEJ components can leave spontaneously arising DSBs unrepaired; the increased frequency of breakage genome wide may then promote translocations.

2.9 How Model Systems Recapitulate Patient Breakpoints

Numerous studies have reported breakpoint junctions for various translocations found in tumors. Oftentimes, only the junction sequences for the oncogenic translocation have been reported, although several publications include both junctions from the reciprocal (balanced) translocation. Concerning deletions, while most of the breakpoints are accompanied by deletions, the median deletion length remains short (e.g., 1 bp, [37]; 5 bp, [40]; 14 bp, [19]). Larger deletions (>1000 bp) are also observed at lower frequency, arising either by resection or possibly as the result of several breaks. Notably, perfect joining of ends has also been reported (e.g., 37% of junctions [37]). Microhomologies have been observed in 20–40% of translocation breakpoint junctions from tumors [34, 37, 59]. The definition of microhomology can differ, however, as authors have sometimes considered there to be microhomologies when short repeated sequences are found at insertions which do not exactly correspond to the original two breakpoint sequences [34, 37], making it difficult to compare various studies. Part of these insertions can arise from a template switching mechanism using microhomology to prime synthesis at the insertion site.

ZFN-induction of the common Ewing translocation has been shown to give rise to breakpoint junctions that fully recapitulate those found in patient tumor cells with a comparable proportion of each type of junction (deletions, insertions, microhomology) [38] [66]. More complex junctions, albeit happening to a lesser extent, can also be recovered in this model and are likely to arise by similar repair mechanisms as in tumor cells from patients. In some cases, a plausible mechanism for their formation is replication primed by one of the DNA ends using microhomology, reminiscent of the template switching mechanism described in patient cells [34, 37]. Of note the use of nCas9 provides more flexibility in DNA end structures, potentially leading to more complex rearrangements found in certain type of tumors, e.g., duplications [18, 41].

In summary, the induction of chromosomal translocations in human cells with programmable nucleases provides a relevant model for deciphering repair mechanisms leading to this type of genome rearrangement and holds promise for deciphering the early events leading to oncogenesis.

Acknowledgments MSK research is supported by NIH/NCI Cancer Center support grant P30 CA008748. This work was supported in part by an Alex's Lemonade Stand Innovation Award and NIH R01CA185660 and R35GM118175 (M.J.). E.B. research is supported by ANR-12-JSV6-0005, the Canceropole IdF, the Institut National du Cancer and la Ligue Contre le Cancer (Equipe de Villartay, Labelisée La Ligue Contre Le Cancer).

References

1. Audebert M, Salles B, Calsou P (2004) Involvement of poly(ADP-ribose) polymerase-1 and XRCC1/DNA ligase III in an alternative route for DNA double-strand breaks rejoining. J Biol Chem 279:55117–55126
2. Barton O, Naumann SC, Diemer-Biehs R, Kunzel J, Steinlage M, Conrad S, Makharashvili N, Wang J, Feng L, Lopez BS et al (2014) Polo-like kinase 3 regulates CtIP during DNA double-strand break repair in G1. J Cell Biol 206:877–894
3. Biehs R, Steinlage M, Barton O, Juhasz S, Kunzel J, Spies J, Shibata A, Jeggo PA, Lobrich M (2017) DNA double-strand break resection occurs during non-homologous end joining in G1 but is distinct from resection during homologous recombination. Mol Cell 65(671–684):e675
4. Blasco RB, Karaca E, Ambrogio C, Cheong TC, Karayol E, Minero VG, Voena C, Chiarle R (2014) Simple and rapid in vivo generation of chromosomal rearrangements using CRISPR/Cas9 technology. Cell Rep 9:1219–1227
5. Breese EH, Buechele C, Dawson C, Cleary ML, Porteus MH (2015) Use of genome engineering to create patient specific MLL translocations in primary human hematopoietic stem and progenitor cells. PLoS One 10:e0136644

6. Brunet E, Simsek D, Tomishima M, DeKelver R, Choi VM, Gregory P, Urnov F, Weinstock DM, Jasin M (2009) Chromosomal translocations induced at specified loci in human stem cells. Proc Natl Acad Sci U S A 106:10620–10625
7. Buechele C, Breese EH, Schneidawind D, Lin CH, Jeong J, Duque-Afonso J, Wong SH, Smith KS, Negrin RS, Porteus M et al (2015) MLL leukemia induction by genome editing of human CD34+ hematopoietic cells. Blood 126:1683–1694
8. Chapman JR, Barral P, Vannier JB, Borel V, Steger M, Tomas-Loba A, Sartori AA, Adams IR, Batista FD, Boulton SJ (2013) RIF1 is essential for 53BP1-dependent nonhomologous end joining and suppression of DNA double-strand break resection. Mol Cell 49:858–871
9. Chen X, Li M, Feng X, Guang S (2015) Targeted chromosomal translocations and essential gene knockout using CRISPR/Cas9 technology in Caenorhabditis elegans. Genetics 201:1295–1306
10. Choi PS, Meyerson M (2014) Targeted genomic rearrangements using CRISPR/Cas technology. Nat Commun 5:3728
11. Colleaux L, d'Auriol L, Gailbert F, Dujon B (1988) Recognition and cleavage site of the intron-encoded *omega* transposase. Proc Natl Acad Sci U S A 85:6022–6026
12. Costantino L, Sotiriou SK, Rantala JK, Magin S, Mladenov E, Helleday T, Haber JE, Iliakis G, Kallioniemi OP, Halazonetis TD (2014) Break-induced replication repair of damaged forks induces genomic duplications in human cells. Science 343:88–91
13. Day TA, Layer JV, Cleary JP, Guha S, Stevenson KE, Tivey T, Kim S, Schinzel AC, Izzo F, Doench J et al (2017) PARP3 is a promoter of chromosomal rearrangements and limits G4 DNA. Nat Commun 8:15110
14. Doyon Y, Vo TD, Mendel MC, Greenberg SG, Wang J, Xia DF, Miller JC, Urnov FD, Gregory PD, Holmes MC (2011) Enhancing zinc-finger-nuclease activity with improved obligate heterodimeric architectures. Nat Methods 8:74–79
15. Elliott B, Richardson C, Jasin M (2005) Chromosomal translocation mechanisms at intronic alu elements in mammalian cells. Mol Cell 17:885–894
16. Elliott B, Richardson C, Winderbaum J, Nickoloff JA, Jasin M (1998) Gene conversion tracts from double-strand break repair in mammalian cells. Mol Cell Biol 18:93–101
17. Ferguson DO, Alt FW (2001) DNA double strand break repair and chromosomal translocation: lessons from animal models. Oncogene 20:5572–5579
18. Ghezraoui H, Piganeau M, Renouf B, Renaud JB, Sallmyr A, Ruis B, Oh S, Tomkinson AE, Hendrickson EA, Giovannangeli C et al (2014) Chromosomal translocations in human cells are generated by canonical nonhomologous end-joining. Mol Cell 55:829–842
19. Gillert E, Leis T, Repp R, Reichel M, Hosch A, Breitenlohner I, Angermuller S, Borkhardt A, Harbott J, Lampert F et al (1999) A DNA damage repair mechanism is involved in the origin of chromosomal translocations t(4;11) in primary leukemic cells. Oncogene 18:4663–4671
20. Harris S, Rudnicki KS, Haber JE (1993) Gene conversions and crossing over during homologous and homeologous ectopic recombination in Saccharomyces cerevisiae. Genetics 135:5–16
21. Jasin M, Haber JE (2016) The democratization of gene editing: insights from site-specific cleavage and double-strand break repair. DNA Repair (Amst) 44:6–16
22. Jasin M, Rothstein R (2013) Repair of strand breaks by homologous recombination. Cold Spring Harb Perspect Biol 5:a012740
23. Jiang J, Zhang L, Zhou X, Chen X, Huang G, Li F, Wang R, Wu N, Yan Y, Tong C et al (2016) Induction of site-specific chromosomal translocations in embryonic stem cells by CRISPR/Cas9. Sci Rep 6:21918
24. Jinek M, Chylinski K, Fonfara I, Hauer M, Doudna JA, Charpentier E (2012) A programmable dual-RNA-guided DNA endonuclease in adaptive bacterial immunity. Science 337:816–821
25. Johnson RD, Jasin M (2000) Sister chromatid gene conversion is a prominent double-strand break repair pathway in mammalian cells. EMBO J 19:3398–3407
26. Kass EM, Jasin M (2010) Collaboration and competition between DNA double-strand break repair pathways. FEBS Lett 584:3703–3708
27. Kim S, Peterson SE, Jasin M, Keeney S (2016) Mechanisms of germ line genome instability. Semin Cell Dev Biol 54:177–187
28. Kim YG, Cha J, Chandrasegaran S (1996) Hybrid restriction enzymes: zinc finger fusions to Fok I cleavage domain. Proc Natl Acad Sci U S A 93:1156–1160
29. Lagutina IV, Valentine V, Picchione F, Harwood F, Valentine MB, Villarejo-Balcells B, Carvajal JJ, Grosveld GC (2015) Modeling of the human alveolar rhabdomyosarcoma Pax3-Foxo1 chromosome translocation in mouse myoblasts using CRISPR-Cas9 nuclease. PLoS Genet 11:e1004951
30. LaRocque JR, Stark JM, Oh J, Bojilova E, Yusa K, Horie K, Takeda J, Jasin M (2011) Interhomolog recombination and loss of heterozygosity in wild-type and Bloom syndrome helicase (BLM)-deficient mammalian cells. Proc Natl Acad Sci U S A 108:11971–11976
31. Maddalo D, Manchado E, Concepcion CP, Bonetti C, Vidigal JA, Han YC, Ogrodowski P, Crippa A, Rekhtman N, de Stanchina E et al (2014) In vivo engineering of oncogenic chromosomal rearrangements with the CRISPR/Cas9 system. Nature 516:423–427
32. Mani RS, Chinnaiyan AM (2010) Triggers for genomic rearrangements: insights into genomic, cellular and environmental influences. Nat Rev Genet 11:819–829
33. Mateos-Gomez PA, Gong F, Nair N, Miller KM, Lazzerini-Denchi E, Sfeir A (2015) Mammalian polymerase theta promotes alternative NHEJ and suppresses recombination. Nature 518:254–257

34. Mattarucchi E, Guerini V, Rambaldi A, Campiotti L, Venco A, Pasquali F, Lo Curto F, Porta G (2008) Microhomologies and interspersed repeat elements at genomic breakpoints in chronic myeloid leukemia. Genes Chromosomes Cancer 47:625–632
35. Miller JC, Tan S, Qiao G, Barlow KA, Wang J, Xia DF, Meng X, Paschon DE, Leung E, Hinkley SJ et al (2011) A TALE nuclease architecture for efficient genome editing. Nat Biotechnol 29:143–148
36. Moynahan ME, Jasin M (2010) Mitotic homologous recombination maintains genomic stability and suppresses tumorigenesis. Nat Rev Mol Cell Biol 11:196–207
37. Nilsson D, Pettersson M, Gustavsson P, Forster A, Hofmeister W, Wincent J, Zachariadis V, Anderlid BM, Nordgren A, Makitie O et al (2017) Whole-genome sequencing of cytogenetically balanced chromosome translocations identifies potentially pathological gene disruptions and highlights the importance of microhomology in the mechanism of formation. Hum Mutat 38:180–192
38. Piganeau M, Ghezraoui H, De Cian A, Guittat L, Tomishima M, Perrouault L, Rene O, Katibah GE, Zhang L, Holmes MC et al (2013) Cancer translocations in human cells induced by zinc finger and TALE nucleases. Genome Res 23:1182–1193
39. Reimer J, Knoess S, Labuhn M, Charpentier EM, Gohring G, Schlegelberger B, Klusmann JH, Heckl D (2017) CRISPR-Cas9-induced t(11;19)/MLL-ENL translocations initiate leukemia in human hematopoietic progenitor cells in vivo. Haematologica 102:1558
40. Reiter A, Saussele S, Grimwade D, Wiemels JL, Segal MR, Lafage-Pochitaloff M, Walz C, Weisser A, Hochhaus A, Willer A et al (2003) Genomic anatomy of the specific reciprocal translocation t(15;17) in acute promyelocytic leukemia. Genes Chromosomes Cancer 36:175–188
41. Renouf B, Piganeau M, Ghezraoui H, Jasin M, Brunet E (2014) Creating cancer translocations in human cells using Cas9 DSBs and nCas9 paired nicks. Methods Enzymol 546:251–271
42. Richardson C, Jasin M (2000a) Coupled homologous and nonhomologous repair of a double-strand break preserves genomic integrity in mammalian cells. Mol Cell Biol 20:9068–9075
43. Richardson C, Jasin M (2000b) Frequent chromosomal translocations induced by DNA double-strand breaks. Nature 405:697–700
44. Richardson C, Moynahan ME, Jasin M (1998) Double-strand break repair by interchromosomal recombination: suppression of chromosomal translocations. Genes Dev 12:3831–3842
45. Scott KP, Mercer DK, Richardson AJ, Melville CM, Glover LA, Flint HJ (2000) Chromosomal integration of the green fluorescent protein gene in lactic acid bacteria and the survival of marked strains in human gut simulations. FEMS Microbiol Lett 182:23–27
46. Sfeir A, Symington LS (2015) Microhomology-mediated end joining: a back-up survival mechanism or dedicated pathway? Trends Biochem Sci 40:701–714
47. Simsek D, Brunet E, Wong SY, Katyal S, Gao Y, McKinnon PJ, Lou J, Zhang L, Li J, Rebar EJ et al (2011) DNA ligase III promotes alternative nonhomologous end-joining during chromosomal translocation formation. PLoS Genet 7:e1002080
48. Simsek D, Jasin M (2010) Alternative end-joining is suppressed by the canonical NHEJ component Xrcc4-ligase IV during chromosomal translocation formation. Nat Struct Mol Biol 17:410–416
49. Soni A, Siemann M, Grabos M, Murmann T, Pantelias GE, Iliakis G (2014) Requirement for Parp-1 and DNA ligases 1 or 3 but not of Xrcc1 in chromosomal translocation formation by backup end joining. Nucleic Acids Res 42:6380–6392
50. Spraggon L, Martelotto LG, Hmeljak J, Hitchman TD, Wang J, Wang L, Slotkin EK, Fan PD, Reis-Filho JS, Ladanyi M (2017) Generation of conditional oncogenic chromosomal translocations using CRISPR-Cas9 genomic editing and homology-directed repair. J Pathol 242:102–112
51. Stark JM, Jasin M (2003) Extensive loss of heterozygosity is suppressed during homologous repair of chromosomal breaks. Mol Cell Biol 23:733–743
52. Stark JM, Pierce AJ, Oh J, Pastink A, Jasin M (2004) Genetic steps of mammalian homologous repair with distinct mutagenic consequences. Mol Cell Biol 24:9305–9316
53. Symington LS, Gautier J (2011) Double-strand break end resection and repair pathway choice. Annu Rev Genet 45:247–271
54. Torres R, Martin MC, Garcia A, Cigudosa JC, Ramirez JC, Rodriguez-Perales S (2014) Engineering human tumour-associated chromosomal translocations with the RNA-guided CRISPR-Cas9 system. Nat Commun 5:3964
55. Torres-Ruiz R, Martinez-Lage M, Martin MC, Garcia A, Bueno C, Castano J, Ramirez JC, Menendez P, Cigudosa JC, Rodriguez-Perales S (2017) Efficient recreation of t(11;22) EWSR1-FLI1+ in human stem cells using CRISPR/Cas9. Stem Cell Reports 8:1408–1420
56. Turner SD, Alexander DR (2005) What have we learnt from mouse models of NPM-ALK-induced lymphomagenesis? Leukemia 19:1128–1134
57. Urnov FD, Miller JC, Lee YL, Beausejour CM, Rock JM, Augustus S, Jamieson AC, Porteus MH, Gregory PD, Holmes MC (2005) Highly efficient endogenous human gene correction using designed zinc-finger nucleases. Nature 435:646–651
58. Vanoli F, Tomishima M, Feng W, Lamribet K, Babin L, Brunet E, Jasin M (2017) CRISPR-Cas9-guided oncogenic chromosomal translocations with conditional fusion protein expression in human mesenchymal cells. Proc Natl Acad Sci U S A 114:3696–3701
59. Weckselblatt B, Hermetz KE, Rudd MK (2015) Unbalanced translocations arise from diverse mutational mechanisms including chromothripsis. Genome Res 25:937–947

60. Weinstock DM, Brunet E, Jasin M (2007) Formation of NHEJ-derived reciprocal chromosomal translocations does not require Ku70. Nat Cell Biol 9:978–981
61. Weinstock DM, Brunet E, Jasin M (2008) Induction of chromosomal translocations in mouse and human cells using site-specific endonucleases. J Natl Cancer Inst Monogr 2008:20–24
62. Weinstock DM, Richardson CA, Elliott B, Jasin M (2006) Modeling oncogenic translocations: distinct roles for double-strand break repair pathways in translocation formation in mammalian cells. DNA Repair (Amst) 5:1065–1074
63. Wray J, Williamson EA, Singh SB, Wu Y, Cogle CR, Weinstock DM, Zhang Y, Lee SH, Zhou D, Shao L et al (2013) PARP1 is required for chromosomal translocations. Blood 121:4359–4365
64. Zhang WW, Lypaczewski P, Matlashewski G (2017) Optimized CRISPR-Cas9 genome editing for Leishmania and its use to target a multigene family, induce chromosomal translocation, and study DNA break repair mechanisms. mSphere 2:e00340
65. Zhang Y, Jasin M (2011) An essential role for CtIP in chromosomal translocation formation through an alternative end-joining pathway. Nat Struct Mol Biol 18:80–84
66. Zucman-Rossi J, Legoix P, Victor JM, Lopez B, Thomas G (1998) Chromosome translocation based on illegitimate recombination in human tumors. Proc Natl Acad Sci U S A 95:11786–11791

Dynamics of Double-Strand Breaks: Implications for the Formation of Chromosome Translocations

3

Henrike Johanna Gothe, Vera Minneker, and Vassilis Roukos

Abstract

Illegitimate joining of chromosome breaks can lead to the formation of chromosome translocations, a catastrophic type of genome rearrangements that often plays key roles in tumorigenesis. Emerging evidence suggests that the mobility of broken DNA loci can be an important determinant in partner search and clustering of individual breaks, events that can influence translocation frequency. We summarize here the recent literature on the mechanisms that regulate chromatin movement, focusing on studies exploring the motion properties of double-strand breaks in the context of chromatin, the functional consequences for DNA repair, and the formation of chromosome fusions.

Keywords

Chromatin motion · Nuclear organization · Double-strand breaks (DSBs) · Repair · Translocations

3.1 Introduction

Approximately a century after the first hypothesis of the German cytologist Theodor Boveri that "malignant tumours might be the consequence of a certain abnormal chromosome constitution" [1], our knowledge regarding the mechanisms that contribute to the formation of chromosome translocations is still limited. Translocations form by the illegitimate joining of chromosome segments that belong to different chromosomes. Translocations are clinically highly relevant as they can be causal to essentially all types of human cancers and account for ≈20% of cancer morbidity [2, 3]. Historically, most of these gene fusions were found in haematological malignancies, such as the prototypical *BCR-ABL* translocation, which was identified as the first fusion to cause cancer [4]. However, a growing number of studies have identified translocations in solid tumours of both mesenchymal and epithelial origin [5]. This is exemplified in prostate cancer, in which ≈50% of cases feature translocations between *transmembrane protease, serine 2* (*TMPRSS2*) and genes encoding ETS transcription factors [6]. Taking advantage of the recurrent nature of specific translocation partners and their association with specific cancer types, translocations are commonly used as diagnostic tools for cancer type stratification, as they can be accurately detected using PCR or cytogenetic methodologies. A translocation can lead to the

Authors Henrike Johanna Gothe and Vera Minneker have equally contributed to this chapter.

H. J. Gothe · V. Minneker · V. Roukos (✉)
Institute of Molecular Biology, Mainz, Germany
e-mail: v.roukos@imb-mainz.de

Y. Zhang (ed.), *Chromosome Translocation*, Advances in Experimental Medicine and Biology 1044,
https://doi.org/10.1007/978-981-13-0593-1_3

formation of chimeric fusion genes or the deregulated expression of oncogenes or tumor suppressor genes. These events may act as drivers of deregulated cell growth in initial stages of cancer and can therefore undergo positive selection during oncogenesis. In addition to this simplistic view, complex chromosome rearrangements have been also identified in patients and are believed to be caused by single catastrophic events during the recently identified phenomenon of chromothrypsis [7].

A textbook view on the events that lead to the formation of a translocation includes three basic, but distinct, steps. First, DNA double-strand breaks (DSBs) must occur in different chromosome loci. Second, these breaks need to be placed in spatial proximity before the physical association and the final joining of the broken partners. Breaks may arise spontaneously during physiological cellular processes, such as DNA replication, or from programmed events during the development of the immune system, such as during V(D)J recombination and immunoglobulin gene class-switch recombination (CSR) [8]. Exogenous conditions, such as oxidative stress, ionizing radiation or the use of chemotherapeutics, are additional sources of DSBs. Whilst the majority of these breaks are repaired *in cis* by cellular DNA repair mechanisms, persistent breaks may physically synapse within the mammalian cell nucleus, increasing the possibility of getting mis-repaired with other chromosome breaks *in trans,* leading to the formation of translocations. The development of novel experimental tools that allow the tracking of individual breaks within the nucleus has recently enabled the investigation of the spatio-temporal regulation of these events in the mammalian cell nucleus. Here we discuss the recent findings on how intact and broken chromatin moves, how breaks find each other, cluster and synapse within the nucleus, and how these events may have functional consequences for DNA repair and the formation of chromosome fusions.

3.2 Chromatin Motion

The early view of chromatin as a static entity of DNA wrapped around histone octamers has been challenged over the years. It is now accepted that continuous changes in nucleosome composition and occupancy, together with changes in post-translational modifications of histones and of the higher order chromatin structure ensure that chromatin is highly dynamic [9, 10]. Seminal work to first demonstrate the highly dynamic nature of chromatin was performed 20 years ago in Sedat's lab, showing that chromatin loci, marked by arrays of fluorescent proteins, move randomly within constraint areas in *S. cerevisiae* [11]. The use of LacO/LacR and TetO/TetR [12–14] operator/repressor system to tag and visualize individual chromosome loci independent of nuclear movement, has allowed the tracking of any chromosome loci of interest in live cells. These experiments revealed that, independently of the organism used to perform the experiments (bacteria, *S. cerevisiae, D. melanogaster*, mammalian cells), the diffusion coefficient of chromatin movement ranges from 10^{-4} to 10^{-3} $\mu m^2/sec$ [11, 15–20]. While early studies suggested that chromatin movement appears to be a constrained random walk [11], it is now clear that genomic loci movement does not fully recapitulate Brownian random walk as constraints appear to restrict chromatin movement [20, 21]. It has been demonstrated in both yeast and mammalian cells that chromatin motion is dependent on ATP levels [11, 15, 17] and depends disproportionately on temperature [18, 20]. Early observations also have shown that chromosome loci mobility changes during the cell cycle and during development, raising the intriguing possibility of a regulated process. In cultured *Drosophila* spermatocytes for example, higher mobility was observed early in differentiation compared to mature spermatocytes during premeiotic development [16]. Experiments using chimaeric versions of core histones fused to photoactivatable or to photobleachable proteins to mark chromatin territories have shown similar mobilities from middle G1 to late G2 [22, 23]. However, a two fold increase in mobility of the chromosome domain was observed in early G1 phase [23]. In support of a cell cycle-dependent chromosome mobility, increased mobility of a fluorescently tagged array has been reported in early G1 in human cells [24] and tagged loci in yeast have shown decreased motion properties during replication [17]. Along the same lines, tracking of decondensed chromatin domains in early S phase revealed higher mobility than condensed heterochromatic domains during mid or late S phase in human cells [25, 26]. Taken together, these observations point to a higher mobility of intact chromosome regions in early G1 phase, which is decreased as cells progress through S phase.

In addition to cell cycle status, changes in transcriptional activity and chromatin state have been associated with changes in the nuclear positioning of a certain locus as a consequence of increased mobility (Fig. 3.1). In mammalian cells it has been shown that tethering of the viral transactivator VP16 to a chromosome site leads to transcription activation and long range directional movement from the nuclear periphery to the interior [27]. In addition, tethering of a VP16 fusion to a non-telomeric locus in yeast increased transcription and mobility [21], whereas targeting to a silent telomeric region repositions it away from the nuclear envelope [28]. Changes in chromatin state can also influence chromatin motion. Decondensation of chromatin without apparent transcriptional changes can mimic the repositioning of different genomic loci during differentiation, indicating that nuclear reorganization is driven by chromatin remodeling rather than transcription [29]. Moreover, targeting of the nucleosome remodeler INO80 to a chromosome locus in *S. cerevisiae* or INO80-dependent eviction of nucleosomes at an endogenous locus, both resulted in an increased movement of the locus [21, 30], arguing that nucleosome remodeling, at least at this specific locus, correlates with locus mobility. Additional factors that influence chromosome dynamics have been recently identified. Association of telomeres with the nuclear envelope and tethering of centromeres to the spindle pole body results in a constrained mobility in *S. cerevisiae* [28, 31, 32]. Likewise, in mammalian cells association of chromosome loci with the nuclear periphery or the nucleolus limits their mobility [15, 22]. Overall, the mobility of a chromatin locus in the interphase nucleus can be considered as an ATP-dependent, non-directed, constrained motion that depends on physical and biological parameters such as the spatial nuclear positioning and chromatin remodeling activities (Fig. 3.1).

3.3 Double-Strand Break Dynamics: Lessons from Yeast and Mammalian Cells

In response to DNA damage, cells have evolved sophisticated response pathways, collectively called the DNA damage response (DDR), which trigger the coordinated recruitment of repair factors at the sites of damage, activate checkpoint pathways to reversibly halt the cell cycle progression and to elicit the repair of DNA lesions. In the presence of DSBs, the local accumulation of repair factors at the sites of breaks results in the formation of microscopically detectable discernible DNA repair foci [33]. Time-lapse microscopy experiments of various DNA repair foci upon different types of DSB-induced DNA damage [25, 34–36] and DNA tagging experiments using the LacO/TetO operator repressor system in the vicinity of endonuclease-induced breaks [17, 19, 37, 38], have been primarily used to probe the motion properties of DSBs. In both yeast and mammalian cells, upon the induction of a single DSB, both intrachromosomal ends remain physically tethered [38–40]. In mammalian cells, both DSBs marked by repair foci and intact chromosome loci undergo similarly constraint motion with a mean squared displacement of ≈1 $\mu m^2 h^{-1}$ [35, 41]. Comparable motion properties have been reported upon challenging the cells with ultrasoft X-rays [36] and when endonuclease induced DSBs were tracked in time by tagging chromosome loci with LacO/LacR [38]. In contrast, large scale DNA damage induced by

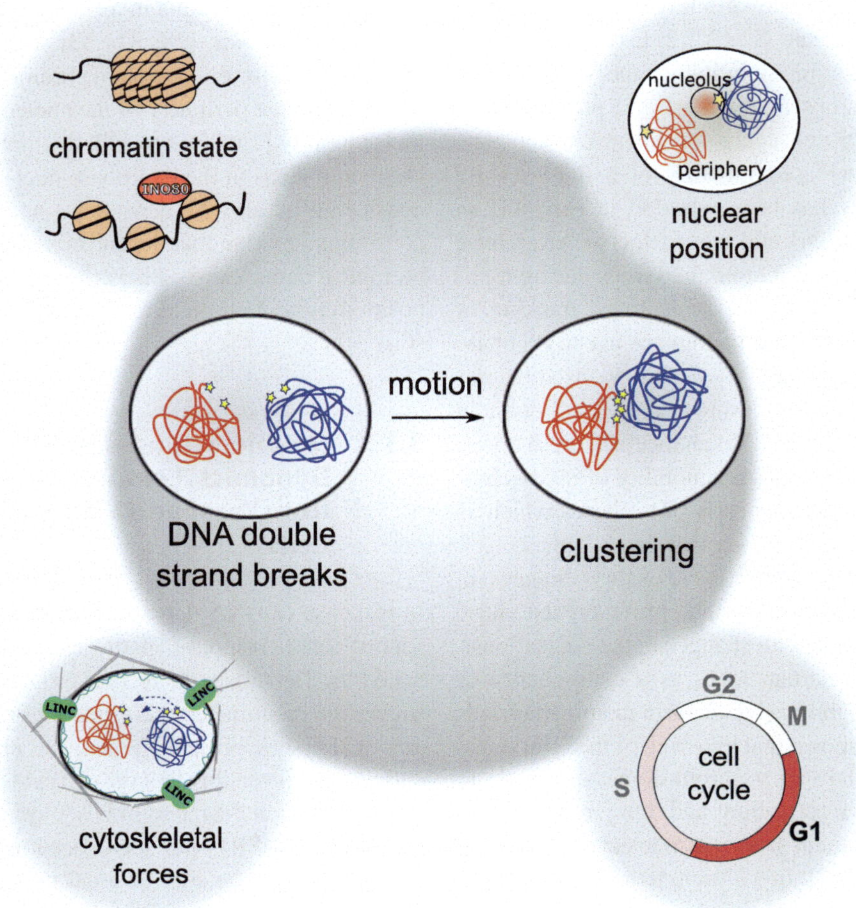

Fig. 3.1 Factors reported to affect chromatin motion may influence clustering and synapsis of DNA double-strand breaks. Cell cycle stage, chromatin status, nuclear positioning and cytoskeleton-generated mechanical forces transferred to the nucleus are all factors that influence the motion of chromosome loci. Chromosome breaks with increased mobility may exhibit a higher probability of clustering, which positively contributes to the formation of translocations

α-particles resulted in extensive motion and clustering of the damaged chromatin domains [34], while increased mobility of DSBs compared to intact loci was observed when 53BP1 foci, which form by the accumulation of 53BP1 molecules at the damaged sites, were tracked in human cells upon ionizing irradiation [25]. However, analysis of motion dynamics of endonuclease ISceI-induced breaks or the corresponding intact chromosome loci demonstrated similar motion properties of broken and intact chromatin in mammalian cells [19] in accordance with previous findings [35, 41]. In yeast cells, induction of DSBs by the endonuclease ISceI leads to an increased motion of undamaged chromosomes and the broken site itself [18, 37, 42], although this does not take place early upon the induction of the breakage [43] or upon induction of different types of breaks such as spontaneous breaks or breaks arising from protein-DNA cross-links [18, 37]. It is therefore intriguing that the increased mobility of DSBs in yeast is associated only with persistent, endonuclease-induced breaks.

It has been suggested that the ability of the DSBs to explore larger volumes of the yeast nucleus facilitates homology search, synapsis and repair and is therefore intriguingly dependent on factors involved in homologous recombination [37, 42]. In yeast, a possible scenario to explain the increased mobility of DSBs has been demonstrated by a recent study showing that disruption of the chromosome anchorage to the spindle pole body is a key event governing the increase in DSB mobility [44]. Moreover, in addition to how the DNA damage was induced, the genomic and chromatin context where breaks have been induced may directly influence the mobility properties (Fig. 3.1). In support of this idea, distinct motion properties have been reported between sub-telomeric and more internal DSBs [45], between breaks introduced in active genes or intergenic loci [46] or breaks within heterochromatin and other chromatin domains [47–49]. In *S. cerevisiae*, breaks within the ribosomal DNA (rDNA) locus led to a transient relocalization to perinucleolar regions in a process that is believed to suppress rDNA hyper-recombination [50]. In a similar manner, in *D. melanogaster*, DSBs in heterochromatic domains rapidly accumulate early markers of DDR and initiate the first steps of HR, followed by their relocalization outside of the domain, apparently to avoid inaccurate recombination [47]. In both mammalian and yeast cells, phosphorylation of H2AX (γH2AX) is found at the periphery of heterochromatic regions, further indicating movement of breaks to the periphery [48, 51]. Importantly, relocation of breaks does not take place in all types of heterochromatin. Breaks within the heterochromatic regions that commonly associate with the nuclear lamina do not relocate [52]. Moreover, DSBs induced by CRISPR-Cas9 in pericentric heterochromatin are positionally stable in G1 and recruit the non-homologous end-joining (NHEJ) repair pathway protein Ku80, while DSBs in S/G2 relocate to the periphery of the heterochromatic domain to be bound by the homologous recombination (HR) repair pathway protein RAD51 [49]. These observations suggest a model in which the commitment to specific DNA repair pathways could regulate DSB mobility and position [49].

Regardless of the observed differences in mobility of DSBs between mammalian cells and yeast, it is important to consider the functional consequences of DSB mobility in both organisms. While 1 μm displacement of a certain locus in a yeast cell is sufficient to roam the entire nucleus, a similar distance travelled in an average 10 μm in diameter mammalian cell nucleus is considered limited [53]. Long range motion of DSBs in mammalian cells is possible, as approximately 5% of DSBs that contribute to translocations are able to move up to 5 μm [19]. In mammalian cells, uncapped telomeres and ionizing radiation-induced repair foci exhibit mobility which is dependent on the presence of the repair factor 53BP1 [54]. These findings together with the observed dependency on HR factors for DSB-mobility in yeast [37, 42, 55], point to a role of major DSB repair factors in mobility of breaks, by yet unidentified mechanisms.

3.4 The Role of Cytoskeletal Components in Motion of Intact Chromosome Loci and DSBs: Random or Directed Motion?

A key open question is whether the mobility of intact or broken chromatin represents an active and directed process, or occurs by passive diffusion. So far, only a handful of studies have demonstrated a directed mobility of a chromatin locus within the nuclear space. Almost a decade ago, Belmont's lab demonstrated migration of an interphase chromosome site from the nuclear periphery to the interior upon targeting a transcriptional activator to this site [27]. This directed mobility was dependent on actin and nuclear myosin I [27]. In similar manner, an actin-dependent motion along a linear trajectory was observed for the U2 small nuclear RNA gene locus towards Cajal bodies in human cells [56]. More recently, directed movement over micrometers was observed for damaged subtelomeric regions, which are repaired or maintained by a recombination-dependent ALT pathway in human cells [45]. It remains to be elucidated, whether the observed actin and myosin-dependent motion is due to a directed, filament-driven mechanism or is merely a consequence of interference with chromatin remodeling activities that utilize actin as a cofactor. Additional studies have demonstrated that cellular cytoskeletal components influence chromatin motion. In the pioneering study from Sedat's lab microtubules' depolymerization increased chromatin movement in *S. cerevisiae* [11]. More recent work in mammalian cells showed that the mobility of uncapped telomeres was reduced upon treatment with microtubule poisons in a reversible manner [57]. A similar reduction in mobility was observed when the LINC complex, which bridges the cytoplasmic cytoskeleton and inner nuclear membrane, was absent, suggesting that the cytoskeleton to nucleus link is required for telomere motion [57]. Importantly, stabilization of microtubules by taxol also led to a decrease in the mobility of ionizing radiation induced repair foci, suggesting that forces transmitted to chromosomes through microtubules influence the motion properties of DSBs [57]. Similarly, in human cells clustering of DSBs seem to require the LINC complex protein SUN2 (but not SUN1) and the FMN2 actin organizer [58]. Similar experiments in yeast cells demonstrated that both cytoplasmic and nuclear actin is required for subtelomeric motion, probably through its action in chromatin-remodeling complexes [30]. In conclusion, these studies suggest that nuclear chromosome motion is influenced by mechanical forces transmitted from cytoplasm to the nucleus (Fig. 3.1). However, additional work is required to fully understand the molecular determinants of their contribution to motion.

3.5 Partner Search, Clustering and Synapsis of DSBs in the Nucleus

Translocations form by the illegitimate fusion of different chromosome ends. An open question is how chromosome ends find their chromosome partners and whether cellular factors promote or inhibit synapsis within the three-dimensional nuclear space. High-throughput sequencing and imaging experiments that have recently been performed to track individual breaks as they synapse with other breaks within the cell nucleus are starting to shed light on this question [19, 42, 58–60]. In *S. cerevisiae,* when the sister chromatid is not available, broken chromosomes search for homologous templates to use for HR in a process that is driven by Rad51 [61–63]. In human cells, it has been proposed that homologous chromosomes also synapse in a process that requires ongoing transcription and ATM activity [46]. In a similar manner, during a recombination-dependent telomere maintenance pathway known as alternative lengthening of telomeres (ALT), DSBs at telomeres trigger long-range movement and clustering between chromosome ends, which is believed to drive homology-directed maintenance of telomere repeats [45]. Interestingly, the damaged telomeres displayed rapid directional movement and association with other telomeres over long-range distances in a Rad51-dependent manner [45].

Factors that influence the clustering and synapsis of DSBs are major candidates as regulators of translocation frequency (Fig. 3.2). In *S. cerevisiae*, multiple DSBs coalesce in common repair centers, in a process that has been suggested to increase repair efficiency and/or to assist the synapsis of DSBs [59]. In mammalian cells, clustering of DSBs has only been observed sporadically, when several DSBs per cell were tracked over time [35, 41]. When up to four individual DSBs were induced concomitantly in mammalian cells, the individual repair foci remain separate in the majority of the cells [19, 38]. During the formation of a translocation however, individual breaks coalesce in a common repair focus, which then resolves over time as the chromosome ends are repaired to form the illegitimate fusion [19]. When multiple (≈100) DSBs were induced in mammalian cells by the use of the endonuclease AsiSI, individual breaks, marked by γH2AX, were shown to coalesce in common repair centers in a process that is dependent on ATM kinase activity [60]. One possibility is that clustering of DSBs is merely a consequence of the random

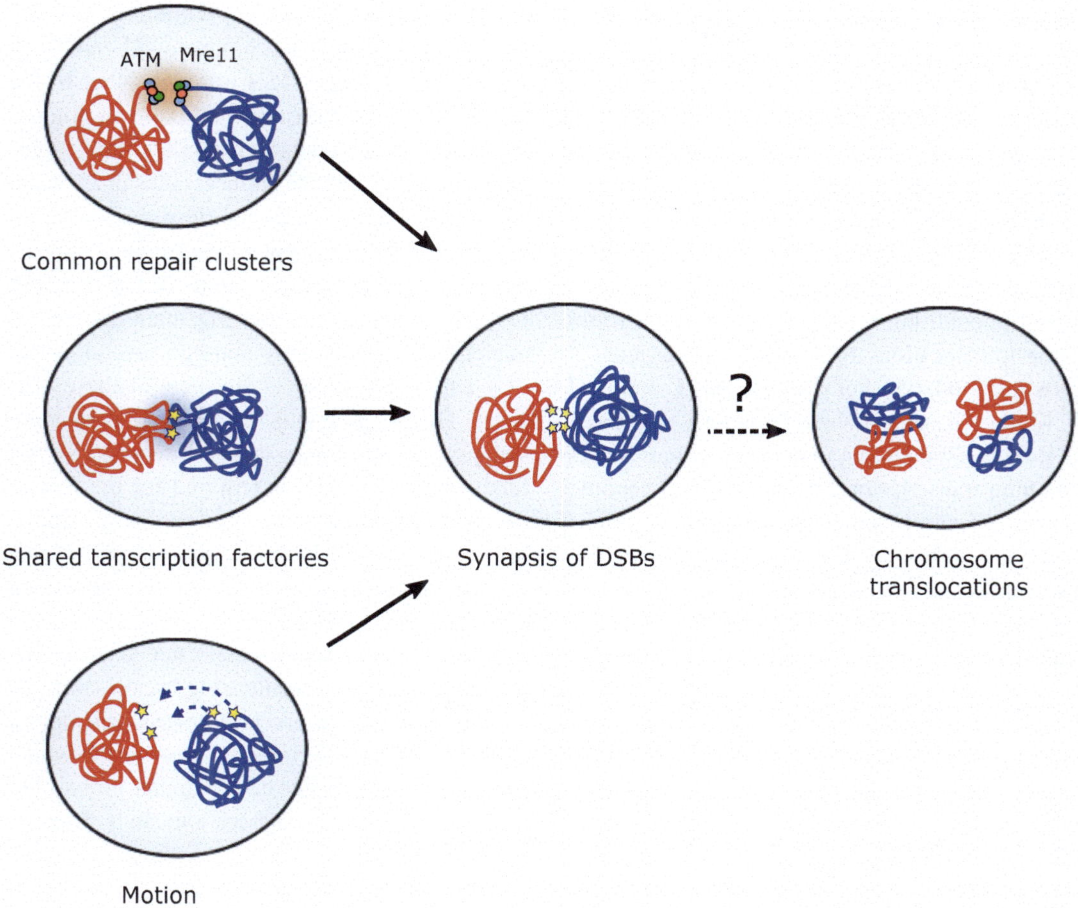

Fig. 3.2 The current view of processes known to influence the clustering of DSBs. Active DDR signaling and repair factor accumulation at damaged sites may contribute to clustering and synapsis of breaks in common repair centers. In a similar fashion, active gene transcription promotes clustering of genes in shared transcription factories. Factors that increase DSB mobility (Fig. 3.1) may also positively contribute to clustering of DSBs and the formation of chromosome translocations

motion of breaks and, therefore, a higher number of induced breaks increases the probability of coalescing in common centers. If not random, congregation of multiple DSBs in repair centers must be tightly controlled as the spatial proximity of the breaks could facilitate illegitimate fusions to form tumorigenic translocations.

Furthermore, the repair protein Mre11 is a well-established factor that mediates the synapsis of DSBs and facilitates the formation of translocations. Using a mammalian cell based system in which individual DSBs can be tracked by high-throughput microscopy and automated image-analysis, Mre11 inhibition by treatment with a small molecule or by knockdown, leads to a substantial decrease in synapsis of DSBs [19]. In line with a role of Mre11 in synapsis of DSBs, clustering of α-particle-induced breaks is altered in cells from patients with ataxia telangiectasia-like disorder (ATLD), which show reduced levels of Mre11 [34]. Moreover, recent studies have also confirmed the role of the Mre11-Rad-Nbs1 (MRN) complex in the synapsis of DSBs, as knockdown of MRE11 or NBS1 led to a decrease in the number of cells with DSB clusters, as shown by γH2AX staining of AsiSI-induced breaks [58]. This elegant study combined AsiSI-induced breakage with high-throughput chromosome conformation capture assay (capture Hi-C) to show that DSBs in human cells indeed cluster, but only when they are induced within transcriptionally active genes [58]. The observed clustering of damaged genes occurred primarily during the G1 phase of the cell cycle and coincided with delayed repair [58]. The correlation of transcriptional activity with clustering of breaks is very intriguing since analogous to the repair center model, coalescing of multiple transcribed genes in common transcription factories has been proposed to contribute to the formation of chromosome translocations by keeping active genes in proximity [64, 65]. For example, MYC and IGH, IGK and IGL genes which frequently translocate in Burkitt's lymphoma in many instances share a common transcription factory [65] and genome-wide studies in lymphocytes have mapped translocation breakpoints near transcription start sites [66, 67]. In an analogous fashion, early replicating fragile sites which are frequently found in translocations that lead to large B cell lymphoma are enriched in genome regions of high transcriptional activity [68]. Similar correlations between high transcriptional activity and the probability to form translocations have been shown for prostate cancer- and anaplastic cell lymphoma-specific translocations [69–71]. Taken together, these observations suggest that DNA repair proteins and active transcription influence the clustering of DSBs and therefore may affect the probability to form translocations by retaining DSBs in close proximity.

3.6 Dynamics of DSBs and Translocation Formation

Since the mobility of DSBs influences the probability of breaks to synapse, altered DSB-dynamics may influence the frequencies by which oncogenic chromosome translocations form. Using an experimental system to track individual IScel induced-DSBs as they move within the 3D mammalian cell nucleus, a recent study has shed light on the timing and sequence of the events leading to the formation of translocations in individual cells [19]. Time-lapse microscopy and single-cell tracking of individual DSBs shows that upon their formation, DSBs undergo a non-directional random motion [19]. Interestingly, both intra-chromosomal ends generated upon a single DSB move together to the area of synapsis with other DSBs. This finding suggests that the separation and illegitimate joining of the chromosome partners takes place when chromosome partners are in close proximity, an observation that can explain the frequent appearance of reciprocal translocations. When different DSBs synapse within the 3D mammalian cell nucleus, the lesions may undergo several cycles of synapsis and dissociation. Few of these synapsed DBSs are finally engaged to permanent synapsis, which makes them susceptible for the final joining step that is required for the formation of chromosome translocations. Surprisingly, breaks that are involved in translocations show a faster component in their motion in comparison with non-translocating breaks [19]. The reason for this apparent difference is not known. The higher mobility of translocating breaks could be a consequence of a directed process or simply a consequence of a selection process during which breaks that move faster have a higher probability to synapse and form translocations. In both cases, the intrinsic higher mobility of some breaks compared to others has unknown etiology, and therefore identifying factors that influence DSB motion is essential to uncover molecular mechanisms of translocation biogenesis. An intriguing possibility foresees that inherent differences in the chromatin environment surrounding individual breaks can directly influence their motion dynamics.

3.7 Conclusions

Chromatin mobility is an intrinsic feature and consequence of the highly dynamic nature of chromatin that has profound effects on genome stability and maintenance. Despite the intense investigation, however, our understanding of the effects of chromatin mobility on the formation of oncogenic translocations is still at an early stage. Systematic efforts to understand how DSBs are generated, move and repaired across the genome in the context of chromatin and nuclear architecture would substantially advance our knowledge on the molecular mechanisms that give rise to oncogenic chromosome translocations. Versatile cell-based systems that are able to track individual DSBs in different chromatin domains and report translocation frequency between different partners across the genome are now possible due to recent advances in recombineering technologies [72, 73]. Moreover, integrative analysis of datasets produced by genome-wide methodologies that are able to probe and quantify DSB occurrence and repair across the genome [74–76] and genome-wide translocation capture methodologies [66, 77] will substantially help in this direction. Elucidating basic principles underpinning the formation of chromosome translocations will unravel fundamental features of cancer etiology.

Acknowledgements We would like to apologize to colleagues whose work could not be cited due to space limitations. We would like to thank Dr. Karen Meaburn for critical reading of the manuscript. This work is supported by the "DFG Major Research Instrumentation Programme" (INST 247/845-1 FUGG).

References

1. Boveri T (2008) Concerning the origin of malignant tumours by Theodor Boveri. Trans Annot Henry Harris J Cell Sci 121(Suppl 1):1–84. https://doi.org/10.1242/jcs.025742
2. Mitelman F, Johansson B, Mertens F (2007) The impact of translocations and gene fusions on cancer causation. Nat Rev Cancer 7(4):233–245. https://doi.org/10.1038/nrc2091
3. Roukos V, Misteli T (2014) The biogenesis of chromosome translocations. Nat Cell Biol 16(4):293–300. https://doi.org/10.1038/ncb2941
4. Rowley JD (1973) Letter: a new consistent chromosomal abnormality in chronic myelogenous leukaemia identified by quinacrine fluorescence and Giemsa staining. Nature 243(5405):290–293
5. Mertens F, Johansson B, Fioretos T, Mitelman F (2015) The emerging complexity of gene fusions in cancer. Nat Rev Cancer 15(6):371–381. https://doi.org/10.1038/nrc3947
6. Tomlins SA, Rhodes DR, Perner S, Dhanasekaran SM, Mehra R, Sun XW, Varambally S, Cao X, Tchinda J, Kuefer R, Lee C, Montie JE, Shah RB, Pienta KJ, Rubin MA, Chinnaiyan AM (2005) Recurrent fusion of TMPRSS2 and ETS transcription factor genes in prostate cancer. Science (New York NY) 310(5748):644–648. https://doi.org/10.1126/science.1117679
7. Stephens PJ, Greenman CD, Fu B, Yang F, Bignell GR, Mudie LJ, Pleasance ED, Lau KW, Beare D, Stebbings LA, McLaren S, Lin ML, McBride DJ, Varela I, Nik-Zainal S, Leroy C, Jia M, Menzies A, Butler AP, Teague JW, Quail MA, Burton J, Swerdlow H, Carter NP, Morsberger LA, Iacobuzio-Donahue C, Follows GA, Green AR, Flanagan AM, Stratton MR, Futreal PA, Campbell PJ (2011) Massive genomic rearrangement acquired in a single catastrophic event during cancer development. Cell 144(1):27–40. https://doi.org/10.1016/j.cell.2010.11.055
8. Alt FW, Zhang Y, Meng FL, Guo C, Schwer B (2013) Mechanisms of programmed DNA lesions and genomic instability in the immune system. Cell 152(3):417–429. https://doi.org/10.1016/j.cell.2013.01.007
9. Campos EI, Reinberg D (2009) Histones: annotating chromatin. Annu Rev Genet 43:559–599. https://doi.org/10.1146/annurev.genet.032608.103928
10. Misteli T (2013) The cell biology of genomes: bringing the double helix to life. Cell 152(6):1209–1212. https://doi.org/10.1016/j.cell.2013.02.048
11. Marshall WF, Straight A, Marko JF, Swedlow J, Dernburg A, Belmont A, Murray AW, Agard DA, Sedat JW (1997) Interphase chromosomes undergo constrained diffusional motion in living cells. Curr Biol CB 7(12):930–939
12. Michaelis C, Ciosk R, Nasmyth K (1997) Cohesins: chromosomal proteins that prevent premature separation of sister chromatids. Cell 91(1):35–45
13. Roukos V, Burgess RC, Misteli T (2014) Generation of cell-based systems to visualize chromosome damage and translocations in living cells. Nat Protoc 9(10):2476–2492. https://doi.org/10.1038/nprot.2014.167
14. Robinett CC, Straight A, Li G, Willhelm C, Sudlow G, Murray A, Belmont AS (1996) In vivo localization of DNA sequences and visualization of large-scale chromatin organization using lac operator/repressor recognition. J Cell Biol 135(6 Pt 2):1685–1700
15. Chubb JR, Boyle S, Perry P, Bickmore WA (2002) Chromatin motion is constrained by association with nuclear compartments in human cells. Curr Biol CB 12(6):439–445
16. Vazquez J, Belmont AS, Sedat JW (2001) Multiple regimes of constrained chromosome motion are regulated in the interphase Drosophila nucleus. Curr Biol CB 11(16):1227–1239
17. Heun P, Laroche T, Shimada K, Furrer P, Gasser SM (2001) Chromosome dynamics in the yeast interphase nucleus. Science (New York, NY) 294(5549):2181–2186. https://doi.org/10.1126/science.1065366
18. Dion V, Kalck V, Seeber A, Schleker T, Gasser SM (2013) Cohesin and the nucleolus constrain the mobil-

ity of spontaneous repair foci. EMBO Rep 14(11):984–991. https://doi.org/10.1038/embor.2013.142
19. Roukos V, Voss TC, Schmidt CK, Lee S, Wangsa D, Misteli T (2013) Spatial dynamics of chromosome translocations in living cells. Science (New York, NY) 341(6146):660–664. https://doi.org/10.1126/science.1237150
20. Weber SC, Spakowitz AJ, Theriot JA (2012) Nonthermal ATP-dependent fluctuations contribute to the in vivo motion of chromosomal loci. Proc Natl Acad Sci U S A 109(19):7338–7343. https://doi.org/10.1073/pnas.1119505109
21. Neumann FR, Dion V, Gehlen LR, Tsai-Pflugfelder M, Schmid R, Taddei A, Gasser SM (2012) Targeted INO80 enhances subnuclear chromatin movement and ectopic homologous recombination. Genes Dev 26(4):369–383. https://doi.org/10.1101/gad.176156.111
22. Wiesmeijer K, Krouwels IM, Tanke HJ, Dirks RW (2008) Chromatin movement visualized with photoactivable GFP-labeled histone H4. Differentiation 76(1):83–90. https://doi.org/10.1111/j.1432-0436.2007.00234.x
23. Walter J, Schermelleh L, Cremer M, Tashiro S, Cremer T (2003) Chromosome order in HeLa cells changes during mitosis and early G1, but is stably maintained during subsequent interphase stages. J Cell Biol 160(5):685–697. https://doi.org/10.1083/jcb.200211103
24. Thomson I, Gilchrist S, Bickmore WA, Chubb JR (2004) The radial positioning of chromatin is not inherited through mitosis but is established de novo in early G1. Curr Biol CB 14(2):166–172
25. Krawczyk PM, Borovski T, Stap J, Cijsouw A, Ten Cate R, Medema JP, Kanaar R, Franken NA, Aten JA (2012) Chromatin mobility is increased at sites of DNA double-strand breaks. J Cell Sci. https://doi.org/10.1242/jcs.089847
26. Pliss A, Malyavantham K, Bhattacharya S, Zeitz M, Berezney R (2009) Chromatin dynamics is correlated with replication timing. Chromosoma 118(4):459–470. https://doi.org/10.1007/s00412-009-0208-6
27. Chuang CH, Carpenter AE, Fuchsova B, Johnson T, de Lanerolle P, Belmont AS (2006) Long-range directional movement of an interphase chromosome site. Curr Biol CB 16(8):825–831. https://doi.org/10.1016/j.cub.2006.03.059
28. Hediger F, Neumann FR, Van Houwe G, Dubrana K, Gasser SM (2002) Live imaging of telomeres: yKu and sir proteins define redundant telomere-anchoring pathways in yeast. Curr Biol CB 12 (24):2076–2089.
29. Therizols P, Illingworth RS, Courilleau C, Boyle S, Wood AJ, Bickmore WA (2014) Chromatin decondensation is sufficient to alter nuclear organization in embryonic stem cells. Science (New York, NY) 346(6214):1238–1242. https://doi.org/10.1126/science.1259587
30. Spichal M, Brion A, Herbert S, Cournac A, Marbouty M, Zimmer C, Koszul R, Fabre E (2016) Evidence for a dual role of actin in regulating chromosome organization and dynamics in yeast. J Cell Sci 129(4):681–692. https://doi.org/10.1242/jcs.175745
31. Zimmer C, Fabre E (2011) Principles of chromosomal organization: lessons from yeast. J Cell Biol 192(5):723–733. https://doi.org/10.1083/jcb.201010058
32. Taddei A, Schober H, Gasser SM (2010) The budding yeast nucleus. Cold Spring Harb Perspect Biol 2(8):a000612. https://doi.org/10.1101/cshperspect.a000612
33. Lukas J, Lukas C, Bartek J (2011) More than just a focus: the chromatin response to DNA damage and its role in genome integrity maintenance. Nat Cell Biol 13(10):1161–1169. https://doi.org/10.1038/ncb2344
34. Aten JA, Stap J, Krawczyk PM, van Oven CH, Hoebe RA, Essers J, Kanaar R (2004) Dynamics of DNA double-strand breaks revealed by clustering of damaged chromosome domains. Science (New York, NY) 303(5654):92–95. https://doi.org/10.1126/science.1088845
35. Jakob B, Splinter J, Durante M, Taucher-Scholz G (2009) Live cell microscopy analysis of radiation-induced DNA double-strand break motion. Proc Natl Acad Sci U S A 106(9):3172–3177. https://doi.org/10.1073/pnas.0810987106
36. Nelms BE, Maser RS, MacKay JF, Lagally MG, Petrini JH (1998) In situ visualization of DNA double-strand break repair in human fibroblasts. Science (New York, NY) 280(5363):590–592
37. Dion V, Kalck V, Horigome C, Towbin BD, Gasser SM (2012) Increased mobility of double-strand breaks requires Mec1, Rad9 and the homologous recombination machinery. Nat Cell Biol 14(5):502–509. https://doi.org/10.1038/ncb2465
38. Soutoglou E, Dorn JF, Sengupta K, Jasin M, Nussenzweig A, Ried T, Danuser G, Misteli T (2007) Positional stability of single double-strand breaks in mammalian cells. Nat Cell Biol 9(6):675–682. https://doi.org/10.1038/ncb1591
39. Kato L, Begum NA, Burroughs AM, Doi T, Kawai J, Daub CO, Kawaguchi T, Matsuda F, Hayashizaki Y, Honjo T (2012) Nonimmunoglobulin target loci of activation-induced cytidine deaminase (AID) share unique features with immunoglobulin genes. Proc Natl Acad Sci U S A 109(7):2479–2484. https://doi.org/10.1073/pnas.1120791109
40. Lobachev K, Vitriol E, Stemple J, Resnick MA, Bloom K (2004) Chromosome fragmentation after induction of a double-strand break is an active process prevented by the RMX repair complex. Curr Biol CB 14(23):2107–2112. https://doi.org/10.1016/j.cub.2004.11.051
41. Kruhlak MJ, Celeste A, Dellaire G, Fernandez-Capetillo O, Muller WG, McNally JG, Bazett-Jones DP, Nussenzweig A (2006) Changes in chromatin structure and mobility in living cells at sites of DNA double-strand breaks. J Cell Biol 172(6):823–834. https://doi.org/10.1083/jcb.200510015
42. Mine-Hattab J, Rothstein R (2012) Increased chromosome mobility facilitates homology search during

recombination. Nat Cell Biol 14(5):510–517. https://doi.org/10.1038/ncb2472

43. Saad H, Gallardo F, Dalvai M, Tanguy-le-Gac N, Lane D, Bystricky K (2014) DNA dynamics during early double-strand break processing revealed by non-intrusive imaging of living cells. PLoS Genet 10(3):e1004187. https://doi.org/10.1371/journal.pgen.1004187
44. Strecker J, Gupta GD, Zhang W, Bashkurov M, Landry MC, Pelletier L, Durocher D (2016) DNA damage signalling targets the kinetochore to promote chromatin mobility. Nat Cell Biol 18(3):281–290. https://doi.org/10.1038/ncb3308
45. Cho NW, Dilley RL, Lampson MA, Greenberg RA (2014) Interchromosomal homology searches drive directional ALT telomere movement and synapsis. Cell 159(1):108–121. https://doi.org/10.1016/j.cell.2014.08.030
46. Gandhi M, Evdokimova VN, TC K, Nikiforova MN, Kelly LM, Stringer JR, Bakkenist CJ, Nikiforov YE (2012) Homologous chromosomes make contact at the sites of double-strand breaks in genes in somatic G0/G1-phase human cells. Proc Natl Acad Sci U S A 109(24):9454–9459. https://doi.org/10.1073/pnas.1205759109
47. Chiolo I, Minoda A, Colmenares SU, Polyzos A, Costes SV, Karpen GH (2011) Double-strand breaks in heterochromatin move outside of a dynamic HP1a domain to complete recombinational repair. Cell 144(5):732–744. https://doi.org/10.1016/j.cell.2011.02.012
48. Jakob B, Splinter J, Conrad S, Voss KO, Zink D, Durante M, Lobrich M, Taucher-Scholz G (2011) DNA double-strand breaks in heterochromatin elicit fast repair protein recruitment, histone H2AX phosphorylation and relocation to euchromatin. Nucleic Acids Res 39(15):6489–6499. https://doi.org/10.1093/nar/gkr230
49. Tsouroula K, Furst A, Rogier M, Heyer V, Maglott-Roth A, Ferrand A, Reina-San-Martin B, Soutoglou E (2016) Temporal and spatial uncoupling of DNA double strand break repair pathways within mammalian heterochromatin. Mol Cell 63(2):293–305. https://doi.org/10.1016/j.molcel.2016.06.002
50. Torres-Rosell J, Sunjevaric I, De Piccoli G, Sacher M, Eckert-Boulet N, Reid R, Jentsch S, Rothstein R, Aragon L, Lisby M (2007) The Smc5-Smc6 complex and SUMO modification of Rad52 regulates recombinational repair at the ribosomal gene locus. Nat Cell Biol 9(8):923–931. https://doi.org/10.1038/ncb1619
51. Kim JA, Kruhlak M, Dotiwala F, Nussenzweig A, Haber JE (2007) Heterochromatin is refractory to gamma-H2AX modification in yeast and mammals. J Cell Biol 178(2):209–218. https://doi.org/10.1083/jcb.200612031
52. Lemaitre C, Grabarz A, Tsouroula K, Andronov L, Furst A, Pankotai T, Heyer V, Rogier M, Attwood KM, Kessler P, Dellaire G, Klaholz B, Reina-San-Martin B, Soutoglou E (2014) Nuclear position dictates DNA repair pathway choice. Genes Dev 28(22):2450–2463. https://doi.org/10.1101/gad.248369.114
53. Misteli T, Soutoglou E (2009) The emerging role of nuclear architecture in DNA repair and genome maintenance. Nat Rev Mol Cell Biol 10(4):243–254. https://doi.org/10.1038/nrm2651
54. Dimitrova N, Chen YC, Spector DL, de Lange T (2008) 53BP1 promotes non-homologous end joining of telomeres by increasing chromatin mobility. Nature 456(7221):524–528. https://doi.org/10.1038/nature07433
55. Seeber A, Dion V, Gasser SM (2013) Checkpoint kinases and the INO80 nucleosome remodeling complex enhance global chromatin mobility in response to DNA damage. Genes Dev 27(18):1999–2008. https://doi.org/10.1101/gad.222992.113
56. Dundr M, Ospina JK, Sung MH, John S, Upender M, Ried T, Hager GL, Matera AG (2007) Actin-dependent intranuclear repositioning of an active gene locus in vivo. J Cell Biol 179(6):1095–1103. https://doi.org/10.1083/jcb.200710058
57. Lottersberger F, Karssemeijer RA, Dimitrova N, de Lange T (2015) 53BP1 and the LINC complex promote microtubule-dependent DSB mobility and DNA repair. Cell 163(4):880–893. https://doi.org/10.1016/j.cell.2015.09.057
58. Aymard F, Aguirrebengoa M, Guillou E, Javierre BM, Bugler B, Arnould C, Rocher V, Iacovoni JS, Biernacka A, Skrzypczak M, Ginalski K, Rowicka M, Fraser P, Legube G (2017) Genome-wide mapping of long-range contacts unveils clustering of DNA double-strand breaks at damaged active genes. Nat Struct Mol Biol 24(4):353–361. https://doi.org/10.1038/nsmb.3387
59. Lisby M, Mortensen UH, Rothstein R (2003) Colocalization of multiple DNA double-strand breaks at a single Rad52 repair Centre. Nat Cell Biol 5(6):572–577. https://doi.org/10.1038/ncb997
60. Caron P, Choudjaye J, Clouaire T, Bugler B, Daburon V, Aguirrebengoa M, Mangeat T, Iacovoni JS, Alvarez-Quilon A, Cortes-Ledesma F, Legube G (2015) Non-redundant functions of ATM and DNA-PKcs in response to DNA double-strand breaks. Cell Rep 13(8):1598–1609. https://doi.org/10.1016/j.celrep.2015.10.024
61. Forget AL, Kowalczykowski SC (2012) Single-molecule imaging of DNA pairing by RecA reveals a three-dimensional homology search. Nature 482(7385):423–427. https://doi.org/10.1038/nature10782
62. Ragunathan K, Liu C, Ha T (2012) RecA filament sliding on DNA facilitates homology search. eLife 1:e00067. https://doi.org/10.7554/eLife.00067
63. Renkawitz J, Lademann CA, Kalocsay M, Jentsch S (2013) Monitoring homology search during DNA double-strand break repair in vivo. Mol Cell 50(2):261–272. https://doi.org/10.1016/j.molcel.2013.02.020

64. Osborne CS, Chakalova L, Brown KE, Carter D, Horton A, Debrand E, Goyenechea B, Mitchell JA, Lopes S, Reik W, Fraser P (2004) Active genes dynamically colocalize to shared sites of ongoing transcription. Nat Genet 36(10):1065–1071. https://doi.org/10.1038/ng1423
65. Osborne CS, Chakalova L, Mitchell JA, Horton A, Wood AL, Bolland DJ, Corcoran AE, Fraser P (2007) Myc dynamically and preferentially relocates to a transcription factory occupied by Igh. PLoS Biol 5(8):e192. https://doi.org/10.1371/journal.pbio.0050192
66. Chiarle R, Zhang Y, Frock RL, Lewis SM, Molinie B, Ho YJ, Myers DR, Choi VW, Compagno M, Malkin DJ, Neuberg D, Monti S, Giallourakis CC, Gostissa M, Alt FW (2011) Genome-wide translocation sequencing reveals mechanisms of chromosome breaks and rearrangements in B cells. Cell 147(1):107–119. https://doi.org/10.1016/j.cell.2011.07.049
67. Klein IA, Resch W, Jankovic M, Oliveira T, Yamane A, Nakahashi H, Di Virgilio M, Bothmer A, Nussenzweig A, Robbiani DF, Casellas R, Nussenzweig MC (2011) Translocation-capture sequencing reveals the extent and nature of chromosomal rearrangements in B lymphocytes. Cell 147(1):95–106. https://doi.org/10.1016/j.cell.2011.07.048
68. Barlow JH, Faryabi RB, Callen E, Wong N, Malhowski A, Chen HT, Gutierrez-Cruz G, Sun HW, McKinnon P, Wright G, Casellas R, Robbiani DF, Staudt L, Fernandez-Capetillo O, Nussenzweig A (2013) Identification of early replicating fragile sites that contribute to genome instability. Cell 152(3):620–632. https://doi.org/10.1016/j.cell.2013.01.006
69. Lin C, Yang L, Tanasa B, Hutt K, Ju BG, Ohgi K, Zhang J, Rose DW, Fu XD, Glass CK, Rosenfeld MG (2009) Nuclear receptor-induced chromosomal proximity and DNA breaks underlie specific translocations in cancer. Cell 139(6):1069–1083. https://doi.org/10.1016/j.cell.2009.11.030
70. Mathas S, Kreher S, Meaburn KJ, Johrens K, Lamprecht B, Assaf C, Sterry W, Kadin ME, Daibata M, Joos S, Hummel M, Stein H, Janz M, Anagnostopoulos I, Schrock E, Misteli T, Dorken B (2009) Gene deregulation and spatial genome reorganization near breakpoints prior to formation of translocations in anaplastic large cell lymphoma. Proc Natl Acad Sci U S A 106(14):5831–5836. https://doi.org/10.1073/pnas.0900912106
71. Roukos V, Mathas S (2015) The origins of ALK translocations. Front Biosci 7:260–268
72. Chen B, Gilbert LA, Cimini BA, Schnitzbauer J, Zhang W, Li GW, Park J, Blackburn EH, Weissman JS, Qi LS, Huang B (2013) Dynamic imaging of genomic loci in living human cells by an optimized CRISPR/Cas system. Cell 155(7):1479–1491. https://doi.org/10.1016/j.cell.2013.12.001
73. Ma H, Tu LC, Naseri A, Huisman M, Zhang S, Grunwald D, Pederson T (2016) Multiplexed labeling of genomic loci with dCas9 and engineered sgRNAs using CRISPRainbow. Nat Biotechnol 34(5):528–530. https://doi.org/10.1038/nbt.3526
74. Canela A, Sridharan S, Sciascia N, Tubbs A, Meltzer P, Sleckman BP, Nussenzweig A (2016) DNA breaks and end resection measured genome-wide by end sequencing. Mol Cell 63(5):898–911. https://doi.org/10.1016/j.molcel.2016.06.034
75. Crosetto N, Mitra A, Silva MJ, Bienko M, Dojer N, Wang Q, Karaca E, Chiarle R, Skrzypczak M, Ginalski K, Pasero P, Rowicka M, Dikic I (2013) Nucleotide-resolution DNA double-strand break mapping by next-generation sequencing. Nat Methods 10(4):361–365. https://doi.org/10.1038/nmeth.2408
76. Tsai SQ, Zheng Z, Nguyen NT, Liebers M, Topkar VV, Thapar V, Wyvekens N, Khayter C, Iafrate AJ, Le LP, Aryee MJ, Joung JK (2015) GUIDE-seq enables genome-wide profiling of off-target cleavage by CRISPR-Cas nucleases. Nat Biotechnol 33(2):187–197. https://doi.org/10.1038/nbt.3117
77. Hu J, Meyers RM, Dong J, Panchakshari RA, Alt FW, Frock RL (2016) Detecting DNA double-stranded breaks in mammalian genomes by linear amplification-mediated high-throughput genome-wide translocation sequencing. Nat Protoc 11(5):853–871. https://doi.org/10.1038/nprot.2016.043

4 The CRISPR/Cas9 System as a Tool to Engineer Chromosomal Translocation In Vivo

Taek-Chin Cheong, Rafael B. Blasco, and Roberto Chiarle

Abstract

The CRISPR/Cas9 system has emerged as a powerful tool to edit the genome. Among many applications, the system generates the exciting possibility of engineering small and large portions of chromosomes to induce a variety of structural alterations such as deletions, inversions, insertions and inter-chromosomal translocations. Furthermore, the availability of viral vectors that express Cas9 has been critical to deliver the CRISPR/Cas9 system directly *in vivo* to induce chromosomal rearrangements. This review provides an overview of the state-of-the-art CRISPR/Cas9 technology to model a variety of rearrangements *in vivo* in animal models.

Keywords

CRISPR/Cas9 · Chromosomal translocations · Cancer · Mouse models · In vivo delivery

4.1 Introduction of CRISPR/Cas9 System

The clustered regularly interspaced short palindromic repeats (CRISPR)/CRISPR-associated (Cas) system was developed by many bacteria and archaea as an adaptive immune system that can protect themselves against invading genetic elements such as viruses or plasmids [1]. Type II CRISPR system adapted from *Streptococcus pyogenes* (Sp) has been used for genome editing and sequence-specific DNA double-strand breaks (DSBs) [1, 2]. For this system, two key components are essentially required: a Cas9 endonuclease and a guide RNA (gRNA) fused with a CRISPR RNA (crRNA) and a transactivating CRISPR RNA (tracrRNA) [2, 3]. In this system, the Cas9 endonuclease forms a complex with a 20-nt gRNA and binds to a target DNA sequence using standard RNA-DNA complementarity base-pairing rules to generate blunt end DSB at 3-nt upstream of a protospacer-adjacent motif (PAM) sequence, 3'-NGG [2, 4, 5]. While the simple 3'-NGG PAM sequence of the SpCas9 occurs on average every 8–12 bp in the human genome, there is a modest limitation of specific PAM requirement [2, 3]. This limitation has been overcome by developing Type II CRISPR system from other species of bacteria because each bacterial adaptive immune system recognizes alternative PAM sequences and utilizes different crRNA and tracrRNA sequences. For instance,

Authors Taek-Chin Cheong and Rafael B. Blasco have equally contributed to this chapter.

T.-C. Cheong · R. B. Blasco · R. Chiarle (✉)
Department of Pathology, Boston Children's Hospital, Boston, MA, USA

Harvard Medical School, Boston, MA, USA
e-mail: roberto.chiarle@childrens.harvard.edu

Y. Zhang (ed.), *Chromosome Translocation*, Advances in Experimental Medicine and Biology 1044,
https://doi.org/10.1007/978-981-13-0593-1_4

Staphylococcus aureus (Sa) Cas9 protein requires 3'-NNGRRT PAM and the *Neisseria meningitides* (Nm) Cas9 nuclease prefers 3'-NNNNGATT PAM [6, 7]. The Cas9 endonucleases found in these type II CRISPR systems generate blunt-end DSBs, but the recently discovered nuclease Cpf1, adapted from the *Francisella novicida* (Fn) type V CRISPR system, has been shown to cleave DNA in a staggered pattern, creating a 4 or 5-nt 5′ overhang 18–23 bases away from a PAM sequence, 5'-TTN [8]. These nuclease-induced DSBs can be repaired by non-homologous end-joining (NHEJ) pathway, which can cause the introduction of insertion/deletion mutations (indels) [5]. Thus, the CRISPR/Cas9 genome editing systems have great potentials to be used in many applications that require engineering of the genome, including structural rearrangements. The topic of this Chapter is a comprehensive review of the applications that exploit the CRISPR/Cas9 system to introduce specific structural rearrangements in the genome, with particular emphasis on its applications *in vivo* in mouse models.

4.2 Applications of Nucleases to Engineer Chromosomal Rearrangements

During the last few years, several methods have been developed for engineering structural rearrangements in the genome by exploiting the capability of nucleases to induce DSBs in the genome, including zinc finger nucleases (ZFNs), transcription activator-like effector nucleases (TALENs), I-SceI meganucleases, and the CRISPR/Cas9 systems [9–11]. Each platform has its own advantages and disadvantages, but this Chapter mostly focuses on the CRISPR/Cas9 system because this system offers several advantages over the ZFNs and TALENs to induce chromosomal rearrangements in terms of target design simplicity and efficiency. The CRISPR/Cas9 genome editing systems have widely applied to RNA-guided targeted genome editing, transactivation and silencing module factors, gene knockout, chromosomal rearrangements, and genome-wide screenings [12–14]. Importantly, the CRISPR/Cas9 system has a unique advantage compared to other meganucleases to engineer structural rearrangements due to its capability of introducing DSBs at multiple sites with an easy design. By expressing the Cas9 nuclease and multiple gRNAs in cells, a variety of small and large deletions, inversions, and chromosomal translocations have been obtained between two DSB sites [2, 15–18]. Thus, this system is one of the most suitable and powerful tools for studying chromosomal rearrangement because of its simple and efficient generation of DNA DSBs at loci between two non-homologous genes of interest. Even though there are many different types of CRISPR/Cas systems, this Chapter focuses on the type II CRISPR/Cas9 system adapted from the *S. pyogenes* because it is the most commonly used and extensively characterized system so far.

4.3 Model Genomic Rearrangements Engineered with the CRISPR/Cas9 System *In Vitro*

Chromosomal translocations are one of the most common types of genomic rearrangements in the genome and are initiated by DNA DSBs at two loci of non-homologous chromosomes, followed by the illegitimate joining between DSBs to form abnormal chromosomes [19] (Fig. 4.1). Since the CRISPR/Cas9 system efficiently, rapidly and synchronously induces DSBs in target genomic loci, chromosomal translocations can be easily obtained. The literature on CRISPR/Cas9 applications *in vitro* for this purposes is extensive and partially reviewed in other chapters, thus we will briefly discuss few examples of chromosomal translocations as well as other types of genomic rearrangements including chromosomal deletions and inversions induced by CRISPR/Cas9 system *in vitro*.

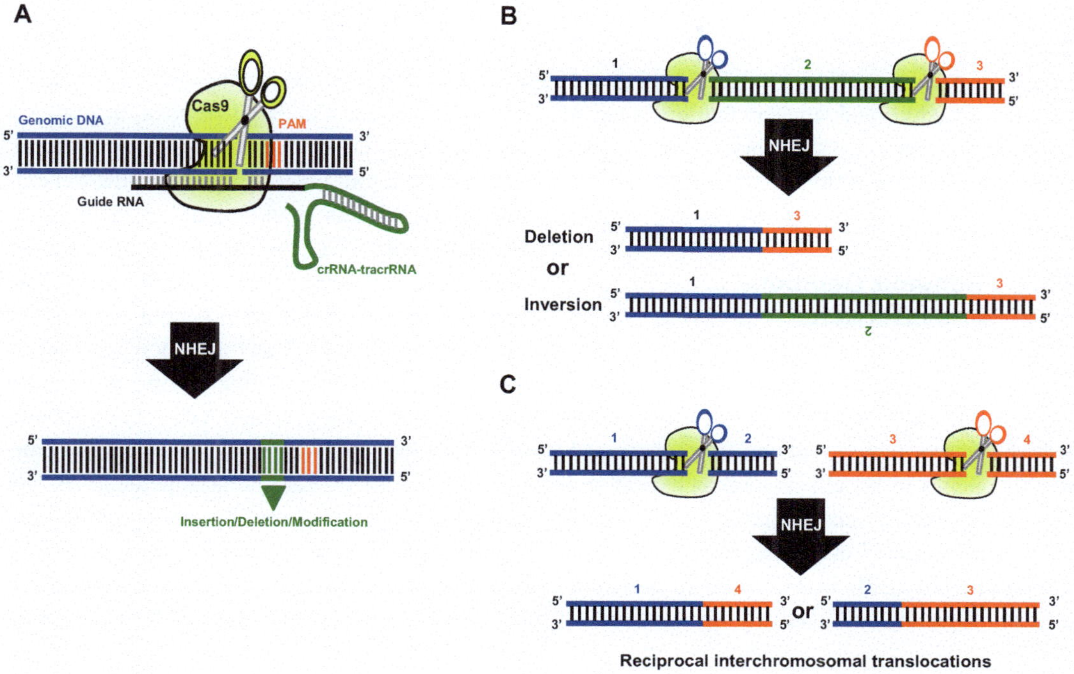

Fig. 4.1 Overview of chromosomal rearrangement using CRISPR-Cas9 system. (**a**) Cas9 nuclease generates double-strand break (DSB) at a DNA target site with complementarity to the 5′ end of a guide RNA (gRNA) at 3-nt upstream of a protospacer-adjacent motif (PAM). gRNA-directed Cas9 nuclease can induce indel mutations. (**b**) Pairs of gRNA-directed Cas9 nucleases in the same chromosome can induce chromosomal rearrangements (e.g., small or large deletions or inversions). (**c**) Pairs of gRNA-directed Cas9 nucleases in two different chromosomes can induce chromosomal translocations. NHEJ: Non-homologous end joining

4.3.1 Chromosomal Translocations

Since several human cancers are driven by chromosomal translocations, one of the most extensive applications of the CRISPR/Cas9 system in this field is to reproduce oncogenic translocations in target cells *in vitro*. Several examples are published but virtually any chromosomal translocation can potentially be engineered *in vitro* provided a good specificity and efficacy of the gRNA sequences and Cas9 expression. As an example, the t(8;21)(q22;q22) is frequently found in acute myeloid leukemia (AML) and generates a fusion between the *ETO* gene on chromosome 8 and the *RUNX1* gene on chromosome 21 [18]. This translocation is an in-frame fusion of almost the entire *ETO* gene with the 5′ region of the *RUNX1* gene [20]. By designing gRNA targeting the introns often involved by DSBs that originate such translocation, the CRISPR/Cas9 system induced chromosomal translocations with an efficiency of 0.96 to 4.07% transduced cells [20]. Together with the t(8;21), a reciprocal t(21;8) translocation was also detected, similar to human AML, thus indicating that the introduction of DSBs at high frequency by the CRISPR/Cas9 system is sufficient to recapitulate reciprocal genomic rearrangements found in human cancers [18]. Indeed, several translocations can be easily engineered in mouse and human cells *in vitro*, including the *CD74-ROS1*, *EWSR1–FLI1, EWSR1-WT1,* and *Pax3/Foxo1* translocations [18, 21–24]. To detect CRISPR/Cas9 induced translocations, several methods are available including polymerase chain reaction (PCR), fluorescence *in situ* hybridization (FISH), and high-throughput, genome-wide, translocation sequencing (HTGTS) [10, 11, 18, 25].

Taken together, the simple and efficient CRISPR/Cas9 system is easily exploited for the generation of chromosomal rearrangements *in vitro* in cell lines with an efficiency varying from

0.1% to 6–8%, likely depending on additional factors including cell type, transduction efficiency, karyotype, efficacy and specificity of gRNA, and additional nuclear features such as distance between two DSBs and chromatin conformation of the target loci.

4.3.2 Genomic Deletions and Insertions

A classical example of the feasibility of engineering genomic deletions by CRISPR/Cas9 system is the editing of mouse and human immunoglobulin (Ig) genes in B cells as well as hybridomas [17]. Class-switch recombination (CSR) in Ig genes is a deletional event that requires the generation of DNA DSBs in the repetitive switch (S) regions, typically initiated by the activation-induced cytidine deaminase (AID) enzyme [26]. The CRISPR/Cas9 system can be used to mimic these events by targeting two desired S regions with specific gRNAs. As a consequence, the DNA segment between the gRNAs is deleted and joined by NHEJ pathway resulting in a switch of the Ig heavy (IgH) chain [17]. Different sizes of deletion can be obtained by the CRISPR/Cas9 system, from few bases up to large deletions of as much as 10–12 Mb [16]. Interestingly, Chen and colleagues recently applied the nickase $Cas9^{D10A}$, a D10A mutation in the catalytic domain of Cas9 that produces DNA single-strand break, to target rearranged genes [27]. They targeted the breakpoints of fusions of *TMEM135-CCDC67*, a fusion gene between intron 13 of transmembrane protein 135 (*TMEM135*) and intron 9 of coiled-coil domain containing 67 (*CCDC67*), and *MAN2A1-FER*, a fusion gene between intron 13 of mannosidase α class 2A member 1 gene (*MAN2A1*) and intron 14 of FER tyrosine kinase (FER), with two adenoviruses to deliver both the nickase $Cas9^{D10A}$ and gRNAs targeting the breakpoint sequences, and an EGFP-thymidine kinase [27]. By this approach, they could specifically target chromosomal translocations by inserting a suicide gene into the genomic breakpoint of chromosomal rearrangements.

4.3.3 Genomic Inversions

Inversions are genomic events where segments of genomic DNA are inverted in position within a chromosome, either pericentric or paracentric. Because genomic inversions require the generation of two DSBs flanking the segment to be inverted, again the CRISPR/Cas9 system showed high efficiency in engineering such events. The very first *in vitro* examples of genomic inversions obtained by CRISPR/Cas9 are oncogenic inversions that generate abnormal fusion proteins in lung cancer. The echinoderm microtubule-associated protein like 4 (*EML4*)- anaplastic lymphoma kinase (*ALK*) rearrangement is an inversion on chromosome 2 in human lung cancer that generates a transcript that fuses the *EML4* and *ALK* genes. In mouse cells, this inversion can be engineered by designing gRNAs targeting intron 13 of the *Eml4* gene and intron 19 of the mouse *Alk* gene on mouse chromosome 17. By introducing DSBs in these two positions, which are approximately 10 megabases (Mb) apart on the same chromosome, the flanked genomic segment can be either inverted or deleted [15, 16]. Similarly, the paracentric *EML4-ALK* inversion on chromosome 2p or the pericentric *KIF5B-RET* inversion on chromosome 10 can be easily obtained in human cells by designing gRNA in their respective breakpoint regions [18, 21].

4.4 Model Chromosomal Rearrangements Engineered by the CRISPR/Cas9 System *In Vivo*

Chromosomal rearrangements are the molecular basis of multiple human diseases and mouse models that recapitulate these events are essential for basic and translational experiments. The generation of Genetic Engineered Mouse Models (GEMMs) reproducing chromosomal rearrangements in the endogenous locus by knock-in strategy is a long and an expensive process. The CRISPR/Cas9 system has emerged as a flexible and relatively easy technology, enabling to simplify the generation of mouse models and par-

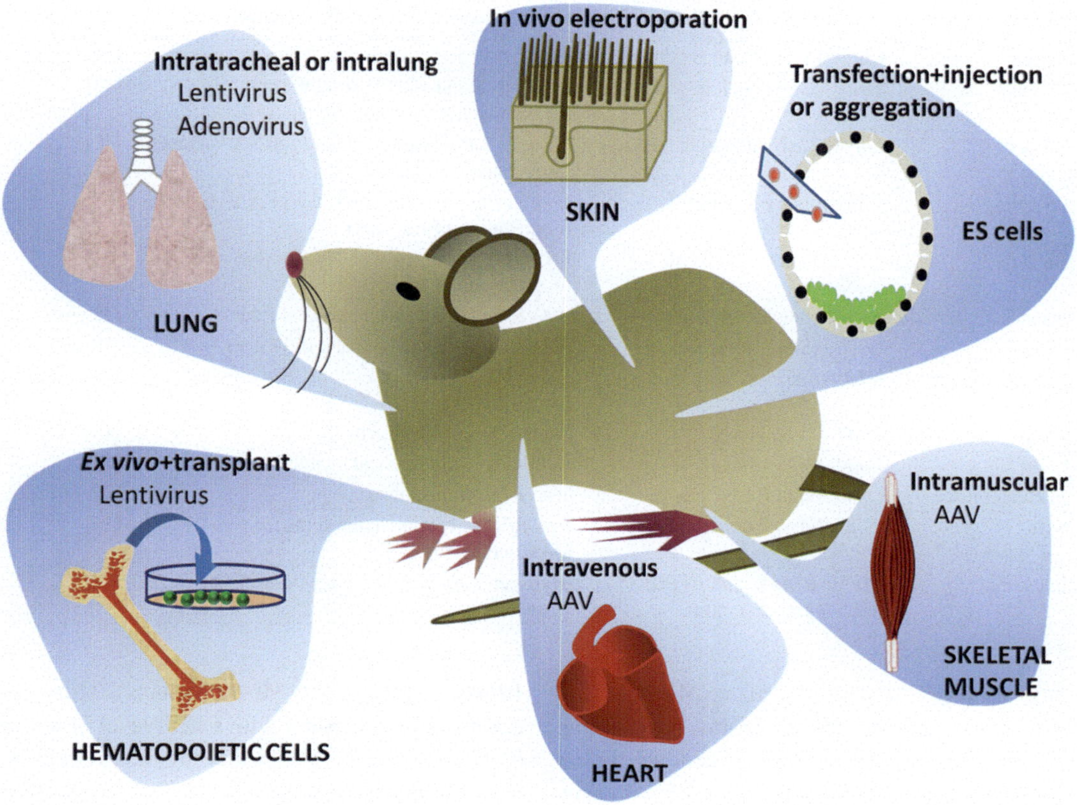

Fig. 4.2 Schematic representation of different approaches to deliver the CRISPR/Cas9 system in mice *ex vivo* or directly *in vivo*

ticularly chromosomal-rearranged mouse models. Three main different approaches have been exploited to generate chromosomal rearrangements in animal models by the CRISPR/Cas9 system: engineering of embryonic stem (ES) cells, *ex vivo* cell engineering with subsequent re-introduction in mice, and *in vivo* direct engineering of mouse tissues typically mediated by delivery of viral particles encoding the CRISPR/Cas9 system. An overview of the successful methods to deliver the components of the CRISPR/Cas9 system to mice *in vivo* is schematically presented in Fig. 4.2 and summarized in Table 4.1.

4.4.1 Mouse Models of Cancer and Genetic Diseases

Multiple hematologic and solid tumors arise or progress owing to fusion oncogenic proteins that are a result of chromosomal translocations. Generation of mouse models that faithfully recapitulate these malignancies is necessary for a better understanding of the molecular process involved, as well as, for testing innovative treatments in a preclinical stage. In spite of the above-described successes in reproducing several of these translocations in human cell lines and mouse primary cells by the CRISRPR-Cas9 system *in vitro*, this technology has been only successfully applied *in vivo* to two mouse models, both reproducing the *EML4-ALK* fusion involved in non-small cell lung cancer (NSCLC) [28]. Simultaneously, two independent groups obtained the *Eml4-Alk* fusion gene as a result of an inversion in the chromosome 17 by delivering viral particles into the lung of mice [15, 16]. Blasco and colleagues inoculated two different lentiviruses encoding for the Cas9 and the corresponding gRNA targeting *Alk* or *Eml4* into the

Table 4.1 Delivery in vivo of the CRISPR/Cas9 system to engineer chromosomal rearrangements in mouse models

Tissue	Chromosomal rearrangement	Delivery	References
ES cells	Titin, Mex5 deletion	Cas9 RNA and gRNAs *in vitro* transcribed (microinjection in 1-day zygotes)	[43]
ES cells	*Laf4*, 353kb intragenic deletion	ES cells transfected with Cas9/gRNA vectors (aggregation of ES cells)	[40]
ES cells	Chromosome 1, 1.67Mb deletion	ES cells transfected with Cas9/gRNA vectors (aggregation of ES cells)	[41]
ES cells	Chromosome 1, 1.06Mb inversion	ES cells transfected with Cas9/gRNA vectors (aggregation of ES cells)	[41]
Heart	Dystrophin, Ex23 deletion	AAV8 (intraperitoneal, neonatal mice)	[35, 36]
Heart	Dystrophin, Ex23 deletion	AAV8 (intravenous, 6 week-old mice)	[35, 36]
$CD34^+$ Human cells	*MLL1-ENL1* translocation	Lentivirus (*ex vivo* infection, cells injected intravenously)	[34]
Lung	*Eml4-Alk* inversion	Adenovirus5 (intratracheal)	[15]
Lung	*Eml4-Alk* inversion	Lentivirus (intratracheal or intralung)	[16]
Skeletal muscle	Dystrophin, Ex23 deletion	AAV8 (intramuscular)	[35, 36]
Skin	*Col7a1*, Ex80 deletion	Cas9/gRNA ribonucleoproteins (*in vivo* electroporation)	[33]
Intestine	EIF3E-RSPO2 and PTPRK-RSPO3 translocations	Transgenic mice with tetracycline-inducible Cas9	[44]

lungs of $p53^{+/-}$ and $p53^{-/-}$ mice. Maddalo and colleagues reported similar results inoculating an adenovirus serotype 5 (Ad5) encoding for the Cas9 and both gRNAs into lungs of wild-type mice. As a result of the expression of EML4-ALK in the alveolar cells, both groups observed the onset of lung adenocarcinomas within 3 months, recapitulating the human histopathology. Moreover, Maddalo and colleagues showed that tumors were sensitive to the ALK-inhibitor crizotinib. Even though both approaches were successful, there are remarkable differences to be considered. The DNA of the adenovirus does not integrate into the host genomic DNA, leading to a transient expression of Cas9 that is sufficient to induce DSBs and chromosomal translocations. In contrast, lentivirus typically integrates into the genome with long lasting Cas9 expression, increasing its potential on-target and off-target activity. Together with the desired *Eml4-Alk* rearrangement generated by the inversion, this approach produced additional genomic rearrangements in alveolar cells, including large 10 Mb deletions and various inter-chromosomal rearrangements between the two chromosomes 17.

In the case of *Eml4-Alk* rearrangement, the two genomic loci are separated approximately by 10 Mb in mice (12.5 Mb in humans). This distance is still consistent with a high level of proximity between these loci that favors a relatively high frequency of these events [29]. A positive correlation between the physical proximity of the target loci and the frequency of translocations obtained by CRISPR/Cas9 was also shown in models of *Pax3-Foxo1* translocations where the frequency of translocation events in a hindlimb or forelimb myoblasts directly correlated with the different proximity of the two genes in these two cell types [22, 30, 31]. However, Boroviak and colleagues showed in mouse zygotes that inversion events are frequent when the sequences targeted by the sgRNAs are separated by 0.155–1.5 Mb [32]. Shin and colleague detected a big frequency of deletions, but no inversion events, by targeting regions separated by <25 kb possibly because small segments of DNA are quickly degraded by exonucleases, and are therefore less likely to be re-integrated into the genome as an inversion [33].

The frequency of translocations events also directly correlates to the fraction of cells transduced with both gRNAs. Therefore, *ex vivo* transduction of cells and their subsequent injection in mice is a valid alternative to direct *in vivo* transduction. This approach is particularly attractive with hematopoietic stem cells. Reimer and colleagues transduced CD34$^+$ human hematopoietic stem cells with a single lentivirus encoding for the Cas9 and the gRNAs for *MLL1* and *ENL1* before transferring them to immunocompromised mice. By this approach, they engineered the *MLL1-ENL1* translocation which is frequent in pediatric acute leukemias, with an efficacy of at least 1.6×10^{-3} cells [34]. Freshly infected CD34$^+$ cells were injected intravenously into immunodeficient mice. Interestingly, the MLL-ENL1 translocation induced the development of a monocytic leukemia like-disease that evolved into a B-cell acute lymphoblastic leukemia (B-ALL) after additional *in vivo* passages, likely due to the acquisition of additional genetic events [34]. Thus, CRISPR/Cas9-based models are a useful tool to study leukemia initiation and progression because the expression of oncogenic fusions is from the endogenous locus in contrast to previous approaches based on overexpression of the fusion oncogenes in stem cells by retroviral transduction.

The high efficiency of the CRISPR/Cas9 system to produce small genomic deletions can also have therapeutic applications *in vivo*. Duchenne muscular dystrophy (DMD) is a degenerative disease caused by mutations or deletions in the dystrophin gene that shift the reading frame, leading to a dysfunctional protein [35]. Following this concept, two independent groups simultaneously developed a CRISPR/Cas9-based strategy to delete exon 23 in a mouse model that faithfully recapitulates DMD [36, 37]. The strategy of delivery was similar in both cases, injection of an adeno-associated virus serotype 8 (AAV8) encoding Cas9 and the gRNAs flanking exon 23 into the tibialis anterior muscle. Nelson and colleagues reported 2% of deletion in the whole muscle lysate by droplet digital PCR (ddPCR). Both groups showed restoration in the expression of dystrophin, and more importantly an increase in the muscular function. Additionally, when the injection of the CRISPR-Cas9 machinery was performed in early life, by intraperitoneal injection, both studies showed recovery of dystrophin expression in abdominal muscles, diaphragm, and heart.

The same idea has been recently applied to a mouse model recapitulating dystrophic epidermolysis bullosa (RDEB). Patients develop this syndrome owing to mutations in the collagen VII protein. Wu and colleagues applied the exon-skipping approach to restoring the expression of a functional collagen VII in the skin stem cells by injecting the components of the CRISPR/Cas9 system in the skin followed by *in vivo* electroporation. By this approach, they could efficiently excise exon 80, which covers the point mutation in the RDEB mouse model and thus restores the correct localization of the collagen VII protein *in vivo* [38].

4.4.2 Mouse Models by Embryonic Stem (ES) Cells Engineering

Several genetic diseases are caused by chromosomal rearrangements or other structural variations of the genome. Duplications, deletions and translocations are involved in the development of Mendelian diseases or complex diseases such as autism or schizophrenia [39]. The delivery of the CRISPR/Cas9 system into ES cells by transfection facilitates the generation of ES clones bearing the desired chromosomal alterations and, therefore, the development of mouse models for these diseases. Many of these genomic alterations affect coding regions with a pathogenicity typically explained by the altered dosage of the genes affected. Several structural variants including deletions, inversions, and duplications can be engineered in mice by manipulating ES cells. Remarkably, Kraft and colleagues engineered deletions, inversions, and also duplications at six different genomic loci ranging from 1.1 kb to 1.6 Mb with efficiencies up to 42% [40]. In similar applications, the CRISPR/Cas9 system can be used to generate structural rearrangements in long non-coding DNA regions. Lupiañez and colleagues recapitulated such events in mice that model rare limb malformations [41]. These particularly inherited malformations are present in families with chromosomal rearrangements affecting three topologically associated domains (TADs) in the extended *WNT6/IHH/EPH4/PAX3* region: one gene-dense region containing *WNT6* and *IHH*, one extended region containing *EPHA4* gene and a third region containing *PAX3* [42]. TADs are structural partitions of the genome thought to work as regulatory units, where promoter and enhancers interact, separated by boundaries [42]. Disruption of these organized structures might connect enhancers with different promoters leading to misexpression of certain genes. The authors reported three unrelated families, with dominantly inherited brachydactyly, presenting different heterozygous deletions (1.7–1.9 Mb) including the *EPHA4* gene along with a big portion of his TAD and extend into the noncoding part of the *PAX3* gene TAD. They reproduced this deletion in ES cells by cotransfecting pairs of gRNAs in order to analyze the mechanism of this genetic disease. The so-called DelB mouse model carried this deletion in heterozygosis and homozygosis. Homozygotes DelB mice developed a phenotype similar to the one observed in human, with a milder version in the case of heterozygotes. In a similar way, the authors reproduced another rare limb malformation, so-called F-syndrome, caused by a 1.1 Mb heterozygous inversion in the same region. The authors conclude that all these pathological syndromes are caused by the disruption of the boundaries between TADs and the reorganization of this genomic region in new TADs, with a subsequent rewiring of the pattern of expression of these genes. These examples show how the CRISPR/Cas9 system can be exploited to engineer complex structural rearrangements that would be technically overwhelming with traditional targeting systems in ES cells.

In similar applications, the CRISPR/Cas9 technology is particularly attractive to efficiently generate small rearrangements such as alternative splicing or exon skipping. An example of this approach has been shown to manipulate the splicing variants that encode for the is7 domain of the titin protein in the mouse skeletal muscles and heart. Is7 domain is encoded by the penultimate exon, so-called Mex5, of the titin gene, an exon alternatively spliced in the M-line of the sarcomere. Charton and colleagues generated a mouse model - by co-microinjecting the spCas9 mRNA and in vitro transcribed gRNAs in one-day zygote - where the expression of alternatively spliced variant(s) carrying the corresponding domain was stably prevented and showed that the phenotype of mice was related to this exon skipping [43].

A critical limitation for modeling cancer-related chromosomal rearrangements *in vivo* is the accessibility of the tissue to be targeted. As an example, modeling the EIF3E-RSPO2 and PTPRK-RSPO3 chromosome rearrangements *in vivo* by direct CRISPR/Cas9 administration is challenging. To overcome this limitation, Han and colleagues generated a transgenic mouse model with a construct expressing Cas9 under the control of a tetracycline response element (TRE^{3G} promoter) and the U6 promoter driving the expression of gRNAs for the corresponding genes [44]. Upon crossing with the R26rtTA mouse strain, they generated transgenic lines that induced the EIF3E-RSPO2 and PTPRK-RSPO3 translocations upon doxycycline treatment to show the pathogenic role of these fusions in tumor initiation and maintenance in the intestine [44].

4.5 Conclusions

The CRISPR/Cas9 system has dramatically empowered and extended the applications to engineering genomic rearrangements in the genome either *in vitro* or directly *in vivo* in mammals. A variety of genomic events including chromosomal translocations, large deletions and microdeletions, inversions, and exon skipping can be efficiently achieved in mice by *ex vivo* or *in vivo* delivery of the components of the CRISPR/Cas9 system. Several factors influence the efficacy of gene rearrangement engineering, including an efficient delivery of the CRISPR/Cas9 components, the specificity of gRNA, the abundance of off-target sites, the levels of expression of the Cas9 nuclease in the target tissue, the proximity, accessibility and chromatin conformation of the genomic loci. Several of these factors are recurrent themes discussed in the CRISPR/Cas9 field, including the specificity and efficacy of the system. Other factors, such as proximity of the target loci and chromatin conformation, are more specifically relevant in the field of chromosomal translocations and genomic rearrangements and are also reviewed in other chapters of this book. A critical factor for the success of CRISPR/Cas9 editing is an effective delivery of the CRISPR/Cas9 components. So far, the most obvious success has been obtained in organs where a direct delivery is straightforward, such as the skin, the muscles, and the lung. To this end, the development of smaller forms of the Cas9 nuclease, such as the Cas9 from *Staphylococcus aureus* (SaCas9) [6], will likely improve the efficacy of delivery together with improved viral or particle-mediated delivery system. With an efficient delivery system, the CRISPR/Cas9 system could further expand its potential from preclinical studies to therapeutic applications that require structural modifications of the genome.

Acknowledgments This work was supported by National Institute of Health grants R01 CA196703-01 and R01 CA222598-01 to R.C.; Associazione Italiana per la Ricerca sul Cancro (AIRC) grant IG-12023 to R.C.; Worldwide Cancer Research (former AICR) grant 12-0216 to R.C.; National Research Foundation of Korea (NRF) fellowship 2016R1A6A3A03006840 to T.C.C.

Author Information The authors declare no competing financial interests. Correspondence and requests for materials should be addressed to RC. (roberto.chiarle@childrens.harvard.edu).

References

1. Wiedenheft B, Sternberg SH, Doudna JA (2012) RNA-guided genetic silencing systems in bacteria and archaea. Nature 482:331–338. https://doi.org/10.1038/nature10886
2. Cong L et al (2013) Multiplex genome engineering using CRISPR/Cas systems. Science 339:819–823. https://doi.org/10.1126/science.1231143
3. Hsu PD et al (2013) DNA targeting specificity of RNA-guided Cas9 nucleases. Nat Biotechnol 31:827–832. https://doi.org/10.1038/nbt.2647
4. Jinek M et al (2012) A programmable dual-RNA-guided DNA endonuclease in adaptive bacterial immunity. Science 337:816–821. https://doi.org/10.1126/science.1225829
5. Sander JD, Joung JK (2014) CRISPR-Cas systems for editing, regulating and targeting genomes. Nat Biotechnol 32:347–355. https://doi.org/10.1038/nbt.2842
6. Ran FA et al (2015) In vivo genome editing using Staphylococcus aureus Cas9. Nature 520:186–191. https://doi.org/10.1038/nature14299
7. Hou Z et al (2013) Efficient genome engineering in human pluripotent stem cells using Cas9 from Neisseria meningitidis. Proc Natl Acad Sci U S A 110:15644–15649. https://doi.org/10.1073/pnas.1313587110
8. Zetsche B et al (2015) Cpf1 is a single RNA-guided endonuclease of a class 2 CRISPR-Cas system. Cell 163:759–771. https://doi.org/10.1016/j.cell.2015.09.038
9. Piganeau M et al (2013) Cancer translocations in human cells induced by zinc finger and TALE nucleases. Genome Res 23:1182–1193. https://doi.org/10.1101/gr.147314.112
10. Chiarle R et al (2011) Genome-wide translocation sequencing reveals mechanisms of chromosome breaks and rearrangements in B cells. Cell 147:107–119. https://doi.org/10.1016/j.cell.2011.07.049
11. Frock RL et al (2015) Genome-wide detection of DNA double-stranded breaks induced by engineered nucleases. Nat Biotechnol 33:179–186. https://doi.org/10.1038/nbt.3101
12. Hsu PD, Lander ES, Zhang F (2014) Development and applications of CRISPR-Cas9 for genome engineering. Cell 157:1262–1278. https://doi.org/10.1016/j.cell.2014.05.010
13. Shalem O, Sanjana NE, Zhang F (2015) High-throughput functional genomics using CRISPR-Cas9. Nat Rev Genet 16:299–311. https://doi.org/10.1038/nrg3899
14. Doudna, J. A. & Charpentier, E. (2014) Genome editing. The new frontier of genome engineering with CRISPR-Cas9. Science 346:1258096. https://doi.org/10.1126/science.1258096
15. Maddalo D et al (2014) In vivo engineering of oncogenic chromosomal rearrangements with the CRISPR/Cas9 system. Nature 516:423–427. https://doi.org/10.1038/nature13902
16. Blasco RB et al (2014) Simple and rapid in vivo generation of chromosomal rearrangements using CRISPR/Cas9 technology. Cell Rep 9:1219–1227. https://doi.org/10.1016/j.celrep.2014.10.051
17. Cheong TC, Compagno M, Chiarle R (2016) Editing of mouse and human immunoglobulin genes by CRISPR-Cas9 system. Nat Commun 7:10934. https://doi.org/10.1038/ncomms10934
18. Torres R et al (2014) Engineering human tumour-associated chromosomal translocations with the RNA-guided CRISPR-Cas9 system. Nat Commun 5:3964. https://doi.org/10.1038/ncomms4964
19. Richardson C, Jasin M (2000) Frequent chromosomal translocations induced by DNA double-strand breaks. Nature 405:697–700. https://doi.org/10.1038/35015097

20. Yuan Y et al (2001) AML1-ETO expression is directly involved in the development of acute myeloid leukemia in the presence of additional mutations. Proc Natl Acad Sci U S A 98:10398–10403. https://doi.org/10.1073/pnas.171321298
21. Choi PS, Meyerson M (2014) Targeted genomic rearrangements using CRISPR/Cas technology. Nat Commun 5:3728. https://doi.org/10.1038/ncomms4728
22. Lagutina IV et al (2015) Modeling of the human alveolar rhabdomyosarcoma Pax3-Foxo1 chromosome translocation in mouse myoblasts using CRISPR-Cas9 nuclease. PLoS Genet 11:e1004951. https://doi.org/10.1371/journal.pgen.1004951
23. Torres-Ruiz R et al (2017) Efficient recreation of t(11;22) EWSR1-FLI1(+) in human stem cells using CRISPR/Cas9. Stem Cell Reports 8:1408–1420. https://doi.org/10.1016/j.stemcr.2017.04.014
24. Vanoli F et al (2017) CRISPR-Cas9-guided oncogenic chromosomal translocations with conditional fusion protein expression in human mesenchymal cells. Proc Natl Acad Sci U S A 114:3696–3701. https://doi.org/10.1073/pnas.1700622114
25. Semrau S et al (2014) FuseFISH: robust detection of transcribed gene fusions in single cells. Cell Rep 6:18–23. https://doi.org/10.1016/j.celrep.2013.12.002
26. Gostissa M, Alt FW, Chiarle R (2011) Mechanisms that promote and suppress chromosomal translocations in lymphocytes. Annu Rev Immunol 29:319–350. https://doi.org/10.1146/annurev-immunol-031210-101329
27. Chen ZH et al (2017) Targeting genomic rearrangements in tumor cells through Cas9-mediated insertion of a suicide gene. Nat Biotechnol 35:543–550. https://doi.org/10.1038/nbt.3843
28. Soda M et al (2007) Identification of the transforming EML4-ALK fusion gene in non-small-cell lung cancer. Nature 448:561–566. https://doi.org/10.1038/nature05945
29. Zhang Y et al (2012) Spatial organization of the mouse genome and its role in recurrent chromosomal translocations. Cell 148:908–921. https://doi.org/10.1016/j.cell.2012.02.002
30. Mani RS et al (2009) Induced chromosomal proximity and gene fusions in prostate cancer. Science 326:1230. https://doi.org/10.1126/science.1178124
31. Lin C et al (2009) Nuclear receptor-induced chromosomal proximity and DNA breaks underlie specific translocations in cancer. Cell 139:1069–1083. https://doi.org/10.1016/j.cell.2009.11.030
32. Boroviak K, Doe B, Banerjee R, Yang F, Bradley A (2016) Chromosome engineering in zygotes with CRISPR/Cas9. Genesis 54:78–85. https://doi.org/10.1002/dvg.22915
33. Shin HY et al (2017) CRISPR/Cas9 targeting events cause complex deletions and insertions at 17 sites in the mouse genome. Nat Commun 8:15464. https://doi.org/10.1038/ncomms15464
34. Reimer J et al (2017) CRISPR-Cas9-induced t(11;19)/MLL-ENL translocations initiate leukemia in human hematopoietic progenitor cells in vivo. Haematologica 102:1558–1566. https://doi.org/10.3324/haematol.2017.164046
35. Hoffman EP, Brown RH Jr, Kunkel LM (1987) Dystrophin: the protein product of the Duchenne muscular dystrophy locus. Cell 51:919–928
36. Nelson CE et al (2016) In vivo genome editing improves muscle function in a mouse model of Duchenne muscular dystrophy. Science 351:403–407. https://doi.org/10.1126/science.aad5143
37. Tabebordbar M et al (2016) In vivo gene editing in dystrophic mouse muscle and muscle stem cells. Science 351:407–411. https://doi.org/10.1126/science.aad5177
38. Wu W et al (2017) Efficient in vivo gene editing using ribonucleoproteins in skin stem cells of recessive dystrophic epidermolysis bullosa mouse model. Proc Natl Acad Sci U S A 114:1660–1665. https://doi.org/10.1073/pnas.1614775114
39. Stankiewicz P, Lupski JR (2010) Structural variation in the human genome and its role in disease. Annu Rev Med 61:437–455. https://doi.org/10.1146/annurev-med-100708-204735
40. Kraft, K. et al. (2015) Deletions, inversions, duplications: engineering of structural variants using CRISPR/Cas in Mice. Cell Reports. https://doi.org/10.1016/j.celrep.2015.01.016
41. Lupianez DG et al (2015) Disruptions of topological chromatin domains cause pathogenic rewiring of gene-enhancer interactions. Cell 161:1012–1025. https://doi.org/10.1016/j.cell.2015.04.004
42. Dixon JR et al (2012) Topological domains in mammalian genomes identified by analysis of chromatin interactions. Nature 485:376–380. https://doi.org/10.1038/nature11082
43. Charton K et al (2016) Exploiting the CRISPR/Cas9 system to study alternative splicing in vivo: application to titin. Hum Mol Genet 25:4518–4532. https://doi.org/10.1093/hmg/ddw280
44. Han T et al (2017) R-Spondin chromosome rearrangements drive Wnt-dependent tumour initiation and maintenance in the intestine. Nat Commun 8:15945. https://doi.org/10.1038/ncomms15945

Generation of Genomic Alteration from Cytidine Deamination

5

Xiaojing Liu and Fei-Long Meng

Abstract

The sources of genome instability can be attributed to many extra- and exo- cellular factors accompanying various biological processes. In leukemia and lymphomas, the collateral effect of programmed DNA alterations during immune diversification is the major source of genome instability. Cytidine deamination from cytidine (C) to uridine (U) at immunoglobulin (Ig) gene loci is required for initiation of antibody diversification, while the same process also contributes to recurrent translocation or mutations outside of Ig loci in lymphocyte-origin tumors. Furthermore, genome sequencing of cancer cells from many tissue origins revealed a significant enrichment of cytidine deaminase mutagenesis signature in human cancers. Thus, cytidine deamination, which can intensively happen in an enzyme-dependent fashion at specific genomic regions, is a widespread genome instability source across many tumor types. AID/APOBEC superfamily proteins are the main single-stranded DNA deaminases in eukaryotes, which play vital roles in adaptive and innate immunity. Their deamination products can be channeled into mutations, insertions and deletions (indels), clusters of mutations called kaetagis, or chromosomal rearrangements/translocations. Here, we review the generation of genome instability from AID/APOBEC-dependent cytidine deamination with emphasis on the most studied enzyme, AID.

Keywords

Genome instability · Cytidine deamination · Activation-induced cytidine deaminase · APOBEC · Base-editing

5.1 AID/APOBEC Deaminase Family

AID/APOBEC proteins represent a group of single-stranded nucleotide cytidine deaminases, which include AID, APOBEC1, APOBEC2, APOBEC3 and APOBEC4 [23]. The first identified member APOBEC1 is responsible for apolipoprotein B (ApoB) pre-mRNA editing and hence was named apolipoprotein B editing complex 1 (APOBEC1) [51, 79, 130]. Subsequently, other members followed the same nomenclature even though none were proven to have RNA editing activity. AID was discovered by Tasuko Honjo group through cDNA subtraction in cytokine-activated B cell lines, hence named

X. Liu · F.-L. Meng (✉)
Shanghai Institute of Biochemistry and Cell Biology, Chinese Academy of Sciences, Shanghai, China
e-mail: feilong.meng@sibcb.ac.cn

Y. Zhang (ed.), *Chromosome Translocation*, Advances in Experimental Medicine and Biology 1044,
https://doi.org/10.1007/978-981-13-0593-1_5

activation-induced deaminase (AID, the coding gene was named as *AICDA*) [75, 76].

5.1.1 RNA or DNA Deamination Model

AID/APOBEC deaminases preferentially target single-stranded (ss) nucleic acids. The RNA editing activity of APOBEC1 led to the assumption that other enzymes in the AID/APOBEC family are also ssRNA deaminases, although so far, APOBEC1 is the only enzyme possessing RNA deamination activity *in vivo*. APOBEC1 targets the 3′ UTR of many cellular mRNAs, which fine-tunes the gene expression by modifying miRNA-binding sites [110]. AID and APOBEC3s targets genomic or viral DNA, which play vital roles in adaptive and innate immunity [132]. Despite the debates on whether AID deaminates RNA or DNA substrates, lines of evidence point to a DNA deamination model [15, 19, 94, 96, 103]. It is worth noting that APOBEC1 also deaminates ssDNA substrate [43, 95], which makes it part of a popular "base editor" tool [54]. Thus, an important question that emerges is how APOBEC1 obtains RNA editing activity during evolution. The deamination activity of APOBEC2 and APOBEC4 has not been fully tested; neither does their physiological role [99, 108].

5.1.2 Cytidine Deaminase Domain and Structure

The deaminase catalytic site in AID/APOBEC proteins lies in the conserved cytidine deaminase (CDA) domain. Like other deaminase family enzymes, the core structure of AID/APOBEC CDA domain contains a zinc-dependent deaminase motif (ZDD). All the AID/APOBEC members contain either one or two CDA domains [23]. AID, APOBEC1, APOBEC2 and APOBEC4 are single CDA domain proteins, while some of the APOBEC3 proteins have dual CDA domains [23]. Different from the sole murine APOBEC3 ortholog, human APOBEC3 has seven members named APOBEC3A, APOBEC3B, APOBEC3C, APOBEC3DE, APOBEC3F, APOBEC3G and APOBEC3H (abbreviated as A3A, A3B, A3C, A3DE, A3F, A3G and A3H in this chapter) through an anthropoid-specific gene expansion [23]. Many efforts had been done to solve the structure of AID/APOBEC proteins in the past years. The structures have given rich information about the deamination catalytic process and also implied potential single-stranded nucleic acid binding surfaces on the proteins (Reviewed in Ref. [114]). Recently, the fine structure of A3A with its ssDNA substrate revealed a U-shaped DNA conformation [120], while AID prefers to recognize structured DNA substrates [101]. Thus, the AID/APOBEC members may utilize distinct substrate recognition models.

5.1.3 DNA Sequence Motif Preference and Deamination Signature

At nucleotide sequence level, AID/APOBEC deaminases have intrinsic preference for local sequence motifs. For example, AID is prone to deaminate C in WRC (W: A/T, R: G/A) motif [53, 109], while A3A and A3B prefer C in the TCA motif context [30]. The motif preference was explained by the A3A-ssDNA structures, as the adjacent nucleotide residues also are involved in DNA-protein interaction [120]. A single amino acid change (D131A) on A3A protein can alter its motif preference [120], and similar change on A3G protein can also change its preference from CCC to TCC [105]. Similarly, biochemical and genetic study had revealed a substrate recognition loop in AID protein [136]. Sequence swapping of AID substrate recognition loop with A3's substrate recognition loop results in AID-3C/F/G mutants that prefer A3C/F/G signatures [136]. The motif preference may be highly linked to its biology function, which could be a result of co-evolution of deaminase and its substrate sequence. The deamination activity of specific motifs leaves footprint of the corresponding enzyme, allowing the identification of the responsible enzyme by mutation signature. Biochemical and DNA sequencing studies have revealed a collection of dinucleotide motif signatures for murine and human AID/APOBEC proteins. For different AID/APOBEC proteins, the adjacent positions around C plays distinct roles, as -1,-2 positions are important for AID/A3G deaminase and -1,+1 positions are important for A3A/B deaminase. Furthermore, biochemical study discovered unique motif preferences for several other AID orthologs [102], implicating that there is a rich pool of deamination dinucleotide signatures for AID/APOBEC proteins.

5.2 Programmed DNA Lesion in Immunity

To fight against the numerous pathogens, B cells can produce a large number of B cell receptors (BCRs, membrane-bound form) or antibodies (Abs, secreted form), which is one of the hallmarks of adaptive immunity [1]. A typical antibody is composed of two immunoglobulin (Ig) heavy chains and two Ig light (IgL) chain, and each Ig chain can be further divided into variable (V) region and constant (C) region. The V region exon recognizes different antigens, while C region exons contain Ig domains that are responsible for activating downstream effectors. V region is assembled by V, D and J gene segments on chromosome through V(D)J recombination. DNA endonuclease complex, RAG1/2, initiates V(D)J recombination to form primary immune (B and T) cell receptor repertoire [1]. The primary B cell receptor repertoire mainly contains low affinity IgM antibodies. To generate highly potent antibody, the antibody gene has to undergo two additional diversification processes, called somatic hypermutation (SHM) and Ig heavy chain (IgH) class switch recombination (CSR) [1] (Fig. 5.1a). SHM introduces mutations and/or insertions and deletions (indels) in the V exon, which allow the selection of B cells that produce high affinity antibodies ([29]), while IgH CSR switches its C regions from IgM to other Ig class, i.e. IgG, IgE, IgA [17] (Fig. 5.1a). Both SHM and CSR are initiated by the same enzyme, AID [46, 75, 106].

5.2.1 Secondary Immune Cell Receptor Diversification Initiated by AID

Naïve mature B cells express IgM and IgD receptors, and circulate to the secondary lymphoid tissues including spleen, Peyer's patch etc. Upon antigen stimulation, B cells can form germinal center (GC) structures with the help of T and other immune cells [134]. GC can be further divided into dark and light zones by immunohistochemistry features. In the GC dark zone, B cells actively divide and undergo SHM/CSR, while in the light zone, B cells compete for T follicular helper cells (T_{FH}) for survival [122, 134]. B cells expressing high affinity BCR are selected and undergo clonal expansion, a process termed affinity maturation. Eventually, B cells exit GC and become plasma or memory B cells, which secrete antibody or gain immune memory, respectively.

SHM and CSR are two distinct processes with different outcomes. Early studies on single B cells derived hybridoma have revealed clones that secrete IgM antibody that carry mutations in the V exon, and also clones that secrete switched IgG antibody with un-mutated V exons. The IgH C genes, e.g. murine IgH C genes, span ~200kb on chromosome 12 in an order of V(D)J-Cμ-Cδ-Cγ3-Cγ1-Cγ2b-Cγ2a-Cε-Cα [17, 37]. During SHM, AID is specifically recruited to the V(D)J exons to generate nucleotide mutations or small indels, while during CSR, AID targets switch (S) regions preceding each C gene to generate DNA double-strand breaks (DSBs). S regions are long repetitive GC-rich introns that can form secondary DNA structures such as G4 or R-loop, which contribute to optimal CSR efficiency [37]. A long-sought question in the field is how AID differentially targets the V(D)J exon and S regions within a short genomic region as the V(D)J exon and the first S region, Sμ, are less than 2kb apart. Until recently, SHM study armed with next-generation sequencing revealed significant SHM events in *ex vivo* cytokine-activated B cells that were considered to be CSR-only cells [143]. Thus, similar AID targeting mechanisms might

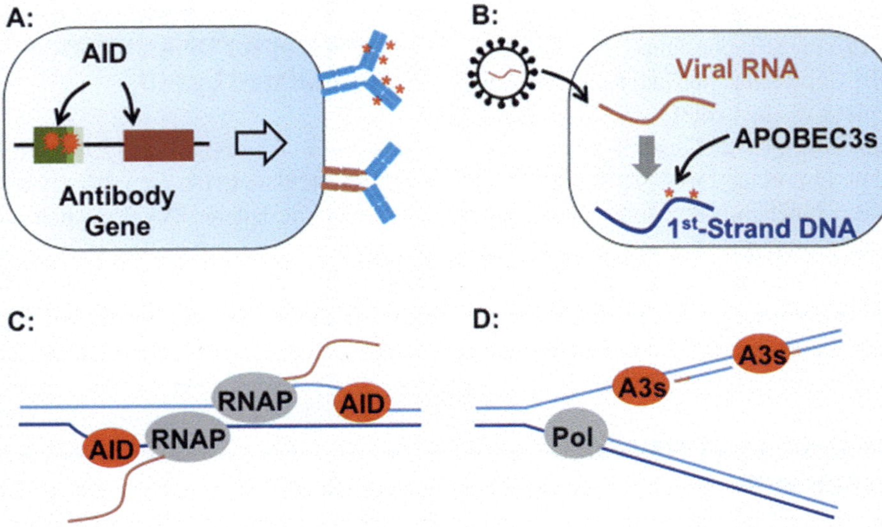

Fig. 5.1 AID/APOBEC-dependent genomic alterations in immunity and cancer genome. (**a**) AID initiates antibody gene somatic hypermutation and class switch recombination in adaptive immunity to generate high affinity antibodies with diverse effector functions. (**b**) A3s can deaminate C in the first-strand viral cDNA, playing anti-viral roles in innate immunity. (**c**) AID preferentially targets to strong convergent transcription regions in the genome. (**d**) A3s target ssDNA on the lagging strand of DNA replication fork. RNAP stands for RNA polymerase together with associated factors, and Pol indicates DNA polymerases (See text for more details)

be employed during SHM and CSR. The sequence intrinsic features of V(D)J exon and S region may underline the distinct outputs of these two processes as the S regions, which contain abundant AID-preferred motifs densely packed together, may generate DSBs more efficiently [37, 39].

General DNA repair proteins are involved in AID-initiated immune diversifications in an error-prone fashion. Cytidine deamination happens in many cell types either naturally or specifically by deaminase activity. To ensure the integrity of genomic information, cells use base excision repair (BER) and/or mismatch repair (MMR) factors to recognize and repair Us on genomic DNA, the ssDNA nicks or gaps are filled and ligated. B cells use the same set of factors to initiate the repair process but channel the Us into mutations or gene rearrangements ([29]). Among the well-defined BER and MMR factors, UNG or MSH gene deficiency results in defective CSR or SHM, suggesting a requirement of these factors to generate the intermediate abasic site or single-stranded nicks [31, 68, 103, 119, 140]. However, genetic aberration of the downstream BER/MMR factor XRCC1 has no obvious CSR defect [38]. Thus, B cell employs error-prone repair pathways to further process the intermediate alterations. In SHM, mutations are generated through translesion synthesis [80]; while in CSR, DSBs are generated via extensive ssDNA nicks [17]. Deficiencies of general DNA repair genes in mouse model and human patients usually result in immunodeficiency that is accompanied by genome instability.

5.2.2 Innate Anti-viral Activity of APOBEC3s

Through homologous protein searching, an anthropoid-specific expansion of APOBEC3 genes was identified [50]. However, the biological function of *APOBEC3s* in human seems to be a big puzzle. At the same time, Michael Malim group compared permissive and non-permissive cell lines that support or not support the production of fully infectious HIV-1 *Δvif* virions by using cDNA subtraction (the same technique used to identify AID), and identified CEM15 (now known as APOBEC3G) as the innate anti-viral factor [117]. In the absence of Vif protein, A3G can be packaged into viral particles and subsequently associates with the viral reverse transcription complex [117]. A3G deaminates C in the nascent minus-strand viral DNA, resulting in degradation and/or non-functional coding of viral DNA [41, 67, 145]. Anti-viral studies of APOBEC3s quickly extended to other A3s besides A3G and to other retroviruses. A3F, A3B, A3C and murine A3 were also found to be packaged into viral particles and inhibiting viral proliferation by similar mechanisms [9, 42] (Fig. 5.1b). A large portion of human genome is composed by retrotransposon elements. Human long interspersed nuclear element (LINE-1 or L1) is the only autonomous non-LTR retrotransposon, which codes L1 proteins that can also mobilize other non-autonomouse non-LTR retrotransposons like *Alu*s. A3 proteins, A3A/B/G/F were identified by different groups as potential retrotransposition inhibitors [10, 11, 125].

Some DNA viruses were later found being inhibited by A3 proteins. Hepatitis B virus (HBV) has to go through a reverse transcription step to replicate its genome, leading to the hypothesis that A3s may target to the minus-strand cDNA [84, 111, 126]. Later study revealed A3A and A3B deaminate the HBV covalently closed circular DNA in nucleus, providing another way of A3 anti-viral activity [62]. Human papillomaviruses (HPVs) infection is also associated with A3B expression, and might be inhibited by various A3s [133]. By which step A3s inhibit DNA virus propagation still need more work.

5.3 Collateral Effects of Cytidine Deamination

The DNA mutator function of AID/APOBEC proteins is like a double-edge sword. AID is required for secondary antibody gene diversity, and its deficiency causes Hyper-IgM syndrome [106]. On the other hand, mis-targeting of AID results in chromosomal translocations between proto-oncogenes and Ig loci that lead to GC-origin B cell lymphomas [107, 146].

5.3.1 Discovery of AID-dependent Translocations and Mutations

Leukemia and lymphoma were characterized by reciprocal chromosomal translocations involving the Ig loci and a variety of partners, often proto-oncogenes. This recurrent feature arises from the aberrant DNA breaks during immune diversity processes. AID-initiated DNA breaks during CSR and SHM are the major sources of genomic instability in mature B cell lymphomas of GC origin [107, 146]. DSBs at proto-oncogenes juxtapose proto-oncogenes next to the Ig loci, bringing the proto-oncogene closer to Ig super-enhancers (SEs) [28, 61]. The strong Ig SEs activate the oncogene expression and results in malignant transformation [61]. In B cell lymphomas, chromosomal translocation junctions were mapped to the Ig gene regions where programmed DNA lesion happens via RAG during V(D)J recombination or AID during antibody CSR/SHM [56].

The aberrant or unrepaired Ig breaks can fuse to breaks arising from many endogenous cellular activities. In activated B cells, AID off-targets beyond Ig genes turn out to be the major translocation partners of Ig break. The DNA mutator function makes AID a threat to genomic DNA. AID frequently targets many off-targets in the genome to generate mutations or DSBs, which eventually contribute to B cell lymphomas. Even before the identification of AID, BCL6, an oncogene, was already found to undergo somatic hypermutation in normal GCB as side effects of GC reaction [89, 118]. Later, more and more genes, such as CD95, CD79a and CD79b, were identified as AID off-targets in normal GCBs [36, 77]. Meanwhile, some proto-oncogenes were found mutated in B cell lymphomas [74, 90].

5.3.2 Systematic Identification of AID Off-targets

The case-by-case studies and emerging numbers of AID off-targets made systematic identification essential. Biochemical and genetic studies revealed that transcription is required for AID-initiated CSR [19, 124, 142]. The Schatz group sequenced hundreds of transcribed genes in GC B cells by Sanger sequencing methods and found a group of AID off-target genes, offering the first large-scale characterization of AID off-targets [60]. Later, AID ChIP-Seq experiments revealed thousands of potential AID targets in the genome [141].

The development of high-throughput sequencing tools has greatly pushed the field forward. The Alt lab has developed an approach termed high-throughput, genome-wide translocation sequencing (HTGTS) to isolate genome wide translocation junctions from large populations of cells [21]. HTGTS approach involves generating bait DSB and allowing the bait to fuse to endogenous breaks in a few cell cycles to avoid oncogenic selection [21]. Massively parallel sequencing of the translocation junctions revealed many AID-initiated breaks in the genome [21, 73]. TC-Seq used a similar approach to identify AID-initiated breaks [52]. Combination of AID over-expression and DNA repair response deficiency further increase the numbers of identified AID off-targets [22, 52, 73]. Single-stranded DNA binding protein RPA ChIP-Seq in 53BP1-deficient cells was also used to identify AID off-targets [100], as in 53BP1$^{-/-}$ cells DNA breaks undergo extensive resection and marked with RPA binding [141]. In summary, any method that can detect endogenous DNA breaks can be used to mark AID off-targets through AID proficient and deficient comparison. Oncogenic chromosomal translocation favoring cell proliferation is selected during tumor transformation and progression. In HTGTS or other methods, cells only proliferate for a few cell cycles, thus the method revealed an unbiased pool of AID off-target [21]. Surprisingly, more than 25% of AID off-targets identified in mouse B cells have orthologs translocated or mutated in human B cell lymphomas [73]. The puzzle was later elaborated by the preferential targeting of intragenic super-enhancer regions by AID (see below sections) [73] and super-enhancers are usually associated with key cell identity and tumor pathogenesis genes [45, 138].

5.3.3 Pervasive APOBEC3 Signature in Cancer Genomes

The role of AID-initiated genomic instability is well recognized in lymphomas of GC B cell origin and leukemia that have aberrant AID expression. The expression of AID depends on transcription factors that also are involved in inflammatory response, leading to the hypothesis that genomic instability in inflammation-associated cancers may also be attributed to AID [69]. Aberrant AID expression was detected in hepatocytes with chronic inflammation caused by HCV infection, cultured cancer lines, etc [128]. How and to what extent AID contributes to cancer development arising from epithelial cells needs more detailed work.

Member of APOBEC3 proteins is expressed in many cell types. Whether they could contribute to cancer genomic instability remains a big mystery, as anthropoid-specific expansion of APOBEC3 genes makes it difficult to study them in other animal models like mouse model. The "21 breast cancer genome" sequencing revealed many unknown mutational signatures, with one group of them showing a clear APOBEC3 signature [81]. Combining gene expression data and APOBEC3 deamination signature, the Harris lab identified A3B as the main enzyme responsible for APOBEC mutational signature (mutagenesis of C in T<u>C</u>A context) [13]. With many other types of cancer genome sequencing data, it was found that this mutation signature is widespread in human cancers [14, 55, 59, 129]. Additional A3 deaminase activity may also account for this mutation signature, as carriers of germline copy number polymorphism of A3A-A3B gene (i.e. A3B deletion) also possess APOBEC3 mutations [82]. The APOBEC signature was mainly found on the lagging-strand template during DNA replication, suggesting a model where A3s may hijack the replication machinery to attack ssDNA genomic widely [40, 47, 116] (Fig.5.1d). Different from AID deamination, so far there has been no evidence that the APOBEC3 signature associates with recurrent translocations.

5.4 Multilayer Regulation of AID Deamination

AID/APOBEC deamination activity must be tightly controlled to ensure genome integrity while playing their roles in immunity. Amongst the AID/APOBEC protein family, AID is the most intensively studied. Here we use AID as an example to review the multilayer regulation of deaminase activity:

5.4.1 Regulation of AID Expression

Although AID was reported to be expressed in pluripotent tissues in many studies [8, 98], its function in DNA demethylation in early development is challenged by the fact that AID-deficient mouse show no sign of developmental defect [75]. Thus, the role of AID in those tissues remains controversial. One of the reasons came from the extreme low expression levels in these cells compared to the level in activated B cells. Early studies with *AICDA* BAC constructs and luciferase assays have discovered four conserved regions of cis-elements around *AICDA* locus [25, 131]. Deletion of the regulatory region abolishes AID-mediated CSR and these regions are responsible for AID expression upon B cell activation [25]. With the development of genomic approaches such as H3K27ac ChIP-Seq and GRO-Seq, a detailed map of regulatory elements was revealed for *AICDA* locus, and more cis regulatory regions were revealed [73]. The cis-elements in these regions offers binding site for many transcription factors including but not restricted to NF-κB, STAT6, Smad4, Pax5, E-boxes protein, BATF, etc [35, 49, 115, 131]. Transcription of AID is also negatively regulated through repressive cis-elements in Intron 1 that contain putative binding sites of Myb and E2f [131], trans-factors like inhibitory E-box protein ID2 [35], or PI3K-Akt-Foxo signaling axis [87]. Inhibition of repressive regulatory factors would turn on the expression of AID and cause genome instability, as shown in the chronic lymphocytic leukemia (CLL) cells and normal B cells treated with PI3K inhibitors [22]. The tight regulation of AID transcription restricts its activity to activated B cells undergoing SHM and CSR.

5.4.2 Intracellular Localization of AID Protein

Once translated, AID protein folds with the assistance from chaperones Hsp90/DnaJA1 [88], and further loads to a large cytoplasmic complex including eEF1A [44]. The cytoplasmic complex protects AID from degradation [44]. To perform its cellular function, AID protein must be imported to the nucleus from cytoplasm. AID has a N-terminal nuclear localization signal (NLS) and C-terminal nuclear export signal (NES) sequence to regulate its trafficking between nucleus and cytoplasm [121]. Mutations of critical amino acid residues in the NLS sequence lost SHM activity but have normal CSR level, i.e. AID^{G23S} knock-in mice show defective SHM but normal Ig levels in serum and intestinal secretions [121, 137]. To strictly control its activity, nucleus AID is actively excluded by CRM1-depedent pathway through the C-terminal NES sequence [70]. More than 90% of the AID proteins were localized in cytoplasm revealed either by immunofluorescent or biochemical fractionation methods [70]. Patients harboring AID C-terminal truncation or mutations display dominant-negative effects in CSR but SHM levels is less affected [3, 48, 127]. How SHM and CSR were differentially regulated by AID still need more work. Many labs have shown that replacement of AID NES with other NESs can restore the nuclear export function but cannot fully support CSR [32, 34]. Later on, the AID C-terminal peptide was suggested to recruit DNA repair factor to Ig loci through a yet-to-be-characterized pathway [104, 144].

5.4.3 Post-translation Modification of AID Protein

Post-translation modification adds another layer of complexity to AID activity regulation. By comparing AID activity purified from activated B cells and 293T cells, the Alt lab found AID^{Bcell} specifically interacts with RPA and possess deamination activity on transcribed dsDNA substrate [18]. The major difference between AID^{Bcell} and AID^{293T} was later identified as Serine 38 phosphorylation through PKA kinase in activated B cells [4]. Mutation of S38 to alanine abolishes the deamination activity *in vitro*, decrease SHM/CSR level in *ex vivo* cultured B cells and in *in vivo* GC B cells [6, 16, 20, 71, 72, 97]. A working model was proposed that PKA is specifically recruited to the S regions to promote active AID complex formation with RPA during CSR [135]. Along with S38, many other phosphorylation residues were also identified on AID protein including T27, Y184, which are also required for optimal CSR [4]. AID contains 198 a.a. residues and is a relative small protein, and many AID mutations were reported in human Type II Hyper-IgM syndrome patients [48, 66]. However, no mutation at AID phosphorylation site has yet to be identified in human patients. It would be of great interest to know whether AID phosphorylation play important roles in human immunodeficiency. Besides phosphorylation, AID was also reported to be poly-ubiquitinated in the nucleus [2], which may offer negative regulation of AID activity through degradation of nuclear AID. Although no specific lysine was identified for AID ubiquitination, genetic screening revealed RNF126 as the E3 ligase for AID [27].

5.4.4 Specific Genomic Targeting of AID

How AID specifically recognize Ig loci is a long-sought question during CSR and SHM. Before AID was identified, genetic evidence suggested transcription is absolutely required for IgH CSR and SHM [33, 63, 64, 112, 113, 123, 142]. Both RNA Polymerase II (Pol II) and Pol I initiated transcription are sufficient to support SHM [33]. Immediately after AID was proven to be the enzyme responsible for CSR and SHM, it was found that AID associates with Pol II transcription machinery [78]. The link between transcription and AID deamination activity strongly suggest that AID targets ssDNA generated during transcription. However, early studies with individual genes revealed that not all transcribed genes are targeted by AID nor the highest-transcribed genes are bona fide AID targets [60]. What makes Ig genes so special as AID targets became an intensively studied question.

Next-generation sequencing based approaches strongly suggested pervasive transcription feature of eukaryotic genome. ChIP-Seq reveals the genome wide localization of a particular protein; RNA-Seq reveals many cellular RNA species at particular subcellular compartment including chromatin associated RNAs [7]; Global Nuclear Run-on sequencing (GRO-Seq) and its related methods uncover genome wide transcription machinery dynamics and nascent RNA generation [24, 58]. Combination of these approaches together with the identification of AID off-targets has given definitive answers for AID targeting associated features [73, 100]. The Nussenzweig lab used genetic screening to find Spt5 as AID interaction protein and suggested AID targets to regions with paused Pol II [91]. The Alt lab used GRO-Seq to check the transcription machinery in *ex vivo* activated B cells and found AID off-targets are enriched in sense-antisense convergent transcription regions and AID-dependent translocation numbers correlates with the convergent transcription levels (defined by the geometric mean of sense and antisense transcription reads) [73] (Fig. 5.1c). The sense transcription comes from gene transcription, while the origin of antisense transcript remained a mystery. Similarly, AID off-targets were found enriched in some divergent transcribed promoters [92]. In eukaryotic genome, transcription outside of gene region mainly comes from enhancers [26]. The Alt lab used H3K27Ac ChIP-Seq to identify typical enhancers and super-enhancers in *ex vivo* activated B cells and found specific enrichment of AID activity in SE regions [73]. Similar conclusion was drawn by an independent study from the Casellas lab [100]. Further bioinformatics examination concludes AID preferentially targets to the intragenic SEs (overlap region between SE and gene pair) with strong convergent transcription [73] (Fig. 5.1c).

GC B cells only take up a small fraction (~5-10%) of total splenic B cells after immunization, which was considered to be impossible for genomic studies like ChIP-Seq. With optimized protocols, the Alt lab further extended the study to *in vivo* GC B cells, and found most of the identified AID off-target regions localized in intragenic SEs with convergent transcription [73]. With human tonsil germinal center SE data, recurrent translocation and kataegis (cluster of mutations in cancer genome) were found frequently located in intragenic SEs [73, 100]. However, most of the conclusions were drawn from non-Ig AID off-targets. The repetitive nature of Ig gene sequences presents a big obstacle to study AID recruitment on Ig genes (Discussed in Refs. [73, 91]). Hopefully, new mouse models and long sequencing reads can solve the problem.

5.4.5 AID Cofactors and Activity Regulation

Many AID cofactors were identified in the last two decades, including the above-mentioned Pol II, CRM1, RPA, Spt5, eEF1A, etc [18, 44, 70, 78, 91]. AID protein has a large portion of negative-charged surface, which contribute to its single-stranded nucleotide acid binding. Purified AID binds to many non-specific ssRNAs, which intrigued to study the role of ssRNA/DNA binding/processing factors during AID activity regulation [12]. IgH S region repetitive DNA sequence contains AGCT motifs, and biochemical experiment isolated 14-3-3 proteins as AGCT motifs binding protein [139]. 14-3-3 proteins were also suggested to directly recruit AID to the S regions, and 14-3-3 gene deficiency led to defective CSR in B cells [139]. In another study, RNA splicing factor PTBP2 was identified as AID binding protein, which was suggested to recruit AID to the S region and required for optimal CSR [85].

AID deamination activity was successfully reconstituted *in vitro* with ssDNA and transcribed dsDNA substrates [4, 18, 19]. However, it remained a big puzzle at the time why AID can only deaminate the non-template strand of transcribed dsDNA substrate *in vitro* [4, 18, 19] as it is well documented that AID can deaminate both strands *in vivo* [86]. Biochemical study eventually discovered RNA exosome as the regulator that can target AID activity to both strands [5]. Since RNA exosome is the major cellular RNA processing machinery, it aroused researcher's curiosity whether RNA or non-coding RNA plays a role in antibody diversification. Later, roles of lncRNA and G4 RNA were reported by the Basu lab [93] and the Chaudhuri lab [147], respectively.

5.5 Future

Nearly two decades have passed since the discovery of AID/APOBEC DNA mutator enzymes. Now we know that multiple layers of regulation are evolved to restrict their deamination activity in immunity. However, many aspects of AID/APOBEC regulation are still unclear, and the recent development of base-editing (BE) tools with AID/APOBEC enzymes [54, 57, 65, 83] raises additional questions and concerns. Whether the engineered deaminases in BE tools still retain dinucleotide motif preference or off-target activity? How error-free and error-prone DNA repair pathways compete for the BE outcomes in different cell types? Answers to these questions will also benefit the understanding of programed AID/APOBEC deamination in both immunity and cancer genome integrity.

Acknowledgements Research in the authors' lab is supported by grants from Science and Technology Commission of Shanghai Municipality (16ZR1441600), National Natural Science Foundation of China (8162200437, 31670929).

References

1. Alt FW, Zhang Y, Meng FL, Guo C, Schwer B (2013) Mechanisms of programmed DNA lesions and genomic instability in the immune system. Cell 152(3):417–429. https://doi.org/10.1016/j.cell.2013.01.007
2. Aoufouchi S, Faili A, Zober C, D'Orlando O, Weller S, Weill JC, Reynaud CA (2008) Proteasomal

degradation restricts the nuclear lifespan of AID. J Exp Med 205(6):1357–1368. https://doi.org/10.1084/jem.20070950
3. Barreto V, Reina-San-Martin B, Ramiro AR, McBride KM, Nussenzweig MC (2003) C-terminal deletion of AID uncouples class switch recombination from somatic hypermutation and gene conversion. Mol Cell 12(2):501–508
4. Basu U, Chaudhuri J, Alpert C, Dutt S, Ranganath S, Li G, Schrum JP, Manis JP, Alt FW (2005) The AID antibody diversification enzyme is regulated by protein kinase A phosphorylation. Nature 438(7067):508–511. https://doi.org/10.1038/nature04255
5. Basu U, Meng FL, Keim C, Grinstein V, Pefanis E, Eccleston J, Zhang T, Myers D, Wasserman CR, Wesemann DR, Januszyk K, Gregory RI, Deng H, Lima CD, Alt FW (2011) The RNA exosome targets the AID cytidine deaminase to both strands of transcribed duplex DNA substrates. Cell 144(3):353–363. https://doi.org/10.1016/j.cell.2011.01.001
6. Basu U, Wang Y, Alt FW (2008) Evolution of phosphorylation-dependent regulation of activation-induced cytidine deaminase. Mol Cell 32(2):285–291. https://doi.org/10.1016/j.molcel.2008.08.019
7. Bhatt DM, Pandya-Jones A, Tong AJ, Barozzi I, Lissner MM, Natoli G, Black DL, Smale ST (2012) Transcript dynamics of proinflammatory genes revealed by sequence analysis of subcellular RNA fractions. Cell 150(2):279–290. https://doi.org/10.1016/j.cell.2012.05.043
8. Bhutani N, Brady JJ, Damian M, Sacco A, Corbel SY, Blau HM (2010) Reprogramming towards pluripotency requires AID-dependent DNA demethylation. Nature 463(7284):1042–1047. https://doi.org/10.1038/nature08752
9. Bishop KN, Holmes RK, Sheehy AM, Davidson NO, Cho SJ, Malim MH (2004) Cytidine deamination of retroviral DNA by diverse APOBEC proteins. Current biology : CB 14(15):1392–1396. https://doi.org/10.1016/j.cub.2004.06.057
10. Bogerd HP, Wiegand HL, Doehle BP, Lueders KK, Cullen BR (2006a) APOBEC3A and APOBEC3B are potent inhibitors of LTR-retrotransposon function in human cells. Nucleic Acids Res 34(1):89–95. https://doi.org/10.1093/nar/gkj416
11. Bogerd HP, Wiegand HL, Hulme AE, Garcia-Perez JL, O'Shea KS, Moran JV, Cullen BR (2006b) Cellular inhibitors of long interspersed element 1 and Alu retrotransposition. Proc Natl Acad Sci U S A 103(23):8780–8785. https://doi.org/10.1073/pnas.0603313103
12. Bransteitter R, Pham P, Scharff MD, Goodman MF (2003) Activation-induced cytidine deaminase deaminates deoxycytidine on single-stranded DNA but requires the action of RNase. Proc Natl Acad Sci U S A 100(7):4102–4107. https://doi.org/10.1073/pnas.0730835100
13. Burns MB, Lackey L, Carpenter MA, Rathore A, Land AM, Leonard B, Refsland EW, Kotandeniya D, Tretyakova N, Nikas JB, Yee D, Temiz NA, Donohue DE, McDougle RM, Brown WL, Law EK, Harris RS (2013a) APOBEC3B is an enzymatic source of mutation in breast cancer. Nature 494(7437):366–370. https://doi.org/10.1038/nature11881
14. Burns MB, Temiz NA, Harris RS (2013b) Evidence for APOBEC3B mutagenesis in multiple human cancers. Nat Genet 45(9):977–983. https://doi.org/10.1038/ng.2701
15. Cascalho M, Wong J, Steinberg C, Wabl M (1998) Mismatch repair co-opted by hypermutation. Science 279(5354):1207–1210
16. Chatterji M, Unniraman S, McBride KM, Schatz DG (2007) Role of activation-induced deaminase protein kinase A phosphorylation sites in Ig gene conversion and somatic hypermutation. J Immunol 179(8):5274–5280
17. Chaudhuri J, Basu U, Zarrin A, Yan C, Franco S, Perlot T, Vuong B, Wang J, Phan RT, Datta A, Manis J, Alt FW (2007) Evolution of the immunoglobulin heavy chain class switch recombination mechanism. Adv Immunol 94:157–214. https://doi.org/10.1016/S0065-2776(06)94006-1
18. Chaudhuri J, Khuong C, Alt FW (2004) Replication protein A interacts with AID to promote deamination of somatic hypermutation targets. Nature 430(7003):992–998. https://doi.org/10.1038/nature02821
19. Chaudhuri J, Tian M, Khuong C, Chua K, Pinaud E, Alt FW (2003) Transcription-targeted DNA deamination by the AID antibody diversification enzyme. Nature 422(6933):726–730. https://doi.org/10.1038/nature01574
20. Cheng HL, Vuong BQ, Basu U, Franklin A, Schwer B, Astarita J, Phan RT, Datta A, Manis J, Alt FW, Chaudhuri J (2009) Integrity of the AID serine-38 phosphorylation site is critical for class switch recombination and somatic hypermutation in mice. Proc Natl Acad Sci U S A 106(8):2717–2722. https://doi.org/10.1073/pnas.0812304106
21. Chiarle R, Zhang Y, Frock RL, Lewis SM, Molinie B, Ho YJ, Myers DR, Choi VW, Compagno M, Malkin DJ, Neuberg D, Monti S, Giallourakis CC, Gostissa M, Alt FW (2011) Genome-wide translocation sequencing reveals mechanisms of chromosome breaks and rearrangements in B cells. Cell 147(1):107–119. https://doi.org/10.1016/j.cell.2011.07.049
22. Compagno M, Wang Q, Pighi C, Cheong TC, Meng FL, Poggio T, Yeap LS, Karaca E, Blasco RB, Langellotto F, Ambrogio C, Voena C, Wiestner A, Kasar SN, Brown JR, Sun J, Wu CJ, Gostissa M, Alt FW, Chiarle R (2017) Phosphatidylinositol 3-kinase delta blockade increases genomic instability in B cells. Nature 542(7642):489–493. https://doi.org/10.1038/nature21406
23. Conticello SG (2008) The AID/APOBEC family of nucleic acid mutators. Genome Biol 9(6):229. https://doi.org/10.1186/gb-2008-9-6-229

24. Core LJ, Waterfall JJ, Lis JT (2008) Nascent RNA sequencing reveals widespread pausing and divergent initiation at human promoters. Science 322(5909):1845–1848. https://doi.org/10.1126/science.1162228
25. Crouch EE, Li Z, Takizawa M, Fichtner-Feigl S, Gourzi P, Montano C, Feigenbaum L, Wilson P, Janz S, Papavasiliou FN, Casellas R (2007) Regulation of AID expression in the immune response. J Exp Med 204(5):1145–1156. https://doi.org/10.1084/jem.20061952
26. De Santa F, Barozzi I, Mietton F, Ghisletti S, Polletti S, Tusi BK, Muller H, Ragoussis J, Wei CL, Natoli G (2010) A large fraction of extragenic RNA pol II transcription sites overlap enhancers. PLoS Biol 8(5):e1000384. https://doi.org/10.1371/journal.pbio.1000384
27. Delker RK, Zhou Y, Strikoudis A, Stebbins CE, Papavasiliou FN (2013) Solubility-based genetic screen identifies RING finger protein 126 as an E3 ligase for activation-induced cytidine deaminase. Proc Natl Acad Sci U S A 110(3):1029–1034. https://doi.org/10.1073/pnas.1214538110
28. Delmore JE, Issa GC, Lemieux ME, Rahl PB, Shi J, Jacobs HM, Kastritis E, Gilpatrick T, Paranal RM, Qi J, Chesi M, Schinzel AC, McKeown MR, Heffernan TP, Vakoc CR, Bergsagel PL, Ghobrial IM, Richardson PG, Young RA, Hahn WC, Anderson KC, Kung AL, Bradner JE, Mitsiades CS (2011) BET bromodomain inhibition as a therapeutic strategy to target c-Myc. Cell 146(6):904–917. https://doi.org/10.1016/j.cell.2011.08.017
29. Di Noia JM, Neuberger MS (2007) Molecular mechanisms of antibody somatic hypermutation. Annu Rev Biochem 76:1–22. https://doi.org/10.1146/annurev.biochem.76.061705.090740
30. Doehle BP, Schafer A, Cullen BR (2005) Human APOBEC3B is a potent inhibitor of HIV-1 infectivity and is resistant to HIV-1 Vif. Virology 339(2):281–288. https://doi.org/10.1016/j.virol.2005.06.005
31. Ehrenstein MR, Neuberger MS (1999) Deficiency in Msh2 affects the efficiency and local sequence specificity of immunoglobulin class-switch recombination: parallels with somatic hypermutation. EMBO J 18(12):3484–3490. https://doi.org/10.1093/emboj/18.12.3484
32. Ellyard JI, Benk AS, Taylor B, Rada C, Neuberger MS (2011) The dependence of Ig class-switching on the nuclear export sequence of AID likely reflects interaction with factors additional to Crm1 exportin. Eur J Immunol 41(2):485–490. https://doi.org/10.1002/eji.201041011
33. Fukita Y, Jacobs H, Rajewsky K (1998) Somatic hypermutation in the heavy chain locus correlates with transcription. Immunity 9(1):105–114
34. Geisberger R, Rada C, Neuberger MS (2009) The stability of AID and its function in class-switching are critically sensitive to the identity of its nuclear-export sequence. Proc Natl Acad Sci U S A 106(16):6736–6741. https://doi.org/10.1073/pnas.0810808106
35. Gonda H, Sugai M, Nambu Y, Katakai T, Agata Y, Mori KJ, Yokota Y, Shimizu A (2003) The balance between Pax5 and Id2 activities is the key to AID gene expression. J Exp Med 198(9):1427–1437. https://doi.org/10.1084/jem.20030802
36. Gordon MS, Kanegai CM, Doerr JR, Wall R (2003) Somatic hypermutation of the B cell receptor genes B29 (Igbeta, CD79b) and mb1 (Igalpha, CD79a). Proc Natl Acad Sci U S A 100(7):4126–4131. https://doi.org/10.1073/pnas.0735266100
37. Hackney JA, Misaghi S, Senger K, Garris C, Sun Y, Lorenzo MN, Zarrin AA (2009) DNA targets of AID evolutionary link between antibody somatic hypermutation and class switch recombination. Adv Immunol 101:163–189. https://doi.org/10.1016/S0065-2776(08)01005-5
38. Han L, Mao W, Yu K (2012) X-ray repair cross-complementing protein 1 (XRCC1) deficiency enhances class switch recombination and is permissive for alternative end joining. Proc Natl Acad Sci U S A 109(12):4604–4608. https://doi.org/10.1073/pnas.1120743109
39. Han L, Masani S, Yu K (2011) Overlapping activation-induced cytidine deaminase hotspot motifs in Ig class-switch recombination. Proc Natl Acad Sci U S A 108(28):11584–11589. https://doi.org/10.1073/pnas.1018726108
40. Haradhvala NJ, Polak P, Stojanov P, Covington KR, Shinbrot E, Hess JM, Rheinbay E, Kim J, Maruvka YE, Braunstein LZ, Kamburov A, Hanawalt PC, Wheeler DA, Koren A, Lawrence MS, Getz G (2016) Mutational Strand Asymmetries in Cancer Genomes Reveal Mechanisms of DNA Damage and Repair. Cell 164(3):538–549. https://doi.org/10.1016/j.cell.2015.12.050
41. Harris RS, Bishop KN, Sheehy AM, Craig HM, Petersen-Mahrt SK, Watt IN, Neuberger MS, Malim MH (2003) DNA deamination mediates innate immunity to retroviral infection. Cell 113(6):803–809
42. Harris RS, Liddament MT (2004) Retroviral restriction by APOBEC proteins. Nature reviews Immunology 4(11):868–877. https://doi.org/10.1038/nri1489
43. Harris RS, Petersen-Mahrt SK, Neuberger MS (2002) RNA editing enzyme APOBEC1 and some of its homologs can act as DNA mutators. Mol Cell 10(5):1247–1253
44. Hasler J, Rada C, Neuberger MS (2011) Cytoplasmic activation-induced cytidine deaminase (AID) exists in stoichiometric complex with translation elongation factor 1alpha (eEF1A). Proc Natl Acad Sci U S A 108(45):18366–18371. https://doi.org/10.1073/pnas.1106729108
45. Hnisz D, Abraham BJ, Lee TI, Lau A, Saint-Andre V, Sigova AA, Hoke HA, Young RA (2013) Super-enhancers in the control of cell identity and disease. Cell 155(4):934–947. https://doi.org/10.1016/j.cell.2013.09.053

46. Honjo T, Kinoshita K, Muramatsu M (2002) Molecular mechanism of class switch recombination: linkage with somatic hypermutation. Annu Rev Immunol 20:165–196. https://doi.org/10.1146/annurev.immunol.20.090501.112049
47. Hoopes J, Cortez L, Mertz T, Malc EP, Mieczkowski PA, Roberts SA (2016) APOBEC3A and APOBEC3B Preferentially Deaminate the Lagging Strand Template during DNA Replication. Cell Reports 14(6):1273–1282. https://doi.org/10.1016/j.celrep.2016.01.021
48. Imai K, Zhu Y, Revy P, Morio T, Mizutani S, Fischer A, Nonoyama S, Durandy A (2005) Analysis of class switch recombination and somatic hypermutation in patients affected with autosomal dominant hyper-IgM syndrome type 2. Clin Immunol 115(3):277–285. https://doi.org/10.1016/j.clim.2005.02.003
49. Ise W, Kohyama M, Schraml BU, Zhang T, Schwer B, Basu U, Alt FW, Tang J, Oltz EM, Murphy TL, Murphy KM (2011) The transcription factor BATF controls the global regulators of class-switch recombination in both B cells and T cells. Nat Immunol 12(6):536–543. https://doi.org/10.1038/ni.2037
50. Jarmuz A, Chester A, Bayliss J, Gisbourne J, Dunham I, Scott J, Navaratnam N (2002) An anthropoid-specific locus of orphan C to U RNA-editing enzymes on chromosome 22. Genomics 79(3):285–296. https://doi.org/10.1006/geno.2002.6718
51. Johnson DF, Poksay KS, Innerarity TL (1993) The mechanism for apo-B mRNA editing is deamination. Biochem Biophys Res Commun 195(3):1204–1210. https://doi.org/10.1006/bbrc.1993.2172
52. Klein IA, Resch W, Jankovic M, Oliveira T, Yamane A, Nakahashi H, Di Virgilio M, Bothmer A, Nussenzweig A, Robbiani DF, Casellas R, Nussenzweig MC (2011) Translocation-capture sequencing reveals the extent and nature of chromosomal rearrangements in B lymphocytes. Cell 147(1):95–106. https://doi.org/10.1016/j.cell.2011.07.048
53. Kohli RM, Maul RW, Guminski AF, McClure RL, Gajula KS, Saribasak H, McMahon MA, Siliciano RF, Gearhart PJ, Stivers JT (2010) Local sequence targeting in the AID/APOBEC family differentially impacts retroviral restriction and antibody diversification. J Biol Chem 285(52):40956–40964. https://doi.org/10.1074/jbc.M110.177402
54. Komor AC, Kim YB, Packer MS, Zuris JA, Liu DR (2016) Programmable editing of a target base in genomic DNA without double-stranded DNA cleavage. Nature 533(7603):420–424. https://doi.org/10.1038/nature17946
55. Kuong KJ, Loeb LA (2013) APOBEC3B mutagenesis in cancer. Nature Genet 45(9):964–965. https://doi.org/10.1038/ng.2736
56. Kuppers R, Dalla-Favera R (2001) Mechanisms of chromosomal translocations in B cell lymphomas. Oncogene 20(40):5580–5594. https://doi.org/10.1038/sj.onc.1204640
57. Kuscu C, Adli M (2016) CRISPR-Cas9-AID base editor is a powerful gain-of-function screening tool. Nature methods 13(12):983–984. https://doi.org/10.1038/nmeth.4076
58. Kwak H, Fuda NJ, Core LJ, Lis JT (2013) Precise maps of RNA polymerase reveal how promoters direct initiation and pausing. Science 339(6122):950–953. https://doi.org/10.1126/science.1229386
59. Leonard B, Hart SN, Burns MB, Carpenter MA, Temiz NA, Rathore A, Vogel RI, Nikas JB, Law EK, Brown WL, Li Y, Zhang Y, Maurer MJ, Oberg AL, Cunningham JM, Shridhar V, Bell DA, April C, Bentley D, Bibikova M, Cheetham RK, Fan JB, Grocock R, Humphray S, Kingsbury Z, Peden J, Chien J, Swisher EM, Hartmann LC, Kalli KR, Goode EL, Sicotte H, Kaufmann SH, Harris RS (2013) APOBEC3B upregulation and genomic mutation patterns in serous ovarian carcinoma. Cancer Res 73(24):7222–7231. https://doi.org/10.1158/0008-5472.CAN-13-1753
60. Liu M, Duke JL, Richter DJ, Vinuesa CG, Goodnow CC, Kleinstein SH, Schatz DG (2008) Two levels of protection for the B cell genome during somatic hypermutation. Nature 451(7180):841–845. https://doi.org/10.1038/nature06547
61. Loven J, Hoke HA, Lin CY, Lau A, Orlando DA, Vakoc CR, Bradner JE, Lee TI, Young RA (2013) Selective inhibition of tumor oncogenes by disruption of super-enhancers. Cell 153(2):320–334. https://doi.org/10.1016/j.cell.2013.03.036
62. Lucifora J, Xia Y, Reisinger F, Zhang K, Stadler D, Cheng X, Sprinzl MF, Koppensteiner H, Makowska Z, Volz T, Remouchamps C, Chou WM, Thasler WE, Huser N, Durantel D, Liang TJ, Munk C, Heim MH, Browning JL, Dejardin E, Dandri M, Schindler M, Heikenwalder M, Protzer U (2014) Specific and nonhepatotoxic degradation of nuclear hepatitis B virus cccDNA. Science 343(6176):1221–1228. https://doi.org/10.1126/science.1243462
63. Lutzker S, Alt FW (1988) Structure and expression of germ line immunoglobulin gamma 2b transcripts. Mol Cell Biol 8(4):1849–1852
64. Lutzker S, Rothman P, Pollock R, Coffman R, Alt FW (1988) Mitogen- and IL-4-regulated expression of germ-line Ig gamma 2b transcripts: evidence for directed heavy chain class switching. Cell 53(2):177–184
65. Ma Y, Zhang J, Yin W, Zhang Z, Song Y, Chang X (2016) Targeted AID-mediated mutagenesis (TAM) enables efficient genomic diversification in mammalian cells. Nature methods 13(12):1029–1035. https://doi.org/10.1038/nmeth.4027
66. Mahdaviani SA, Hirbod-Mobarakeh A, Wang N, Aghamohammadi A, Hammarstrom L, Masjedi MR, Pan-Hammarstrom Q, Rezaei N (2012) Novel mutation of the activation-induced cytidine deaminase gene in a Tajik family: special review on hyper-immunoglobulin M syndrome. Expert Rev Clin Immunol 8(6):539–546. https://doi.org/10.1586/eci.12.46

67. Mangeat B, Turelli P, Caron G, Friedli M, Perrin L, Trono D (2003) Broad antiretroviral defence by human APOBEC3G through lethal editing of nascent reverse transcripts. Nature 424(6944):99–103. https://doi.org/10.1038/nature01709
68. Martin A, Li Z, Lin DP, Bardwell PD, Iglesias-Ussel MD, Edelmann W, Scharff MD (2003) Msh2 ATPase activity is essential for somatic hypermutation at a-T basepairs and for efficient class switch recombination. J Exp Med 198(8):1171–1178. https://doi.org/10.1084/jem.20030880
69. Marusawa H, Takai A, Chiba T (2011) Role of activation-induced cytidine deaminase in inflammation-associated cancer development. Adv Immunol 111:109–141. https://doi.org/10.1016/B978-0-12-385991-4.00003-9
70. McBride KM, Barreto V, Ramiro AR, Stavropoulos P, Nussenzweig MC (2004) Somatic hypermutation is limited by CRM1-dependent nuclear export of activation-induced deaminase. J Exp Med 199(9):1235–1244. https://doi.org/10.1084/jem.20040373
71. McBride KM, Gazumyan A, Woo EM, Barreto VM, Robbiani DF, Chait BT, Nussenzweig MC (2006) Regulation of hypermutation by activation-induced cytidine deaminase phosphorylation. Proc Natl Acad Sci U S A 103(23):8798–8803. https://doi.org/10.1073/pnas.0603272103
72. McBride KM, Gazumyan A, Woo EM, Schwickert TA, Chait BT, Nussenzweig MC (2008) Regulation of class switch recombination and somatic mutation by AID phosphorylation. J Exp Med 205(11):2585–2594. https://doi.org/10.1084/jem.20081319
73. Meng FL, Du Z, Federation A, Hu J, Wang Q, Kieffer-Kwon KR, Meyers RM, Amor C, Wasserman CR, Neuberg D, Casellas R, Nussenzweig MC, Bradner JE, Liu XS, Alt FW (2014) Convergent transcription at intragenic super-enhancers targets AID-initiated genomic instability. Cell 159(7):1538–1548. https://doi.org/10.1016/j.cell.2014.11.014
74. Migliazza A, Martinotti S, Chen W, Fusco C, Ye BH, Knowles DM, Offit K, Chaganti RS, Dalla-Favera R (1995) Frequent somatic hypermutation of the 5′ noncoding region of the BCL6 gene in B-cell lymphoma. Proc Natl Acad Sci U S A 92(26):12520–12524
75. Muramatsu M, Kinoshita K, Fagarasan S, Yamada S, Shinkai Y, Honjo T (2000) Class switch recombination and hypermutation require activation-induced cytidine deaminase (AID), a potential RNA editing enzyme. Cell 102(5):553–563
76. Muramatsu M, Sankaranand VS, Anant S, Sugai M, Kinoshita K, Davidson NO, Honjo T (1999) Specific expression of activation-induced cytidine deaminase (AID), a novel member of the RNA-editing deaminase family in germinal center B cells. J Biol Chem 274(26):18470–18476
77. Muschen M, Re D, Jungnickel B, Diehl V, Rajewsky K, Kuppers R (2000) Somatic mutation of the CD95 gene in human B cells as a side-effect of the germinal center reaction. J Exp Med 192(12):1833–1840
78. Nambu Y, Sugai M, Gonda H, Lee CG, Katakai T, Agata Y, Yokota Y, Shimizu A (2003) Transcription-coupled events associating with immunoglobulin switch region chromatin. Science 302(5653):2137–2140. https://doi.org/10.1126/science.1092481
79. Navaratnam N, Morrison JR, Bhattacharya S, Patel D, Funahashi T, Giannoni F, Teng BB, Davidson NO, Scott J (1993) The p27 catalytic subunit of the apolipoprotein B mRNA editing enzyme is a cytidine deaminase. J Biol Chem 268(28):20709–20712
80. Neuberger MS, Rada C (2007) Somatic hypermutation: activation-induced deaminase for C/G followed by polymerase eta for A/T. J Exp Med 204(1):7–10. https://doi.org/10.1084/jem.20062409
81. Nik-Zainal S, Alexandrov LB, Wedge DC, Van Loo P, Greenman CD, Raine K, Jones D, Hinton J, Marshall J, Stebbings LA, Menzies A, Martin S, Leung K, Chen L, Leroy C, Ramakrishna M, Rance R, Lau KW, Mudie LJ, Varela I, DJ MB, Bignell GR, Cooke SL, Shlien A, Gamble J, Whitmore I, Maddison M, Tarpey PS, Davies HR, Papaemmanuil E, Stephens PJ, McLaren S, Butler AP, Teague JW, Jonsson G, Garber JE, Silver D, Miron P, Fatima A, Boyault S, Langerod A, Tutt A, Martens JW, Aparicio SA, Borg A, Salomon AV, Thomas G, Borresen-Dale AL, Richardson AL, Neuberger MS, Futreal PA, Campbell PJ, Stratton MR, Breast Cancer Working Group of the International Cancer Genome C (2012) Mutational processes molding the genomes of 21 breast cancers. Cell 149(5):979–993. https://doi.org/10.1016/j.cell.2012.04.024
82. Nik-Zainal S, Wedge DC, Alexandrov LB, Petljak M, Butler AP, Bolli N, Davies HR, Knappskog S, Martin S, Papaemmanuil E, Ramakrishna M, Shlien A, Simonic I, Xue Y, Tyler-Smith C, Campbell PJ, Stratton MR (2014) Association of a germline copy number polymorphism of APOBEC3A and APOBEC3B with burden of putative APOBEC-dependent mutations in breast cancer. Nature Genet 46(5):487–491. https://doi.org/10.1038/ng.2955
83. Nishida K, Arazoe T, Yachie N, Banno S, Kakimoto M, Tabata M, Mochizuki M, Miyabe A, Araki M, Hara KY, Shimatani Z, Kondo A (2016) Targeted nucleotide editing using hybrid prokaryotic and vertebrate adaptive immune systems. Science 353(6305). https://doi.org/10.1126/science.aaf8729
84. Noguchi C, Ishino H, Tsuge M, Fujimoto Y, Imamura M, Takahashi S, Chayama K (2005) G to A hypermutation of hepatitis B virus. Hepatology 41(3):626–633. https://doi.org/10.1002/hep.20580
85. Nowak U, Matthews AJ, Zheng S, Chaudhuri J (2011) The splicing regulator PTBP2 interacts with the cytidine deaminase AID and promotes binding of AID to switch-region DNA. Nat Immunol 12(2):160–166. https://doi.org/10.1038/ni.1977
86. Ohm-Laursen L, Barington T (2007) Analysis of 6912 unselected somatic hypermutations in human VDJ rearrangements reveals lack of strand

specificity and correlation between phase II substitution rates and distance to the nearest 3′ activation-induced cytidine deaminase target. J Immunol 178(7):4322–4334
87. Omori SA, Cato MH, Anzelon-Mills A, Puri KD, Shapiro-Shelef M, Calame K, Rickert RC (2006) Regulation of class-switch recombination and plasma cell differentiation by phosphatidylinositol 3-kinase signaling. Immunity 25(4):545–557. https://doi.org/10.1016/j.immuni.2006.08.015
88. Orthwein A, Patenaude AM, Affar el B, Lamarre A, Young JC, Di Noia JM (2010) Regulation of activation-induced deaminase stability and antibody gene diversification by Hsp90. J Exp Med 207(12):2751–2765. https://doi.org/10.1084/jem.20101321
89. Pasqualucci L, Migliazza A, Fracchiolla N, William C, Neri A, Baldini L, Chaganti RS, Klein U, Kuppers R, Rajewsky K, Dalla-Favera R (1998) BCL-6 mutations in normal germinal center B cells: evidence of somatic hypermutation acting outside Ig loci. Proc Natl Acad Sci U S A 95(20):11816–11821
90. Pasqualucci L, Neumeister P, Goossens T, Nanjangud G, Chaganti RS, Kuppers R, Dalla-Favera R (2001) Hypermutation of multiple proto-oncogenes in B-cell diffuse large-cell lymphomas. Nature 412(6844):341–346. https://doi.org/10.1038/35085588
91. Pavri R, Gazumyan A, Jankovic M, Di Virgilio M, Klein I, Ansarah-Sobrinho C, Resch W, Yamane A, Reina San-Martin B, Barreto V, Nieland TJ, Root DE, Casellas R, Nussenzweig MC (2010) Activation-induced cytidine deaminase targets DNA at sites of RNA polymerase II stalling by interaction with Spt5. Cell 143(1):122–133. https://doi.org/10.1016/j.cell.2010.09.017
92. Pefanis E, Wang J, Rothschild G, Lim J, Chao J, Rabadan R, Economides AN, Basu U (2014) Noncoding RNA transcription targets AID to divergently transcribed loci in B cells. Nature. https://doi.org/10.1038/nature13580
93. Pefanis E, Wang J, Rothschild G, Lim J, Kazadi D, Sun J, Federation A, Chao J, Elliott O, Liu ZP, Economides AN, Bradner JE, Rabadan R, Basu U (2015) RNA exosome-regulated long non-coding RNA transcription controls super-enhancer activity. Cell 161(4):774–789. https://doi.org/10.1016/j.cell.2015.04.034
94. Petersen-Mahrt SK, Harris RS, Neuberger MS (2002) AID mutates E. coli suggesting a DNA deamination mechanism for antibody diversification. Nature 418(6893):99–103. https://doi.org/10.1038/nature00862
95. Petersen-Mahrt SK, Neuberger MS (2003) In vitro deamination of cytosine to uracil in single-stranded DNA by apolipoprotein B editing complex catalytic subunit 1 (APOBEC1). J Biol Chem 278(22):19583–19586. https://doi.org/10.1074/jbc.C300114200
96. Pham P, Bransteitter R, Petruska J, Goodman MF (2003) Processive AID-catalysed cytosine deamination on single-stranded DNA simulates somatic hypermutation. Nature 424(6944):103–107. https://doi.org/10.1038/nature01760
97. Pham P, Smolka MB, Calabrese P, Landolph A, Zhang K, Zhou H, Goodman MF (2008) Impact of phosphorylation and phosphorylation-null mutants on the activity and deamination specificity of activation-induced cytidine deaminase. J Biol Chem 283(25):17428–17439. https://doi.org/10.1074/jbc.M802121200
98. Popp C, Dean W, Feng S, Cokus SJ, Andrews S, Pellegrini M, Jacobsen SE, Reik W (2010) Genome-wide erasure of DNA methylation in mouse primordial germ cells is affected by AID deficiency. Nature 463(7284):1101–1105. https://doi.org/10.1038/nature08829
99. Prochnow C, Bransteitter R, Klein MG, Goodman MF, Chen XS (2007) The APOBEC-2 crystal structure and functional implications for the deaminase AID. Nature 445(7126):447–451. https://doi.org/10.1038/nature05492
100. Qian J, Wang Q, Dose M, Pruett N, Kieffer-Kwon KR, Resch W, Liang G, Tang Z, Mathe E, Benner C, Dubois W, Nelson S, Vian L, Oliveira TY, Jankovic M, Hakim O, Gazumyan A, Pavri R, Awasthi P, Song B, Liu G, Chen L, Zhu S, Feigenbaum L, Staudt L, Murre C, Ruan Y, Robbiani DF, Pan-Hammarstrom Q, Nussenzweig MC, Casellas R (2014) B cell super-enhancers and regulatory clusters recruit AID tumorigenic activity. Cell 159(7):1524–1537. https://doi.org/10.1016/j.cell.2014.11.013
101. Qiao Q, Wang L, Meng FL, Hwang JK, Alt FW, Wu H (2017) AID Recognizes Structured DNA for Class Switch Recombination. Molecular cell 67(3):361–373. e364. https://doi.org/10.1016/j.molcel.2017.06.034
102. Quinlan EM, King JJ, Amemiya CT, Hsu E, Larijani M (2017) Biochemical regulatory features of AID remain conserved from lamprey to humans. Mol Cell Biol. https://doi.org/10.1128/MCB.00077-17
103. Rada C, Di Noia JM, Neuberger MS (2004) Mismatch recognition and uracil excision provide complementary paths to both Ig switching and the A/T-focused phase of somatic mutation. Mol Cell 16(2):163–171. https://doi.org/10.1016/j.molcel.2004.10.011
104. Ranjit S, Khair L, Linehan EK, Ucher AJ, Chakrabarti M, Schrader CE, Stavnezer J (2011) AID binds cooperatively with UNG and Msh2-Msh6 to Ig switch regions dependent upon the AID C terminus. J Immunol 187(5):2464–2475. https://doi.org/10.4049/jimmunol.1101406
105. Rathore A, Carpenter MA, Demir O, Ikeda T, Li M, Shaban NM, Law EK, Anokhin D, Brown WL, Amaro RE, Harris RS (2013) The local dinucleotide preference of APOBEC3G can be altered from 5′-CC to 5′-TC by a single amino acid substitution. J Mol Biol 425(22):4442–4454. https://doi.org/10.1016/j.jmb.2013.07.040
106. Revy P, Muto T, Levy Y, Geissmann F, Plebani A, Sanal O, Catalan N, Forveille M,

Dufourcq-Labelouse R, Gennery A, Tezcan I, Ersoy F, Kayserili H, Ugazio AG, Brousse N, Muramatsu M, Notarangelo LD, Kinoshita K, Honjo T, Fischer A, Durandy A (2000) Activation-induced cytidine deaminase (AID) deficiency causes the autosomal recessive form of the Hyper-IgM syndrome (HIGM2). Cell 102(5):565–575

107. Robbiani DF, Nussenzweig MC (2013) Chromosome translocation, B cell lymphoma, and activation-induced cytidine deaminase. Annu Rev Pathol 8:79–103. https://doi.org/10.1146/annurev-pathol-020712-164004
108. Rogozin IB, Basu MK, Jordan IK, Pavlov YI, Koonin EV (2005) APOBEC4, a new member of the AID/APOBEC family of polynucleotide (deoxy)cytidine deaminases predicted by computational analysis. Cell cycle 4(9):1281–1285. https://doi.org/10.4161/cc.4.9.1994
109. Rogozin IB, Diaz M (2004) Cutting edge: DGYW/WRCH is a better predictor of mutability at G:C bases in Ig hypermutation than the widely accepted RGYW/WRCY motif and probably reflects a two-step activation-induced cytidine deaminase-triggered process. J Immunol 172(6):3382–3384
110. Rosenberg BR, Hamilton CE, Mwangi MM, Dewell S, Papavasiliou FN (2011) Transcriptome-wide sequencing reveals numerous APOBEC1 mRNA-editing targets in transcript 3′ UTRs. Nat Struct Mol Biol 18 (2):230-236. https://doi.org/10.1038/nsmb.1975
111. Rosler C, Kock J, Kann M, Malim MH, Blum HE, Baumert TF, von Weizsacker F (2005) APOBEC-mediated interference with hepadnavirus production. Hepatology 42(2):301–309. https://doi.org/10.1002/hep.20801
112. Rothman P, Chen YY, Lutzker S, Li SC, Stewart V, Coffman R, Alt FW (1990) Structure and expression of germ line immunoglobulin heavy-chain epsilon transcripts: interleukin-4 plus lipopolysaccharide-directed switching to C epsilon. Mol Cell Biol 10(4):1672–1679
113. Rothman P, Lutzker S, Cook W, Coffman R, Alt FW (1988) Mitogen plus interleukin 4 induction of C epsilon transcripts in B lymphoid cells. J Exp Med 168(6):2385–2389
114. Salter JD, Bennett RP, Smith HC (2016) The APOBEC Protein Family: United by Structure, Divergent in Function. Trends Biochem Sci 41(7):578–594. https://doi.org/10.1016/j.tibs.2016.05.001
115. Sayegh CE, Quong MW, Agata Y, Murre C (2003) E-proteins directly regulate expression of activation-induced deaminase in mature B cells. Nat Immunol 4(6):586–593. https://doi.org/10.1038/ni923
116. Seplyarskiy VB, Soldatov RA, Popadin KY, Antonarakis SE, Bazykin GA, Nikolaev SI (2016) APOBEC-induced mutations in human cancers are strongly enriched on the lagging DNA strand during replication. Genome Res 26(2):174–182. https://doi.org/10.1101/gr.197046.115
117. Sheehy AM, Gaddis NC, Choi JD, Malim MH (2002) Isolation of a human gene that inhibits HIV-1 infection and is suppressed by the viral Vif protein. Nature 418(6898):646–650. https://doi.org/10.1038/nature00939
118. Shen HM, Peters A, Baron B, Zhu X, Storb U (1998) Mutation of BCL-6 gene in normal B cells by the process of somatic hypermutation of Ig genes. Science 280(5370):1750–1752
119. Shen HM, Tanaka A, Bozek G, Nicolae D, Storb U (2006) Somatic hypermutation and class switch recombination in Msh6(-/-)Ung(-/-) double-knockout mice. J Immunol 177(8):5386–5392
120. Shi K, Carpenter MA, Banerjee S, Shaban NM, Kurahashi K, Salamango DJ, McCann JL, Starrett GJ, Duffy JV, Demir O, Amaro RE, Harki DA, Harris RS, Aihara H (2017) Structural basis for targeted DNA cytosine deamination and mutagenesis by APOBEC3A and APOBEC3B. Nature Struct Mol Biol 24(2):131–139. https://doi.org/10.1038/nsmb.3344
121. Shinkura R, Ito S, Begum NA, Nagaoka H, Muramatsu M, Kinoshita K, Sakakibara Y, Hijikata H, Honjo T (2004) Separate domains of AID are required for somatic hypermutation and class-switch recombination. Nat Immunol 5(7):707–712. https://doi.org/10.1038/ni1086
122. Shulman Z, Gitlin AD, Targ S, Jankovic M, Pasqual G, Nussenzweig MC, Victora GD (2013) T follicular helper cell dynamics in germinal centers. Science 341(6146):673–677. https://doi.org/10.1126/science.1241680
123. Stavnezer-Nordgren J, Sirlin S (1986) Specificity of immunoglobulin heavy chain switch correlates with activity of germline heavy chain genes prior to switching. EMBO J 5(1):95–102
124. Stavnezer J (1986) Presence of a polyadenylated RNA fragment encoding the membrane domain for immunoglobulin alpha chain indicates that mRNAs for both secreted and membrane-bound alpha chains can be produced from the same RNA transcript. Nucleic Acids Res 14(15):6129–6144
125. Stenglein MD, Harris RS (2006) APOBEC3B and APOBEC3F inhibit L1 retrotransposition by a DNA deamination-independent mechanism. J Biol Chem 281(25):16837–16841. https://doi.org/10.1074/jbc.M602367200
126. Suspene R, Guetard D, Henry M, Sommer P, Wain-Hobson S, Vartanian JP (2005) Extensive editing of both hepatitis B virus DNA strands by APOBEC3 cytidine deaminases in vitro and in vivo. Proc Natl Acad Sci U S A 102(23):8321–8326. https://doi.org/10.1073/pnas.0408223102
127. Ta VT, Nagaoka H, Catalan N, Durandy A, Fischer A, Imai K, Nonoyama S, Tashiro J, Ikegawa M, Ito S, Kinoshita K, Muramatsu M, Honjo T (2003) AID mutant analyses indicate requirement for class-switch-specific cofactors. Nat Immunol 4(9):843–848. https://doi.org/10.1038/ni964

128. Takai A, Toyoshima T, Uemura M, Kitawaki Y, Marusawa H, Hiai H, Yamada S, Okazaki IM, Honjo T, Chiba T, Kinoshita K (2009) A novel mouse model of hepatocarcinogenesis triggered by AID causing deleterious p53 mutations. Oncogene 28(4):469–478. https://doi.org/10.1038/onc.2008.415
129. Taylor BJ, Nik-Zainal S, Wu YL, Stebbings LA, Raine K, Campbell PJ, Rada C, Stratton MR, Neuberger MS (2013) DNA deaminases induce break-associated mutation showers with implication of APOBEC3B and 3A in breast cancer kataegis. eLife 2:e00534. https://doi.org/10.7554/eLife.00534
130. Teng B, Burant CF, Davidson NO (1993) Molecular cloning of an apolipoprotein B messenger RNA editing protein. Science 260(5115):1816–1819
131. Tran TH, Nakata M, Suzuki K, Begum NA, Shinkura R, Fagarasan S, Honjo T, Nagaoka H (2010) B cell-specific and stimulation-responsive enhancers derepress Aicda by overcoming the effects of silencers. Nat Immunol 11(2):148–154. https://doi.org/10.1038/ni.1829
132. Turelli P, Trono D (2005) Editing at the crossroad of innate and adaptive immunity. Science 307(5712):1061–1065. https://doi.org/10.1126/science.1105964
133. Vartanian JP, Guetard D, Henry M, Wain-Hobson S (2008) Evidence for editing of human papillomavirus DNA by APOBEC3 in benign and precancerous lesions. Science 320(5873):230–233. https://doi.org/10.1126/science.1153201
134. Victora GD, Nussenzweig MC (2012) Germinal centers. Annu Rev Immunol 30:429–457. https://doi.org/10.1146/annurev-immunol-020711-075032
135. Vuong BQ, Lee M, Kabir S, Irimia C, Macchiarulo S, McKnight GS, Chaudhuri J (2009) Specific recruitment of protein kinase A to the immunoglobulin locus regulates class-switch recombination. Nat Immunol 10(4):420–426. https://doi.org/10.1038/ni.1708
136. Wang M, Rada C, Neuberger MS (2010) Altering the spectrum of immunoglobulin V gene somatic hypermutation by modifying the active site of AID. J Exp Med 207(1):141–153. https://doi.org/10.1084/jem.20092238
137. Wei M, Shinkura R, Doi Y, Maruya M, Fagarasan S, Honjo T (2011) Mice carrying a knock-in mutation of Aicda resulting in a defect in somatic hypermutation have impaired gut homeostasis and compromised mucosal defense. Nat Immunol 12(3):264–270. https://doi.org/10.1038/ni.1991
138. Whyte WA, Orlando DA, Hnisz D, Abraham BJ, Lin CY, Kagey MH, Rahl PB, Lee TI, Young RA (2013) Master transcription factors and mediator establish super-enhancers at key cell identity genes. Cell 153(2):307–319. https://doi.org/10.1016/j.cell.2013.03.035
139. Xu Z, Fulop Z, Wu G, Pone EJ, Zhang J, Mai T, Thomas LM, Al-Qahtani A, White CA, Park SR, Steinacker P, Li Z, Yates J 3rd, Herron B, Otto M, Zan H, Fu H, Casali P (2010) 14-3-3 adaptor proteins recruit AID to 5′-AGCT-3′-rich switch regions for class switch recombination. Nat Struct Mol Biol 17(9):1124–1135. https://doi.org/10.1038/nsmb.1884
140. Xue K, Rada C, Neuberger MS (2006) The in vivo pattern of AID targeting to immunoglobulin switch regions deduced from mutation spectra in msh2-/- ung-/- mice. J Exp Med 203(9):2085–2094. https://doi.org/10.1084/jem.20061067
141. Yamane A, Resch W, Kuo N, Kuchen S, Li Z, Sun HW, Robbiani DF, McBride K, Nussenzweig MC, Casellas R (2011) Deep-sequencing identification of the genomic targets of the cytidine deaminase AID and its cofactor RPA in B lymphocytes. Nat Immunol 12(1):62–69. https://doi.org/10.1038/ni.1964
142. Yancopoulos GD, DePinho RA, Zimmerman KA, Lutzker SG, Rosenberg N, Alt FW (1986) Secondary genomic rearrangement events in pre-B cells: VHDJH replacement by a LINE-1 sequence and directed class switching. EMBO J 5(12):3259–3266
143. Yeap LS, Hwang JK, Du Z, Meyers RM, Meng FL, Jakubauskaite A, Liu M, Mani V, Neuberg D, Kepler TB, Wang JH, Alt FW (2015) Sequence-intrinsic mechanisms that target AID mutational outcomes on antibody genes. Cell 163(5):1124–1137
144. Zahn A, Eranki AK, Patenaude AM, Methot SP, Fifield H, Cortizas EM, Foster P, Imai K, Durandy A, Larijani M, Verdun RE, Di Noia JM (2014) Activation induced deaminase C-terminal domain links DNA breaks to end protection and repair during class switch recombination. Proc Natl Acad Sci U S A. https://doi.org/10.1073/pnas.1320486111
145. Zhang H, Yang B, Pomerantz RJ, Zhang C, Arunachalam SC, Gao L (2003) The cytidine deaminase CEM15 induces hypermutation in newly synthesized HIV-1 DNA. Nature 424(6944):94–98. https://doi.org/10.1038/nature01707
146. Zhang Y, Gostissa M, Hildebrand DG, Becker MS, Boboila C, Chiarle R, Lewis S, Alt FW (2010) The role of mechanistic factors in promoting chromosomal translocations found in lymphoid and other cancers. Adv Immunol 106:93–133. https://doi.org/10.1016/S0065-2776(10)06004-9
147. Zheng S, Vuong BQ, Vaidyanathan B, Lin JY, Huang FT, Chaudhuri J (2015) Non-coding RNA generated following Lariat Debranching mediates targeting of AID to DNA. Cell 161(4):762–773. https://doi.org/10.1016/j.cell.2015.03.020

The Role for the DSB Response Pathway in Regulating Chromosome Translocations

6

Rajib Ghosh, Debamitra Das, and Sonia Franco

Abstract

In response to DNA double strand breaks (DSB), mammalian cells activate the DNA Damage Response (DDR), a network of factors that coordinate their detection, signaling and repair. Central to this network is the ATM kinase and its substrates at chromatin surrounding DSBs H2AX, MDC1 and 53BP1. In humans, germline inactivation of ATM causes Ataxia Telangiectasia (A-T), an autosomal recessive syndrome of increased proneness to hematological malignancies driven by clonal chromosomal translocations. Studies of cancers arising in A-T patients and in genetically engineered mouse models (GEMM) deficient for ATM and its substrates have revealed complex, multilayered roles for ATM in translocation suppression and identified functional redundancies between ATM and its substrates in this context. "Programmed" DSBs at antigen receptor loci in developing lymphocytes employ ubiquitous DDR factors for signaling and repair and have been particularly useful for mechanistic studies because they are region-specific and can be monitored *in vitro* and *in vivo*. In this context, murine thymocytes deficient for ATM recapitulate the molecular events that lead to transformation in T cells from A-T patients and provide a widely used model to study the mechanisms that suppress RAG recombinase-dependent translocations. Similarly, analyses of the fate of Activation induced Cytidine Deaminase (AID)-dependent DSBs during mature B cell Class Switch Recombination (CSR) have defined the genetic requirements for end-joining and translocation suppression in this setting. Moreover, a unique role for 53BP1 in the promotion of synapsis of distant DSBs has emerged from these studies.

Keywords

Class switch recombination · V(D)J recombination · AID · RAG · ATM · Chromosomal translocation

6.1 The DNA Damage Response (DDR) at Sites of Double-Strand Breaks (DSBs)

6.1.1 Overview: ATM Orchestrates the DDR in Mammalian Cells

DNA DSBs represent the most deleterious DNA lesion; failure to repair them may lead to genomic instability and cell death or senescence [45]. At

R. Ghosh · D. Das · S. Franco (✉)
Department of Radiation Oncology and Molecular Radiation Sciences, Johns Hopkins University School of Medicine and the Sidney Kimmel Comprehensive Cancer Center, Baltimore, MD, USA
e-mail: sfranco2@jhmi.edu

Y. Zhang (ed.), *Chromosome Translocation*, Advances in Experimental Medicine and Biology 1044,
https://doi.org/10.1007/978-981-13-0593-1_6

the organismal level, defective DSB repair translates into tissue dysfunction and premature aging and promotes tumorigenesis [45]. To prevent these outcomes, mammalian cells have evolved the DDR, a rapid set of responses that coordinates the assembly of repair complexes at the break with the activation of cell cycle checkpoints and the transcriptional networks that ultimately mediate cellular outcomes [153]. Moreover, it is becoming apparent that the DDR is not limited to the nucleus but rather coordinates the nuclear and cytoplasmic responses to DNA damage [62].

This rapid and widespread response is made possible by coordinated posttranslational modifications of DNA repair factors and histones at the break site, including phosphorylation, ubiquitination, SUMOylation, PARylation and others [15, 16, 29, 85, 126, 127, 182, 192]. In the context of phosphorylation, the DDR is regulated by three highly related PI3 kinase-like kinases (PI3KKs): Ataxia Telangiectasia Mutated (ATM), Ataxia-telangiectasia and RAD3 Related protein (ATR) and the catalytic subunit of the DNA protein kinase (DNA-PKcs) [108]. All three factors are activated in response to DSBs and phosphorylate hundreds of substrates at target SQ/TQ motifs [112], often in a redundant manner [159, 174]. Their activity is regulated by multiple mechanisms, including the cell cycle [83] and their mutual interactions [119, 193] and ultimately promotes repair and suppresses chromosomal translocations. In this Chapter, we will focus on key roles for the ATM kinase and its substrates in translocation suppression. Please see Fig. 6.1 for introductory schematic of ATM domains and its main regulatory functions at sites of DSBs.

6.1.2 General Mechanisms of Translocation Suppression by the DDR

DSBs threaten genomic integrity because their repair may introduce mutations at the break site and/or proceed aberrantly to generate chromosomal rearrangements. Indeed, genome-wide mapping of chromosomal translocations arising in primary mouse B lymphocytes harboring traceable DSBs revealed that translocation formation is enhanced by proximity [44, 93], favoring intrachromosomal rearrangements and resulting in frequent deletions *in cis* [44, 93]. Although most of these translocations likely have no consequence to cellular functions, rare events may disable anti-cancer mechanisms via inactivation or overexpression of anti- or pro-oncogenic factors, respectively [64, 135]. Therefore, a key function of the DDR is to suppress pathogenic clonal translocations by promoting the rejoining of DNA ends across the break.

In mammalian cells, DSB rejoining is mediated via either Homologous Recombination (HR), an error-free pathway active in the replicative phases of the cell cycle [87, 134] or NonHomologous End-Joining (NHEJ), a versatile, ligase IV-dependent pathway that re-ligates broken DNA ends across the cell cycle using no homology or micro-homologies [21, 102]. In addition to the canonical NHEJ pathway, cancer cells may activate a back-up or alternative NHEJ pathway that rather employs ligase I/III and may repair persistent breaks with slower kinetics [41]. Regardless of the repair pathway used, DSB repair is slow (minutes to hours) and necessitates a strategy to prevent DNA end dissociation prior to ligation. This end tethering function is provided by the DDR in conjunction with repair factors. The complex formed by MRE11, RAD50 and NBS1 (MRN) ubiquitously binds to broken DNA ends and plays key roles in their sensing and processing throughout the cell cycle [164]. Specifically, the hook domains at the apex of two RAD50 coiled coil domains dimerize to bridge two DNA molecules bound by the RAD50 globular heads [78] and this function is facilitated by MRE11 dimerization [178]. Indirectly, the endonuclease activity of MRE11 initiates end-resection and activates the ATM kinase [129], a main orchestrator of the DDR [112]. This role for MRN in DNA end alignment and bridging is highly conserved and occurs in different chromatin contexts, including canonical and alternative NHEJ [50, 180] and programmed breaks generated during

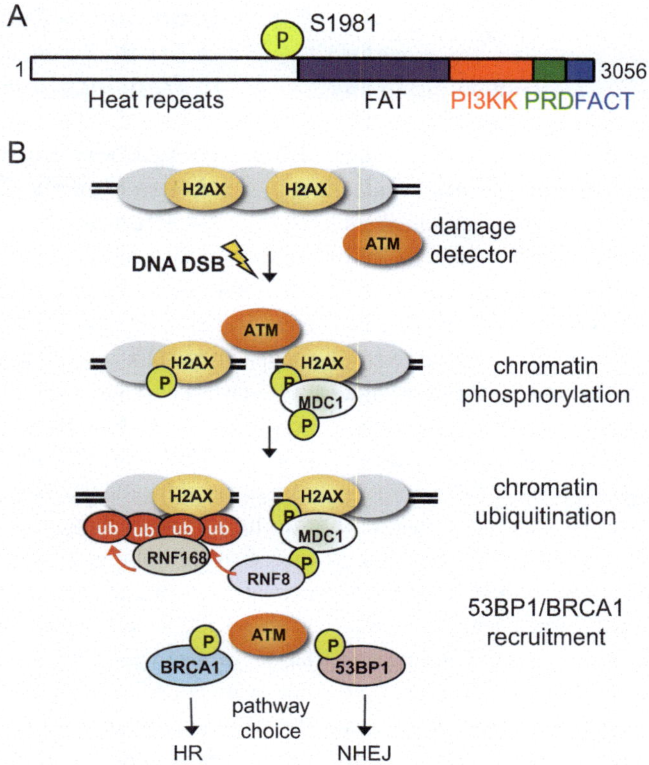

Fig. 6.1 ATM is a kinase with pleiotropic roles in the DNA Damage Response (DDR). (**a**) The carboxi-terminal domain of ATM contains 40–50 alpha-helical repeats that mediate interaction with the MRE11/RAD50/NBS1 (MRN) complex. The carboxi-terminal domain contains a PI3 kinase-like kinase (PI3KK) domain that modifies S/T(Q) motifs in target proteins. The kinase domain is flanked by a FRAP-ATM-TRRAP (FAT) domain and a FAT carboxi-terminal (FACT) domain. ATM auto-phosphorylates at Ser1981 upon induction of double strand breaks (DSB) and the modified protein is commonly used as a biomarker for DDR activation. (**b**) ATM regulates the DDR at multiple steps, including the rapid detection of break-induced alterations in chromosomal structure leading to auto-phosphorylation at Ser1981 and activation; the maintenance of DNA end tethering via modification of MRE11/RAD50/NBS1 complex; the recruitment of ubiquitin ligases via phosphorylation of H2AX and MDC1; and the competition between 53BP1 and BRCA1 for DNA ends, a critical event during DSB repair pathway choice. ATM also phosphorylates many proteins in the nucleoplasm (such as p53) and outside the nucleus (no depicted)

V(D)J recombination [74] and Class Switch Recombination (CSR; see below) [54].

ATM, together with DNA-PKcs and ATR, phosphorylates the amino-terminal tail of histone H2AX at Ser139 to form γ-H2AX [142]. This modification spreads both sides of the break and anchors MDC1 [107] to form a platform for the recruitment of BRCA1, 53BP1 and their effectors [14]. These multiprotein complexes, detected as "foci" by standard immunocytochemistry assays, may themselves function as "glue" to suppress DNA end dissociation [11, 186, 191]. In addition to DDR factors, components of the NHEJ pathway have also been implicated in the formation of the synaptic complex via different mechanisms [28, 68, 157], and their absence eventually results in DNA end dissociation [30, 103]. Therefore, the DDR and the NHEJ pathway normally cooperate to maintain DNA ends aligned and tethered until ligated.

In addition to position, transcriptional status has emerged as a main determinant of translocation proneness. Indeed, unbiased genome-wide translocation sequencing in primary mouse cells revealed that transcribed regions, and in particular active transcription start sites, are prone to translocations [44, 93]. Transcription results in the generation of R loops, short RNA-DNA hybrids that leave the nontemplate DNA strand exposed to mutagenic activities, such as oxidative stress, cellular cytidine deaminases or others [147]. R loops are abundant in human cells [65] and, in some contexts, have been clearly shown to promote DNA DSBs and chromosomal translocations [76, 82]. The best characterized example is perhaps the promotion of chromosomal translocations between C-MYC and the immunoglobulin heavy chain (IgH) locus, a hallmark of many B cell malignancies [140]. R loops form at the at the MYC locus [59] and at the S region of the immunoglobulin heavy chain [23, 188]. AID, a cytidine deaminase expressed in B cells, binds to and modifies these structures, leading to formation of DNA DSBs and IgH-MYC translocations [59]. In support of a mechanistic role for R loops in this setting, loss of TOP3B, which relaxes negative supercoiling and increases R loop formation at the MYC promoter, also increases the frequency of Igh-Myc translocations in mice [183]. Moreover, AID induces translocations involving a heterologous S region and MYC in yeast THO mutants, known to accumulate R loops [144]. Together, these data suggest that the simultaneous formation of R loops in Ig and transcribed MYC (or other transcribed genes) may promote their translocations. Recent evidence indicates that many DDR and DNA repair factors may suppress chromosomal translocations via direct modulation of R loop formation and dissolution ([20, 73]; reviewed in [162]). In this context, ATM, a suppressor of IgH-Myc translocations *in vivo* [135], also suppresses R loop formation in proliferating cells [184] and is activated by R loops at sites of UV-induced DNA damage [169]. Finally, emerging data implicates transcription-independent DSBs generated by topoisomerase 2B (TOP2B) at chromosome loop anchors as a cause of chromosomal fragility [36], another scenario where the DDR may function to limit translocations.

6.2 Mammalian Genetic Models to Study Roles for the DDR in Translocation Suppression

6.2.1 Genetically Engineered Mouse Models (GEMM) of DDR Deficiency

Mice with germline inactivation of DDR factors have provided valuable insights into their requirements in translocation suppression. Embryonic fibroblasts and B and T lymphocytes deficient for ATM [34, 64], histone H2AX [12, 40, 64], MDC1 [107] or 53BP1 [64, 117] all accumulate chromosomal translocations. Mechanistically, these translocations are though to occur as a result of defective end-joining across the break leading to persistent breaks and end dissociation [11, 63]. In addition, DDR factors may play roles in the regulation of pathway choice during translocation. For example, ATM and H2AX not only suppress translocations but also enhance fidelity at translocation breakpoints by promoting rejoining via classical over alternative NHEJ [19]. Although all H2AX [31], MDC1 [81, 96, 109, 110] and 53BP1 [42, 61] are ATM substrates in the DDR, their roles in translocation suppression are not fully epistatic with ATM. In this regard, cytogenetic analysis of primary B and T cells deficient for 53BP1 and ATM revealed increased frequency of chromosomal breaks and translocations in double mutant primary cells associated to a greater defect in end-joining [146]. Similarly, combined deficiency for H2AX and ATM leads to a marked increase in the frequency of chromosomal breaks and translocations in embryonic fibroblasts [190] and *in vitro* cultured T cells [185]. However, mechanistic understanding of these interactions has mostly relied on the analysis of translocations arising at loci undergoing programmed DSBs during lymphocyte development, which can be traced in time and space. In Section 3 below, we discuss roles for ATM and its substrates in translocation suppression using RAG-dependent DSBs during V(D)J recombination as a model system. Furthermore, Section 4 summarizes our understanding on how the ATM network suppresses AID-dependent translocations during Class Switch Recombination (CSR).

6.2.2 Human Genetic Syndromes of DDR Deficiency: Ataxia-Telangiectasia

Mutations in H2AX, MDC1 or 53BP1 have not been observed in humans. In contrast, biallelic mutations in the ATM kinase result in the autosomal recessive syndrome Ataxia-Telangiectasia (A-T; OMIM#208900). The ATM locus at human chromosome 11q22 contains 66 exons and encodes a 350 kDa protein with a C-terminal PI3 kinase-like kinase (PI3KK) domain [149]. Mutations leading to A-T span the entire locus and most patients are compound heterozygous [167]. Approximately 85% of A-T patients harbor biallelic null mutations and display the most severe or "classical" form of the disease, including neurodegeneration, immunodeficiency and increased cancer predisposition [22, 143]. Neurodegeneration is particularly severe in the cerebellum, with progressive loss of Purkinje cells (PC) and, to a lesser extent, granule cells (GCs) [25, 128, 173]. On average, ataxia first manifests in the toddler years and patients become wheelchair bound at a mean age of 8 years [120]. Currently, the mechanisms leading to neurodegeneration remain unclear. In contrast, the phenotypes of immunodeficiency, gonadal atrophy, radiosensitivity, premature aging and cancer proneness are clearly related to ATM functions in DSB repair. In this regard, A-T primary cells show frequent chromosomal breaks and translocations [94, 95]. Moreover, accelerated telomere shortening [114, 125, 155, 170, 179], defective response to oxidative stress [70] and other phenotypes may cooperate with defective DSB repair to augment genomic instability in A-T cells. Lymphocytes from A-T patients harbor chromosomal breaks and clonal translocations that mainly involve T [77, 92, 94] and B [32] cell receptor loci. These translocation are detected in the blood of many A-T patients years prior to the development of malignancy [165] and are discussed in detail in below.

6.3 The DDR Suppresses Chromosomal Translocations During V(D)J Recombination

6.3.1 Mechanisms of V(D)J Recombination

Adaptive immunity relies on the clonal expansion of B and T lymphocytes upon binding of their surface receptors to specific antigens [48, 168]. The generation of both B and T cells involves the expression of the lymphocyte-specific RAG1/RAG2 (RAG) endonuclease to introduce DSBs between V, D and J coding sequences and their flanking recombination signal sequences (RSS) at antigen receptor loci, followed by deletional or inversional recombination [151]. To prevent genomic instability, this process is initiated and completed during the G1 phase of the cell cycle [88, 152]. In the bone marrow, B lineage cells successively rearrange their V to D gene segments in the variable region of the immunoglobulin heavy chain (IgH) and light chain (IgL) loci to form pre-B cells [106]. Additional rearrangement of D to J gene segments yields mature B cells that express the B cell receptor (BCR) and exit the bone marrow to colonize the spleen and lymph nodes [106] (diagrammed in Fig. 6.2). Similarly, developing T cells in the thymus sequentially rearrange V, D and J gene segments to generate T cell receptors TCRα, TCRβ, TCRγ and TCRδ. TCRβ, TCRγ and TCRδ variable region exons are assembled first, at the $CD4^-/CD8^-$ ("double negative") stage. Productive VδDδJδ and VγJγ rearrangements generate TCRδ and TCRγ chains, respectively, which assemble in the surface to form the TCRγδ receptor and induce differentiation along this lineage [18]. Alternatively, a productive VβDβJβ rearrangement generates TCRβ chains that promote differentiation to the $CD4^+/CD8^+$ ("double positive") stage. Here, a productive VαJα rearrangement generates a TCRα chain that associates with TCRβ to promote differentiation to either $CD4^+$ or $CD8^+$ ("single positive") T cells [18].

In all cases, successful recombination requires the rejoining of RAG-liberated DSBs via ubiquitous DDR and NHEJ factors [10, 104]. Therefore, deficiencies for RAG as well as many DDR/NHEJ factors block B and T cell develop-

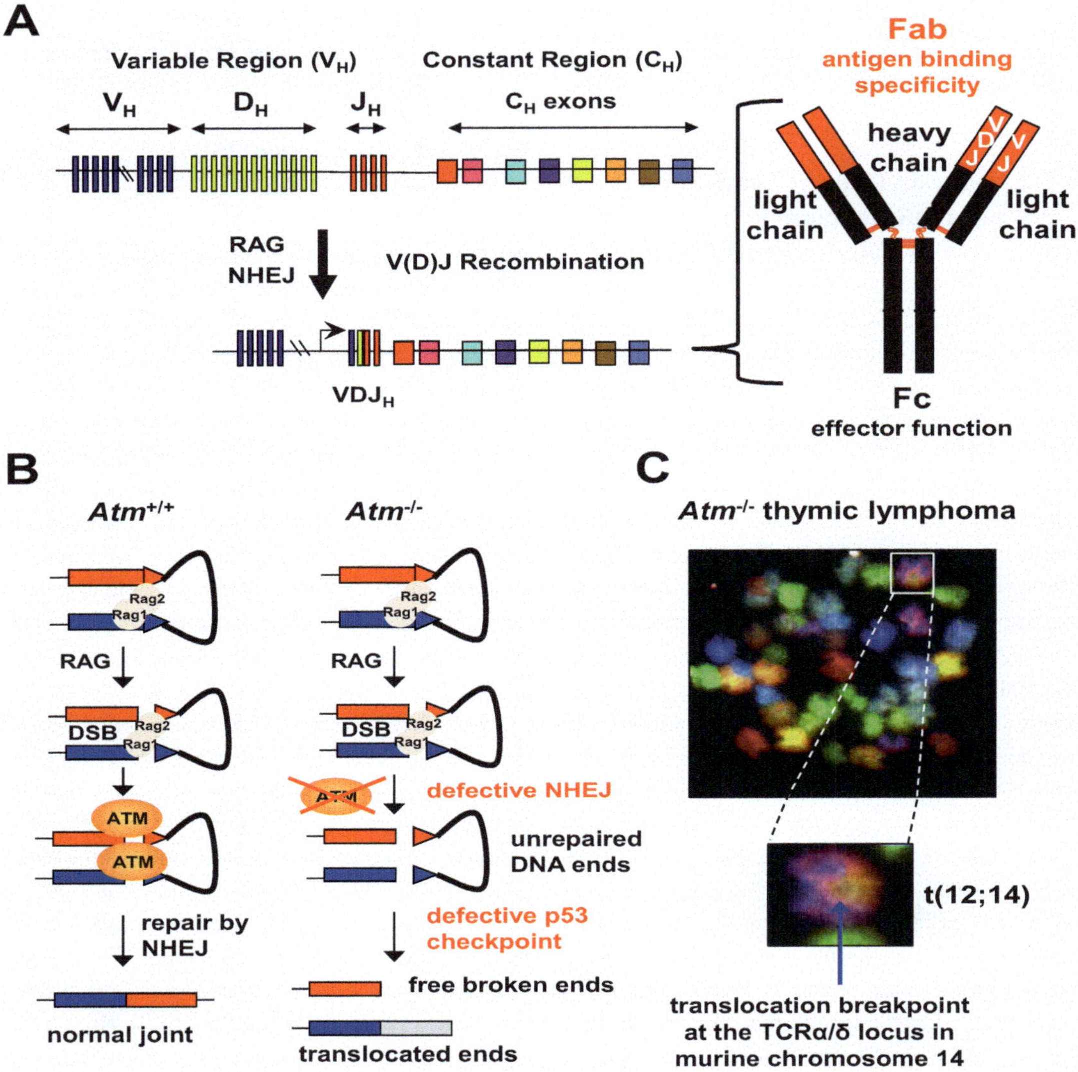

Fig. 6.2 The DDR prevents translocations of RAG-dependent DSBs. (**a**) Schematic of the immunoglobulin heavy chain (IgH) locus at mouse chromosome 12. During V(D)J recombination, the RAG recombinase introduces double-strand breaks (DSBs) adjacent to V_H, D_H and J_H segments and the Nonhomologous End-Joining (NHEJ) pathway of DSB repair ligates the ends to generate a coding VDJ_H exon. The expression of this VDJ_H exon and a downstream constant region (C_H) exon generates the immunoglobulin heavy chain (IgH). The immunoglobulin light chain (IgL) is similarly generated via V_L to J_L recombination at the immunoglobulin light chain kappa (Igκ) or lambda (Igλ) loci. Binding of heavy and light chains results in antibody formation (depicted on the right). (**b**) Schematic of molecular events at the TCRα/δ locus in developing murine thymocytes. In wild-type mice (*Atm*$^{+/+}$ mice), RAG-dependent DSBs are sensed and signaled by ATM and repaired via NHEJ. In mice with a germline deletion of Atm (*Atm*$^{-/-}$ mice), a subset of RAG-dependent DSBs dissociates prior to repair, leading to either free, unrepaired DNA ends or aberrant repair by translocation to another broken DNA ends elsewhere in the genome. (**c**) Spectral karyotyping (SKY) analysis of thymic lymphomas arising in *Atm*$^{-/-}$ mice reveals a clonal translocation involving chromosomes 12 and 14. The breakpoint at chromosome 14 localizes to the TCRα/δ locus

ment at early stages to variable extent [156]. Specifically, loss of RAG results in a complete block or severe combined immunodeficiency (scid) with antigen receptor loci in germline configuration. In contrast, loss of DDR or NHEJ factors results in variable degrees of immunodeficiency as a result of failed recombination of RAG-dependent DSBs. It is only in the latter scenario that unrepaired DSBs signal cell death or engage in chromosomal translocations. Moreover, concomitant defects in DSB repair elsewhere often lead to coexisting nonimmunological manifestations.

In mice and humans, the genes encoding TCRα and TCRδ occupy the same locus in chromosome 14 and their sequential rearrangement is regulated via specific enhancers [98]. Both human and mouse TCRδ loci resided between Vα and Jα segments, and therefore TCRα rearrangement deletes the TCR δ gene and commits cells to the α/β lineage. These rearrangements are driven by specific enhancer elements, Eδ and Eα [154], which play major roles in the generation of translocations (see below).

6.3.2 RAG-Dependent Breaks and Translocations in DDR-Deficient Mice

Recombining thymocytes deficient for NHEJ factors (including Ku70, Ku80, DNA-PKcs, Artemis, ligase IV or XRCC4) fail to ligate RAG-generated DSBs, leading to pro-B and pro-T cell apoptosis and absence or marked depletion of mature lymphocytes [3]. In contrast, deficiencies for DDR factors result in variable, milder defects in T cell maturation, in line with their redundant roles in DSB detection and signaling. Deficiency for components of the MRN complex [74] or ATM [7] results in the most severe defects, while residual recombination and lymphocyte development is observed in thymocytes deficient for ATM substrates H2AX [13, 40], MDC1 [107] or 53BP1 [176]. Of note, deficiencies for specific factors impact B and T cell development to different extent. For example, loss of ATM results in marked depletion of thymocytes and peripheral T cells early in life, while the B cell compartment is less affected.

Atm$^{-/-}$ mice [7, 24, 181, 190] have been particularly valuable to understand how the DDR suppresses the translocation of RAG-induced DSBs during in developing lymphocytes (Fig. 6.2). Cytogenetic analysis of interphase and dividing *Atm*$^{-/-}$ thymocytes and peripheral T cells has been used extensively to quantify chromosomal breaks at the TCRα/δ locus [24, 80, 105]. Moreover, this approach also allows monitoring of their progression to nonclonal chromosomal translocations and, over a period of a few months, clonal selection and full malignant transformation [7, 24, 181, 190]. The rapid clonal progression uniquely observed in this model results from the requirement for ATM in the activation of the p53-dependent G1/S cell cycle checkpoint [37, 91] and apoptosis in response to unrepaired DSBs [58]. Thus, rapid transformation observed in thymocytes deficient for ATM but not its substrates (see below) results from the unique pleomorphic roles for ATM in DSB detection, signaling and repair [34]. Interestingly, breeding into a RAG-deficient background does not prevent lymphomagenesis in *Atm*$^{-/-}$ mice [132, 133]. However, lymphomas in *Atm*$^{-/-}$/*Rag2*$^{-/-}$ mice lack chromosomal translocations involving antigen receptor loci [132, 133], further highlighting the essential role for ATM in promoting repair in this context. Finally, we note that ATM is activated in response cellular stresses other than DSBs [129], a mechanism that may potentially cross-talk with its functions at the DDR to modulate the fate of RAG-dependent DSBs .

Atm$^{-/-}$ thymi show blockade at the DP to SP transition and marked reductions in total cellularity, α/β T cells and SP CD4^{+} and CD8^{+} T cells [7, 24, 181, 190]. Consistent with defective repair of RAG-induced DSBs in the absence of ATM, fluorescence in situ hybridization (FISH) analyses of *Atm*$^{-/-}$ T cells with probes that hybridize to sequences flanking the TCRα/δ locus in chromosome 14 reveal frequent locus-specific chromosomal breaks [34, 105] (see Fig. 6.2 for an example).

Atm$^{-/-}$ mice succumb to T-cell acute ALLs with clonal translocations that typically involve the TCRα/δ locus in chromosome 14, the TCRβ locus on chromosome 6 and the immunoglobulin heavy chain locus in chromosome 12 that recombines D and J segments in thymocytes [33, 105, 189]. Moreover, T cells harboring translocations with a breakpoint at this locus are detected in the peripheral blood of *Atm*$^{-/-}$ mice [34, 105], harboring the onset of malignancy. Mechanistically, elegant experiments by Sleckman and colleagues demonstrated that ATM functions to maintain RAG-generated DNA ends in repair complexes [26], preventing DNA end dissociation and chromosomal translocation.

Using array comparative genomic hybridization (CGH) analysis to map the translocation breakpoints within TCRα/δ, Zha and colleagues demonstrated that the TCRα/δ translocation is associated to defective rearrangement at the TCRδ rather than the TCRα locus [189], suggesting an earlier developmental origin than previously thought [33]. In support of this notion, deletion of Eδ [89] but not Eα [189] rescued clonal translocations *in vivo*. In addition, the CGH data revealed that the T(12;14) is associated to amplification of a set of genes upstream of the TCRα/δ locus [189], suggesting that breakage-fusion-bridge (BFB) cycles may act as intermediaries. Todate, it remains unclear whether the sequences in chromosome 12 represent "passengers" or, alternatively, contribute to tumorigenesis by inactivating a tumor suppressor gene [189]. In this regard, the translocation deletes one copy of Bcl11b [189], a haploinsufficient tumor suppressor in the mouse [90], leading to decreased expression [189]. However, monoallelic deletion of Bcl11b in double negative thymocytes did not accelerate lymphomagenesis in *Atm*$^{-/-}$ mice [60], suggesting an alternative mechanism. Finally, the murine translocation deletes TCL1 [189], a gene that is translocated in the human translocation, even though it is located in the syntenic area.

ATM substrates H2AX, MDC1 and 53BP1 have also been implicated in the suppression of chromosomal translocations in developing lymphocytes. Immuno-FISH using antibodies that recognize γ-H2AX and DNA probes that hybridize to sequences at TCR loci revealed the presence of γ-H2AX foci at chromatin surrounding RAG-dependent DSBs [43]. The functional significance of this ATM- (and likely DNA-PKcs-) dependent modification has been investigated in detail using mice with germline [13, 40] or T cell-specific [185] H2AX inactivation. Collectively, these studies clearly demonstrate a requirement for H2AX for end-joining of RAG-dependent DNA ends at the TCRα/δ locus via its functions in DNA end anchoring prior to ligation [11, 186] and in protection from aberrant CtIP-mediated resection [75].

Interestingly, the presence of persistent, unrepaired RAG-dependent breaks in *H2afx*$^{-/-}$ thymocytes is not sufficient to trigger transformation and *H2afx*$^{-/-}$ mice are not lymphoma prone [13, 40]. This is likely due to the vigorous p53-dependent apoptotic response elicited by DSBs in *H2afx*$^{-/-}$ developing T cells. Indeed, breeding of *H2afx*$^{-/-}$ mice to mice with germline inactivation of p53 (*Trp53*$^{-/-}$ mice), themselves lymphoma prone [55, 72, 84], greatly accelerates lymphomagenesis relative to single mutants [13, 39]. Most significantly, the mechanisms driving transformation in *Trp53*$^{-/-}$ and *H2afx*$^{-/-}$/ *Trp53*$^{-/-}$ thymocytes are distinct. *Trp53*$^{-/-}$ lymphomas are driven by point mutations in *Pten* and other tumor-associated loci [56, 101] while *H2afx*$^{-/-}$/ *Trp53*$^{-/-}$ lymphomas are driven by clonal chromosomal translocations [13, 39]. Interestingly, these translocations do not involve the TCRα/δ locus in chromosome 14 or breakpoints at either TCRβ in chromosome 6 or TCRγ in chromosome 13, even though the TCRβ is rearranged in the tumor cells. Instead, Spectral Karyotyping (SKY) analysis of *H2afx*$^{-/-}$/ *Trp53*$^{-/-}$ revealed clonal translocations originated by rejoining of DSBs that presumably occur at "random" sites during periods of rapid cellular proliferation. Similarly, conditional inactivation of H2AX and p53 in double negative thymocytes using an Lck-Cre transgenic mouse model resulted in thymic lymphomas driven by clonal translocations that did not typically involve antigen receptor loci [187]. Finally, although deletion of an H2AX conditional allele in ATM-deficient thymocytes increased the number of RAG-dependent chromosomal translocations *in vitro* [185], it did not accelerate lymphomagenesis *in vivo* [185]. Altogether, these observations suggest that, unlike ATM, H2AX is mostly dispensable for the repair of RAG-dependent DSBs, but becomes limiting at DSBs that arise via other mechanisms, such as replication, oxidation or others. A potential explanation for these findings is that RAG itself may function to promotes DNA end synapsis, providing an overlapping mechanisms with H2AX (and other foci factors) in this context [11]. This notion also may explain the lack of significant defects in V(D)J recombination in *Mdc1*$^{-/-}$ mice [107].

Similar to *H2afx*$^{-/-}$ mice, *Trp53bp1*$^{-/-}$ mice show decreased thymic size and decreased number of peripheral T cells, including α/β and γ/δ T cells [52], pointing to a defect in maturation. Interestingly, the mechanism driving T cell immunodeficiency in the absence of 53BP1 is unique. While ATM, H2AX and MDC1 function to promote repair across a DSBs, 53BP1 is mainly required for the synapsis of two distant DSBs, a critical step during recombination. Specifically, *Trp53bp1*$^{-/-}$ thymocytes are impaired for V to DJ joining, leading to degradation of persistent coding ends and apoptosis. If the apoptotic response is blunted by breeding to a p53-deficient background, thymic lymphomas are observed [116, 175]. However, most *Trp53bp1*$^{-/-}$/*Trp53*$^{-/-}$ lymphomas are driven by polyploidy or by clonal translocations that spare antigen receptor loci [116, 175].

An epistaxis analysis of ATM and 53BP1 functions at RAG-dependent DSBs was also conducted using a murine model of combined germline deficiency [146]. This work indicated that combined loss aggravates the T cells maturation defect, further reducing thymic output [146]. Moreover, *Atm*$^{-/-}$/*Trp53bp1*$^{-/-}$ mice develop thymic lymphomas earlier in life and with higher penetrance than *Atm*$^{-/-}$ controls [146]. Like *Atm*$^{-/-}$ lymphomas, *Atm*$^{-/-}$/*Trp53bp1*$^{-/-}$ lymphomas are driven by clonal chromosomal translocations involving the TCRα/δ locus [146]. Altogether, these data suggest that H2AX and 53BP1 play modest ATM-independent functions in translocation suppression *in vivo*.

6.3.3 RAG-Dependent Break and Translocations in Ataxia-Telangiectasia Patients

The requirement for ATM in the repair of RAG-dependent DSBs is highly conserved in mice and humans. Approximately two thirds of patients with classical A-T have low lymphocyte counts and immunodeficiency [121]. Although both circulating B and T cells are decreased, the most common observation is low number of CD4^{+} T cells with impaired response to mitogens and antigens and anergy [121]. The degree of immunodeficiency varies significantly from one patient to another, but tends to be stable over time. Immunodeficiency tends to be less common in the variant, milder forms of the disease [172].

In addition to lymphopenia, A-T patients show increased predisposition to hematological malignancies. Specifically, the risk of lymphoid but not myeloid malignancies is markedly increased [166]. T cell cancers are more frequent than B cell tumors and include T cell acute lymphoblastic leukemia (ALL), T cell lymphomas and, in older A-T patients, T cell prolymphocytic leukemia (T-PLL) [166] . Cytogenetic analysis indicates that tumor cells typically harbor clonal chromosomal rearrangement involving antigen receptor loci [27], suggesting that they originate during V(D)J recombination.

Childhood T cell ALL is the most common malignancy in A-T and frequently involves clonal translocations involving TCRA/D locus in chromosome 14 or TCRB in chromosome 7 [166]. In humans, the TCRA/D locus is located in chromosome 14 and translocations in A-T leukemias are typically inv(14)(q11q32), or tandem translocations of chromosome 14 with breakpoints at q11 and q32 and del(14)(q11q32) [27]. Older A-T patients can harbor clonal expansions of peripheral T cells with inv(14)(q11;q32) and, in addition, t(14,14)(q11;32.1) and more rarely t(X;14)(q28;q1); some of these patients will develop T cell PLL. In these translocations, the TCRA/D breakpoint at 14q11 is fused with a breakpoint at the TCL1 oncogene at 14q32.1 and at the MTCP1 oncogene at Xq28 [166] [145] [130] [6, 47]. These translocations can be detected in the blood of asymptomatic patients for years [27] and likely evolve to full malignancy upon the acquisition of additional alterations, such as trisomy of 8q containing *C-MYC* and others [27].

Interestingly, over half of non-A-T patients with T-PLL carry a somatic mutation of ATM

[158] and the same TCRA;TLC1 and TCRA;MTCP1 translocations are also recurrent clonal lesions in this setting [49, 163]. These translocation likely drive transformation in both A-T and non A-T patients by placing the *TCL1* or *MTCP1* under the control of the TCRα transcriptional enhancer (Eα) [27]. Similarly, cytogenetic abnormalities involving antigen receptor loci are often present in T cell ALL in the general population [2], suggesting that uncharacterized defects in the DDR or NHEJ may promote leukemogenesis more broadly.

The improvement in supportive care has increased the life expectancy of A-T patients and also uncovered their predisposition to solid tumors in the second and third decades of life [171, 172]. Moreover, patients with the milder, "variant" form of the disease tend to develop solid tumors rather than leukemias typically observed in the classical form of the disease [136, 171]. Future investigations for the presence of fusion transcripts in tumor DNA will help determine whether the role for ATM in translocation suppression may also be relevant in this scenario.

6.4 The DDR Suppresses Chromosomal Translocations During Class Switch Recombination

6.4.1 Mechanisms of Class Switch Recombination

Upon encounter with antigen, IgM^{+} B cells undergo CSR to diversify their effector functions by expressing the same variable region as a secondary isotype (i.e., IgG, IgA or IgE). Mechanistically, this process involves a deletional recombination reaction at the immunoglobulin heavy (IgH) chain locus constant region (diagrammed in Fig. 6.3). In particular, activation-induced cytidine deaminase (AID; gene symbol, *AICDA*) [118] works in concert with ubiquitous DNA repair pathways (including Base Excision Repair (BER) and Mismatch Repair (MMR) to introduce DSBs at "Switch" (S) regions upstream of Cμ (encoding IgM) and a downstream C_H exon. AID-dependent DSBs are sensed and signaled via ubiquitous DDR factors, brought together across long chromosomal distances ("synapsed") and rejoined via ubiquitous NHEJ. Completion of the recombination reaction results in deletion of Cμ and expression of the variable region together with Cα, Cε or Cγ (to encode IgA, IgE or IgG, respectively). Intervening DNA is sealed into a circle by NHEJ and eventually lost upon division. The general mechanisms of CSR have been the subject of recent excellent reviews [3, 38, 113, 115].

Successful CSR requires that two distant DSBs are repaired by rejoining to each other rather than via rejoining of DNA ends across each individual DSB, exploiting a general cellular response that promotes DSBs repair *in cis* [67]. As diagrammed in Fig. 6.3, AID is thought to introduce numerous DSBs within each of the two recombining S regions. These concurrent DSBs may be rejoined to either DSBs within the same S region (intra-S region recombination, leading to an internal deletions or "shorter" S region) or to DSBs within the recombining S region (inter-S region recombination, leading to CSR). Studies in wild-type B cells indicate that CSR is normally favored over internal deletion. For example, in a typical B cell activation with α-CD40 antibody and Il-4, over half of the cells undergo CSR. In contrast, less than 10% show intra-S deletions (when assayed by Southern blotting which would not detect small deletions). As described below, mutations in specific DDR components impair end-joining during CSR by decreasing the efficiency of synapsis, while others impair repair (i.e., NHEJ) *per se*.

Like V(D)J recombination, CSR is initiated and completed in the G1 phase of the cell cycle [64]. Consistently, defects for AID or the DDR/NHEJ factors that regulate DSB repair during the G1 phase of the cell cycle impair CSR to variable extent. In this context, ATM and its substrates H2AX and 53BP1 are required for efficient CSR in mice and humans, as described in detail below. In contrast, defects in HR or DDR factors that regulate DSB repair in the replicative phases of the cell cycle do not directly interfere with CSR

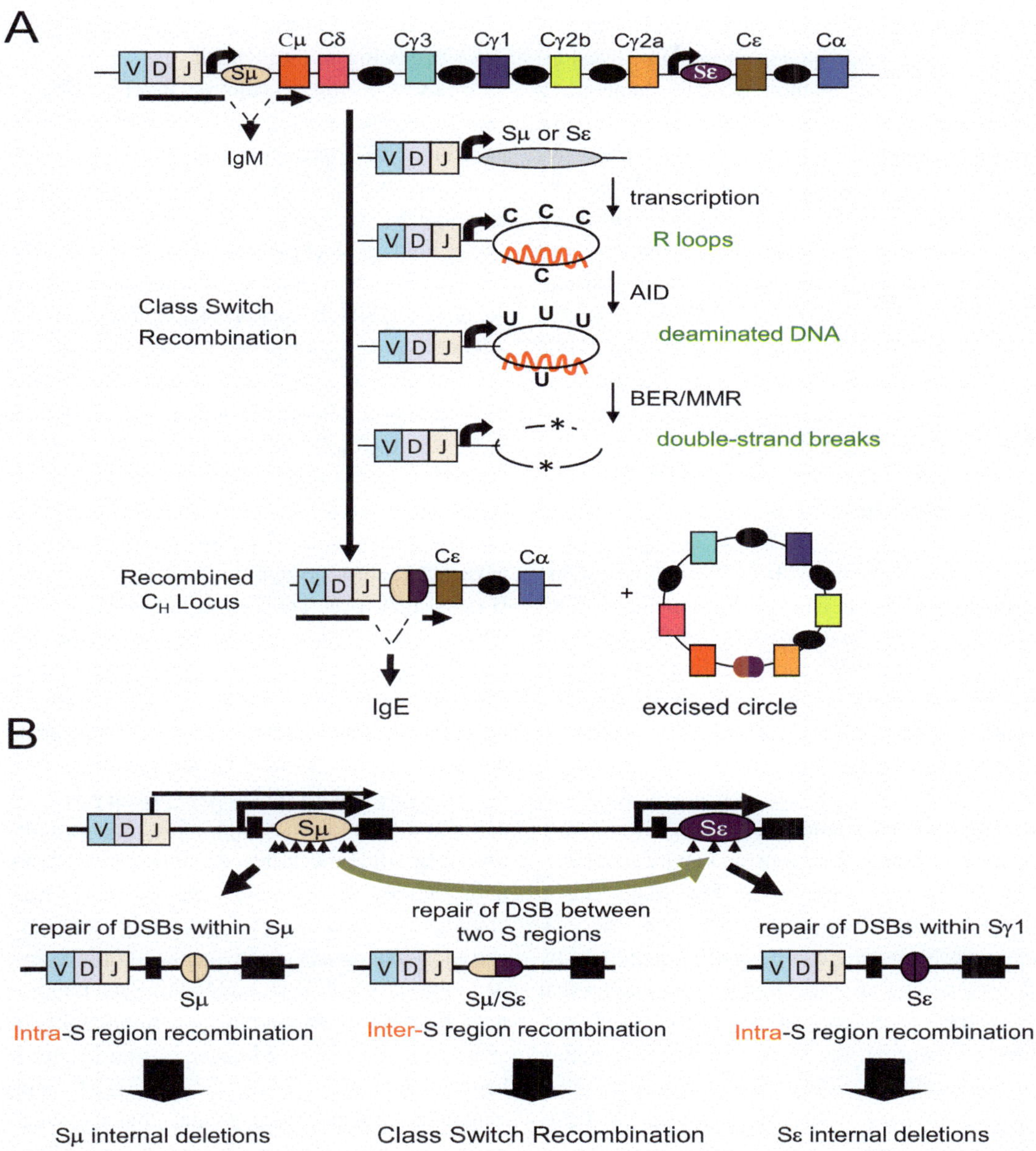

Fig. 6.3 Mechanisms of Class Switch Recombination (CSR). (**a**) In mature B cells, the VDJ_H exon formed during V(D)J recombination of the immunoglobulin heavy chain (IgH) locus at the pro-B cell stage is initially transcribed with the most upstream exon of the IgH constant region, Cμ, to generate IgM. Upon antigen encounter, mature B cells may undergo CSR to a secondary isotype (such as IgE encoded by the Cε exon in the example depicted here). CSR requires sterile transcription of repetitive "switch" (S) regions upstream of the recombining exons, Sμ and Sε. The nontranscribed strand is preferentially deaminated by AID and processed via ubiquitous DNA repair pathways to generate DNA single-strand breaks (SSBs). Two SSBs in opposite strands are sensed as a double-strand break (DSB) and activate the DNA Damage Response (DDR). DSBs at Sμ and Sε are brought together ("synapsed") and rejoined via the ubiquitous nonhomologous end-joining (NHEJ) pathway to effect recombination. After recombination, the VDJ_H exon is transcribed with the Cε exon, to generate IgE. Intervening DNA is rejoined in a circle and is eventually lost upon replication. (**b**) Possible fates for DSBs generated at S regions of recombining exons during CSR. DSBs within S regions can sometimes rejoin to each other (intra-S region recombination) to form an internal deletion. However, during normal CSR, DSBs within an S region preferentially rejoin to DSBs at another S region (inter-S region recombination), resulting in CSR

and will not be discussed here. Finally, the IgH locus is also modified via programmed lesion-repair cycles in mature B cells undergoing Somatic Hypermutation (SHM). Like CSR, this process employs ubiquitous DNA repair pathways to introduce point mutations into the IgH variable region to increase clonal affinity [51]. However, SHM differs from CSR in that it does not proceed through intermediary DSBs [57, 115]. Consistent with this notion, ATM [122, 123, 137], H2AX [139] and 53BP1 [111] are dispensable for SHM.

6.4.2 AID-Dependent Breaks and Translocations in DDR-Deficient Mice

Murine B cells activated for CSR *in vitro* provide a facile system to dissect the genetic requirements for recombination during CSR. The IgH locus heavy chain of B cells residing in the mouse spleen and lymph nodes is mostly in germline configuration (IgM^+). Upon isolation and activation with cytokines that mimic either a T cell-independent or a T cell-dependent response, these IgM^+ cells proliferate, induce AID and transcription through S regions and switch to a secondary isotype within a few days in a quasi-synchronous manner. Moreover, the efficiency of switching can be readily quantified by flow cytometry after labeling with antibodies that recognize secondary isotypes in the B cell surface. In addition, the highly proliferative nature of these cultures makes them amenable to cytogenetic analysis with IgH locus-specific FISH probes, providing a correlate between the switching defect and the frequency of IgH locus-specific chromosomal breaks and translocations (locus-specific genomic instability; see Fig. 6.4 for schematic of the FISH assay and possible outcomes and interpretation). This experimental pipeline has been applied extensively to understand the contribution of ATM and many of its substrates to the repair of AID-induced DSBs. In general, these studies have found that the molecular events upstream of DSBs (i.e., proliferation, AID induction and S region transcription) occur normally in DDR mutants. Rather, the repair of AID-dependent DSBs is compromised, as described below in more detail.

Analysis of $Atm^{-/-}$ B cells activated for CSR *in vitro* has revealed that the efficiency of switching is reduced to approximately half of the ATM-proficient control cultures [64] [137]. This defect is associated to frequent genomic instability at one or both IgH loci [64], revealing a requirement for ATM in the rejoining of a subset of AID-dependent DSBs. In support of this notion, loss of AID rescues most chromosomal instability at IgH in $Atm^{-/-}$ B cells [34]. Some residual IgH breaks observed in B cells deficient for both ATM and AID is thought to reflect on persistent RAG-dependent DSBs in B cells precursors that fail to trigger apoptosis in the absence of ATM-dependent cell cycle checkpoints [34] and/or when masked as dicentric chromosomes [79]. In $Atm^{-/-}$ activated B cells, IgH locus breaks and translocations are observed frequently (in up to 50% of cells in one study) [64]. Indeed, the most common IgH translocation partner is the broken IgH locus on the other chromosome 12 [64]. Murine chromosomes are acrocentric and therefore the majority of *de novo* rearrangements observed in primary $Atm^{-/-}$ B cell cultures are dicentrics [64, 79, 135]. In addition to IgH-IgH dicentrics, dicentrics with breakpoints at the IgH locus and a chromosomal break elsewhere or between two apparently "random" chromosomal breaks are frequent and often coexist in the same cell [64], highlighting the requirement for ATM in maintaining genomic stability in switching B cells. Moreover, translocations between IgH and c-Myc, a hallmark of many human B cell lymphomas, are detected at low frequency in primary activated $Atm^{-/-}$ B cells [135]. Interestingly, unlike TCRα/δ locus translocations arising in $Atm^{-/-}$ thymocytes, IgH-c-myc translocations in $Atm^{-/-}$ B cells are not clonally selected *in vivo* and $Atm^{-/-}$ mice are not prone to B cell lymphomas [79, 99]. Although the mechanisms underlying these lineage-specific differences are not known, they may relate to differential responses downstream of DSBs that ultimately determine cellular outcomes. In this regard, loss of ATM activates type I interferon signaling [71], a pathway that promotes cell death or survival in a context-dependent manner.

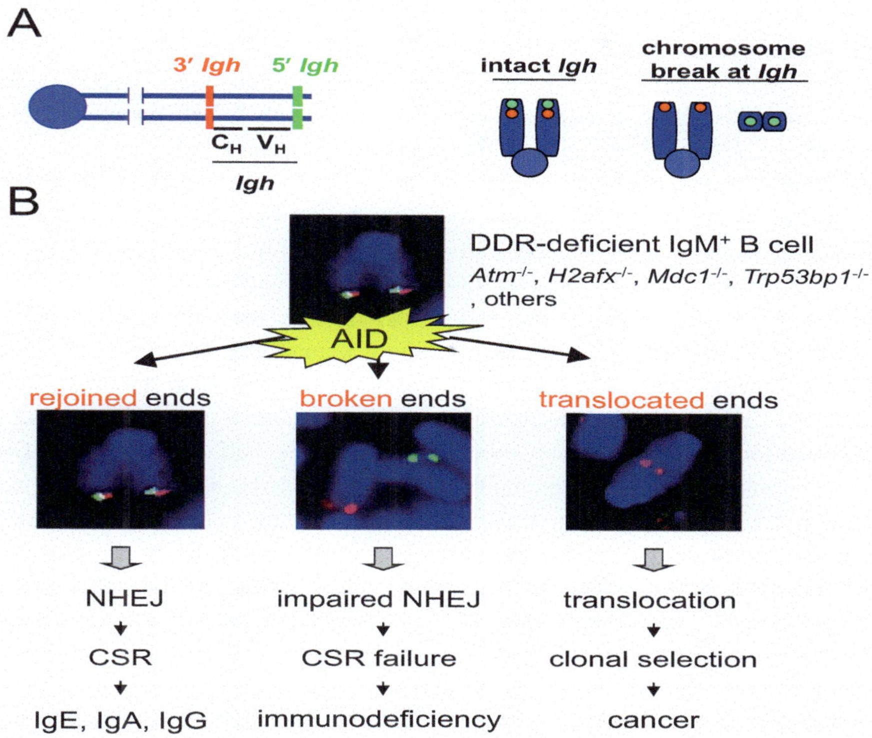

Fig. 6.4 Analysis of the fate of AID-dependent DNA ends using two-color fluorescence in situ hybridization (FISH) on murine B cell metaphases. (**a**) Schematic of the immunoglobulin heavy chain (IgH) locus in murine chromosome 12. The IgH locus localizes to the subtelomeric region, with the variable region (V_H) genes oriented towards the telomere and the constant region (C_H) exons oriented towards the centromere. During efficient CSR, rejoining of AID-dependent double-strand breaks (DSB) results in two close-by FISH signals in the subtelomeric region of the chromosome ("intact IgH"). In contrast, failure to rejoin the ends results in "split signals", or localization of each signal to a distinct chromosome fragment. (**b**) B cells were isolated from the spleen of mice deficient for DDR factors and activated for CSR *in vitro* using cytokines. After about three days, the fate of DNA ends was analyzed on metaphase spreads using two-color IgH locus FISH. Rejoining of DNA ends results in close-by signals and indicates recombination and expression of a secondary immunoglobulin isotype (such as IgG, IgE or IgA). However, rejoining is impaired in a subset of DDR-deficient cells, leading to end dissociation and "split signals" (broken or "free" ends). Defective switching manifests clinically as an immunodeficiency characterized by decreased titers of secondary isotypes. Finally, some breaks are repaired aberrantly by rejoining to a break in another chromosome, generating a chromosomal translocation. Most translocations in primary B cells likely have no consequence, but selection for rare oncogenic translocations may promote B cell transformation

ATM likely mediates the formation of γ-H2AX foci at the IgH locus in B cells undergoing CSR [131]. Although DNA-PKcs may also modify H2AX in this context, the ATM-dependent events may serve an important regulatory function by controlling the spread and density of the modification [148]. Like *Atm*$^{-/-}$ B cells, *H2afx*$^{-/-}$ B cells are impaired for CSR to multiple isotypes [64], although the severity of the defect tends to be lesser. Moreover, activated *H2afx*$^{-/-}$ B cells accumulate IgH locus chromosomal breaks and dicentrics with breakpoints at IgH, clearly pointing to a defect in the end-joining phase of CSR. Breeding into AID-deficient mice completely rescues genomic instability at IgH in *H2afx*$^{-/-}$ B cells [64], indicating that these breaks result from switching and excluding their origin as byproducts of transcription, replication or other processes. Finally, although *H2afx*$^{-/-}$ mice develop B cell lymphomas when bred into a p53-deficient background, T cell malignancies are more frequent [13].

The analysis of CSR in murine B cells deficient for 53BP1 (*Trp53bp1*$^{-/-}$ B cells) was particularly gratifying in that it led to novel mechanistic insights into the unique regulation of this process. Unexpectedly, *Trp53bp1*$^{-/-}$ B cells were found to be impaired for CSR to a much greater extent than *Atm*$^{-/-}$ or *H2afx*$^{-/-}$ B cells [111, 177], despite their lesser defect in the repair of "general" chromosomal breaks. The CSR defect is due to defective end-joining of AID-dependent DSBs because IgH locus-specific analysis of activated *Trp53bp1*$^{-/-}$ B cell metaphases revealed frequent IgH breaks [64, 135] that were completely rescued by breeding into an AID-deficient background [135]. However, the mechanism underlying the end-joining defect uniquely relies on defective synapsis of the two recombining DSBs. As a result, *Trp53bp1*$^{-/-}$ B cells show a dramatic increase in the frequency of intra-S switch region recombination [138], presumably reflecting on increased "local" repair in the absence of synapsis. In contrast, activated *H2afx*$^{-/-}$ B cells do not accumulate internal deletions to a greater extent than wild-type B cells [139], suggesting that they fail at rejoining synapsed S regions. In further support of the unique roles for 53BP1 during CSR, others have shown defective synapsis of V and DJ exons during V(D)J recombination of 53BP1-deficient T cells [52] and defective end-joining of dysfunctional telomeres in 53BP1-deficient cells [53].

6.4.3 AID-Dependent Breaks and Translocations in A-T Patients

The mechanisms and regulation of CSR are generally conserved between mice and humans [124]. Specifically, a conserved role for ATM in CSR has been demonstrated by the analysis of B cells from A-T patients. Consistent with defective CSR, individuals with A-T show variable decreases in secondary immunoglobulins, most commonly IgG4, IgA, IgE and IgG2, isolated or in combination [121]. These abnormalities are clinically relevant, resulting in impaired antibody response to pathogens and frequent sinopulmonary infections that negatively affect the quality of life of A-T children. Despite these defects, individuals with A-T are not prone to B cell lymphomas but rather to T cell leukemias. These lineage-specific differences in the progression from chromosomal breaks to full malignancy are also observed in the murine models and remain incompletely understood.

Finally, we note that somatic ATM inactivation is common in sporadic mature B cell lymphomas [4, 5, 46, 100, 160, 161]. Deletions at 11q22, containing the ATM locus, occur in approximately half of mantle cell lymphomas (MCLs) [4, 161], a mature B cell malignancy characterized by a clonal T(11,14) that fuses the IgH locus to cyclin D1 to drive its overexpression. The 11q22 deletion in MCL typically leads to loss of ATM function due to mutation of the second allele [35, 150, 161] and correlates with poor clinical outcome [46]. However, the T(11,14) translocation is thought to occur in pro-B cells undergoing V(D)J recombination [86], and may precede the ATM mutation during malignant progression.

ATM mutations are also observed at low frequency in cancers of the breast [1], pancreas [141], bladder [69], prostate [17] and other solid tumors. It currently remains unclear whether ATM roles in translocation suppression may contribute to tumor initiation and/or progression in this context, potentially in cooperation with roles in the activation of cell cycle checkpoints, metabolic regulation and others [97]. Finally, roles for DDR factors other than ATM in translocation suppression in human cancers have not been clearly established, although Bartek and colleagues reported that MDC1 and 53BP1 were lost in a subset of human carcinomas [9]. In summary, the DDR represents a main barrier to transformation in a wide range of human cancers [8, 66] and more work is needed to determine whether its functions in translocation suppression may extend beyond its well-documented roles in hematological malignancies.

References

1. Ahmed M, Rahman N (2006) ATM and breast cancer susceptibility. Oncogene 25(43):5906–5911. doi:1209873 [pii] https://doi.org/10.1038/sj.onc.1209873
2. Aifantis I, Raetz E, Buonamici S (2008) Molecular pathogenesis of T-cell leukaemia and lymphoma. Nat Rev Immunol 8(5):380–390. https://doi.org/10.1038/nri2304
3. Alt FW, Zhang Y, Meng FL, Guo C, Schwer B (2013) Mechanisms of programmed DNA lesions and genomic instability in the immune system. Cell 152(3):417–429. https://doi.org/10.1016/j.cell.2013.01.007
4. Au WY, Gascoyne RD, Viswanatha DS, Connors JM, Klasa RJ, Horsman DE (2002) Cytogenetic analysis in mantle cell lymphoma: a review of 214 cases. Leuk Lymphoma 43(4):783–791. https://doi.org/10.1080/10428190290016890
5. Austen B, Barone G, Reiman A, Byrd PJ, Baker C, Starczynski J, Nobbs MC, Murphy RP, Enright H, Chaila E, Quinn J, Stankovic T, Pratt G, Taylor AM (2008) Pathogenic ATM mutations occur rarely in a subset of multiple myeloma patients. Br J Haematol 142(6):925–933. https://doi.org/10.1111/j.1365-2141.2008.07281.x
6. Baer R, Heppell A, Taylor AM, Rabbitts PH, Boullier B, Rabbitts TH (1987) The breakpoint of an inversion of chromosome 14 in a T-cell leukemia: sequences downstream of the immunoglobulin heavy chain locus are implicated in tumorigenesis. Proc Natl Acad Sci U S A 84(24):9069–9073
7. Barlow C, Hirotsune S, Paylor R, Liyanage M, Eckhaus M, Collins F, Shiloh Y, Crawley JN, Ried T, Tagle D, Wynshaw-Boris A (1996) Atm-deficient mice: a paradigm of ataxia telangiectasia. Cell 86(1):159–171
8. Bartkova J, Horejsi Z, Koed K, Kramer A, Tort F, Zieger K, Guldberg P, Sehested M, Nesland JM, Lukas C, Orntoft T, Lukas J, Bartek J (2005) DNA damage response as a candidate anti-cancer barrier in early human tumorigenesis. Nature 434(7035):864–870
9. Bartkova J, Horejsi Z, Sehested M, Nesland JM, Rajpert-De Meyts E, Skakkebaek NE, Stucki M, Jackson S, Lukas J, Bartek J (2007) DNA damage response mediators MDC1 and 53BP1: constitutive activation and aberrant loss in breast and lung cancer, but not in testicular germ cell tumours. Oncogene 26(53):7414–7422. https://doi.org/10.1038/sj.onc.1210553
10. Bassing CH, Alt FW (2004a) The cellular response to general and programmed DNA double strand breaks. DNA Repair (Amst) 3(8-9):781–796
11. Bassing CH, Alt FW (2004b) H2AX may function as an anchor to hold broken chromosomal DNA ends in close proximity. Cell Cycle 3(2):149–153
12. Bassing CH, Chua KF, Sekiguchi J, Suh H, Whitlow SR, Fleming JC, Monroe BC, Ciccone DN, Yan C, Vlasakova K, Livingston DM, Ferguson DO, Scully R, Alt FW (2002) Increased ionizing radiation sensitivity and genomic instability in the absence of histone H2AX. Proc Natl Acad Sci U S A 99(12):8173–8178
13. Bassing CH, Suh H, Ferguson DO, Chua KF, Manis J, Eckersdorff M, Gleason M, Bronson R, Lee C, Alt FW (2003) Histone H2AX: a dosage-dependent suppressor of oncogenic translocations and tumors. Cell 114(3):359–370
14. Bekker-Jensen S, Lukas C, Kitagawa R, Melander F, Kastan MB, Bartek J, Lukas J (2006) Spatial organization of the mammalian genome surveillance machinery in response to DNA strand breaks. J Cell Biol 173(2):195–206. https://doi.org/10.1083/jcb.200510130
15. Bekker-Jensen S, Mailand N (2010) Assembly and function of DNA double-strand break repair foci in mammalian cells. DNA Repair (Amst) 9(12):1219–1228. https://doi.org/10.1016/j.dnarep.2010.09.010
16. Bekker-Jensen S, Mailand N (2011) The ubiquitin- and SUMO-dependent signaling response to DNA double-strand breaks. FEBS Lett 585(18):2914–2919. https://doi.org/10.1016/j.febslet.2011.05.056
17. Beltran H, Yelensky R, Frampton GM, Park K, Downing SR, MacDonald TY, Jarosz M, Lipson D, Tagawa ST, Nanus DM, Stephens PJ, Mosquera JM, Cronin MT, Rubin MA (2013) Targeted next-generation sequencing of advanced prostate cancer identifies potential therapeutic targets and disease heterogeneity. Eur Urol 63(5):920–926. https://doi.org/10.1016/j.eururo.2012.08.053
18. Bhandoola A, von Boehmer H, Petrie HT, Zuniga-Pflucker JC (2007) Commitment and developmental potential of extrathymic and intrathymic T cell precursors: plenty to choose from. Immunity 26(6):678–689. https://doi.org/10.1016/j.immuni.2007.05.009
19. Bhargava R, Carson CR, Lee G, Stark JM (2017) Contribution of canonical nonhomologous end joining to chromosomal rearrangements is enhanced by ATM kinase deficiency. Proc Natl Acad Sci U S A 114(4):728–733. https://doi.org/10.1073/pnas.1612204114
20. Bhatia V, Barroso SI, Garcia-Rubio ML, Tumini E, Herrera-Moyano E, Aguilera A (2014) BRCA2 prevents R-loop accumulation and associates with TREX-2 mRNA export factor PCID2. Nature 511(7509):362–365. https://doi.org/10.1038/nature13374
21. Boboila C, Alt FW, Schwer B (2012) Classical and alternative end-joining pathways for repair of lymphocyte-specific and general DNA double-strand breaks. Adv Immunol 116:1–49. https://doi.org/10.1016/B978-0-12-394300-2.00001-6
22. Boder E, Sedgwick RP (1958) Ataxia-telangiectasia; a familial syndrome of progressive cerebellar ataxia, oculocutaneous telangiectasia and frequent pulmonary infection. Pediatrics 21(4):526–554

23. Boerma EG, Siebert R, Kluin PM, Baudis M (2009) Translocations involving 8q24 in Burkitt lymphoma and other malignant lymphomas: a historical review of cytogenetics in the light of todays knowledge. Leukemia 23(2):225–234. https://doi.org/10.1038/leu.2008.281
24. Borghesani PR, Alt FW, Bottaro A, Davidson L, Aksoy S, Rathbun GA, Roberts TM, Swat W, Segal RA, Gu Y (2000) Abnormal development of Purkinje cells and lymphocytes in Atm mutant mice. Proc Natl Acad Sci U S A 97(7):3336–3341
25. Bottini AR, Gatti RA, Wirenfeldt M, Vinters HV (2012) Heterotopic Purkinje cells in ataxia-telangiectasia. Neuropathology 32(1):23–29. https://doi.org/10.1111/j.1440-1789.2011.01210.x
26. Bredemeyer AL, Sharma GG, Huang CY, Helmink BA, Walker LM, Khor KC, Nuskey B, Sullivan KE, Pandita TK, Bassing CH, Sleckman BP (2006) ATM stabilizes DNA double-strand-break complexes during V(D)J recombination. Nature 442(7101):466–470. doi:nature04866 [pii]. https://doi.org/10.1038/nature04866
27. Brito-Babapulle V, Catovsky D (1991) Inversions and tandem translocations involving chromosome 14q11 and 14q32 in T-prolymphocytic leukemia and T-cell leukemias in patients with ataxia telangiectasia. Cancer Genet Cytogenet 55(1):1–9. doi:0165-4608(91)90228-M [pii]
28. Brouwer I, Sitters G, Candelli A, Heerema SJ, Heller I, de Melo AJ, Zhang H, Normanno D, Modesti M, Peterman EJ, Wuite GJ (2016) Sliding sleeves of XRCC4-XLF bridge DNA and connect fragments of broken DNA. Nature 535(7613):566–569. https://doi.org/10.1038/nature18643
29. Brown JS, Jackson SP (2015) Ubiquitylation, neddylation and the DNA damage response. Open Biol 5(4):150018. https://doi.org/10.1098/rsob.150018
30. Bunting SF, Nussenzweig A (2013) End-joining, translocations and cancer. Nat Rev Cancer 13(7):443–454. https://doi.org/10.1038/nrc3537
31. Burma S, Chen BP, Murphy M, Kurimasa A, Chen DJ (2001) ATM phosphorylates histone H2AX in response to DNA double-strand breaks. J Biol Chem 276(45):42462–42467. https://doi.org/10.1074/jbc.C100466200. C100466200 [pii]
32. Butterworth SV, Taylor AM (1986) A subpopulation of t(2;14)(p11;q32) cells in ataxia telangiectasia B lymphocytes. Hum Genet 73(4):346–349
33. Callen E, Bunting S, Huang CY, Difilippantonio MJ, Wong N, Khor B, Mahowald G, Kruhlak MJ, Ried T, Sleckman BP, Nussenzweig A (2009) Chimeric IgH-TCRalpha/delta translocations in T lymphocytes mediated by RAG. Cell Cycle 8(15):2408–2412. doi:9085 [pii]
34. Callen E, Jankovic M, Difilippantonio S, Daniel JA, Chen HT, Celeste A, Pellegrini M, McBride K, Wangsa D, Bredemeyer AL, Sleckman BP, Ried T, Nussenzweig M, Nussenzweig A (2007) ATM prevents the persistence and propagation of chromosome breaks in lymphocytes. Cell 130(1):63–75. doi:S0092-8674(07)00783-0 [pii]. https://doi.org/10.1016/j.cell.2007.06.016
35. Camacho E, Hernandez L, Hernandez S, Tort F, Bellosillo B, Bea S, Bosch F, Montserrat E, Cardesa A, Fernandez PL, Campo E (2002) ATM gene inactivation in mantle cell lymphoma mainly occurs by truncating mutations and missense mutations involving the phosphatidylinositol-3 kinase domain and is associated with increasing numbers of chromosomal imbalances. Blood 99(1):238–244
36. Canela A, Maman Y, Jung S, Wong N, Callen E, Day A, Kieffer-Kwon KR, Pekowska A, Zhang H, Rao SSP, Huang SC, McKinnon PJ, Aplan PD, Pommier Y, Aiden EL, Casellas R, Nussenzweig A (2017) Genome Organization Drives Chromosome Fragility. Cell 170(3):507–521. e518. https://doi.org/10.1016/j.cell.2017.06.034
37. Canman CE, Lim DS, Cimprich KA, Taya Y, Tamai K, Sakaguchi K, Appella E, Kastan MB, Siliciano JD (1998) Activation of the ATM kinase by ionizing radiation and phosphorylation of p53. Science 281(5383):1677–1679
38. Casellas R, Basu U, Yewdell WT, Chaudhuri J, Robbiani DF, Di Noia JM (2016) Mutations, kataegis and translocations in B cells: understanding AID promiscuous activity. Nat Rev Immunol 16(3):164–176. https://doi.org/10.1038/nri.2016.2
39. Celeste A, Difilippantonio S, Difilippantonio MJ, Fernandez-Capetillo O, Pilch DR, Sedelnikova OA, Eckhaus M, Ried T, Bonner WM, Nussenzweig A (2003) H2AX haploinsufficiency modifies genomic stability and tumor susceptibility. Cell 114(3):371–383
40. Celeste A, Petersen S, Romanienko PJ, Fernandez-Capetillo O, Chen HT, Sedelnikova OA, Reina-San-Martin B, Coppola V, Meffre E, Difilippantonio MJ, Redon C, Pilch DR, Olaru A, Eckhaus M, Camerini-Otero RD, Tessarollo L, Livak F, Manova K, Bonner WM, Nussenzweig MC, Nussenzweig A (2002) Genomic instability in mice lacking histone H2AX. Science 296(5569):922–927
41. Chang HHY, Pannunzio NR, Adachi N, Lieber MR (2017) Non-homologous DNA end joining and alternative pathways to double-strand break repair. Nat Rev Mol Cell Biol. https://doi.org/10.1038/nrm.2017.48
42. Chapman JR, Barral P, Vannier JB, Borel V, Steger M, Tomas-Loba A, Sartori AA, Adams IR, Batista FD, Boulton SJ (2013) RIF1 is essential for 53BP1-dependent nonhomologous end joining and suppression of DNA double-strand break resection. Mol Cell. https://doi.org/10.1016/j.molcel.2013.01.002
43. Chen HT, Bhandoola A, Difilippantonio MJ, Zhu J, Brown MJ, Tai X, Rogakou EP, Brotz TM, Bonner WM, Ried T, Nussenzweig A (2000) Response to RAG-mediated VDJ cleavage by NBS1 and gamma-H2AX. Science 290(5498):1962–1965
44. Chiarle R, Zhang Y, Frock RL, Lewis SM, Molinie B, Ho YJ, Myers DR, Choi VW, Compagno M, Malkin DJ, Neuberg D, Monti S, Giallourakis CC,

Gostissa M, Alt FW (2011) Genome-wide translocation sequencing reveals mechanisms of chromosome breaks and rearrangements in B cells. Cell 147(1):107–119. doi:S0092-8674(11)01001-4 [pii]. https://doi.org/10.1016/j.cell.2011.07.049
45. Ciccia A, Elledge SJ (2010) The DNA damage response: making it safe to play with knives. Mol Cell 40(2):179–204. https://doi.org/10.1016/j.molcel.2010.09.019
46. Cuneo A, Bigoni R, Rigolin GM, Roberti MG, Milani R, Bardi A, Minotto C, Agostini P, De Angeli C, Narducci MG, Sabbioni S, Russo G, Negrini M, Castoldi G (2000) Acquired chromosome 11q deletion involving the ataxia telangiectasia locus in B-cell non-Hodgkin's lymphoma: correlation with clinicobiologic features. J Clin Oncol 18(13):2607–2614
47. Davey MP, Bertness V, Nakahara K, Johnson JP, McBride OW, Waldmann TA, Kirsch IR (1988) Juxtaposition of the T-cell receptor alpha-chain locus (14q11) and a region (14q32) of potential importance in leukemogenesis by a 14;14 translocation in a patient with T-cell chronic lymphocytic leukemia and ataxia-telangiectasia. Proc Natl Acad Sci U S A 85(23):9287–9291
48. Davis MM, Bjorkman PJ (1988) T-cell antigen receptor genes and T-cell recognition. Nature 334(6181):395–402. https://doi.org/10.1038/334395a0
49. Dearden CE (2006) T-cell prolymphocytic leukemia. Med Oncol 23(1):17–22. https://doi.org/10.1385/MO:23:1:17
50. Della-Maria J, Zhou Y, Tsai MS, Kuhnlein J, Carney JP, Paull TT, Tomkinson AE (2011) Human Mre11/human Rad50/Nbs1 and DNA ligase IIIalpha/XRCC1 protein complexes act together in an alternative nonhomologous end joining pathway. J Biol Chem 286(39):33845–33853. https://doi.org/10.1074/jbc.M111.274159
51. Di Noia JM, Neuberger MS (2007) Molecular mechanisms of antibody somatic hypermutation. Annu Rev Biochem 76:1–22. https://doi.org/10.1146/annurev.biochem.76.061705.090740
52. Difilippantonio S, Gapud E, Wong N, Huang CY, Mahowald G, Chen HT, Kruhlak MJ, Callen E, Livak F, Nussenzweig MC, Sleckman BP, Nussenzweig A (2008) 53BP1 facilitates long-range DNA end-joining during V(D)J recombination. Nature 456(7221):529–533. doi:nature07476 [pii]. https://doi.org/10.1038/nature07476
53. Dimitrova N, Chen YC, Spector DL, de Lange T (2008) 53BP1 promotes non-homologous end joining of telomeres by increasing chromatin mobility. Nature 456(7221):524–528. https://doi.org/10.1038/nature07433
54. Dinkelmann M, Spehalski E, Stoneham T, Buis J, Wu Y, Sekiguchi JM, Ferguson DO (2009) Multiple functions of MRN in end-joining pathways during isotype class switching. Nat Struct Mol Biol 16(8):808–813. https://doi.org/10.1038/nsmb.1639
55. Donehower LA, Harvey M, Slagle BL, McArthur MJ, Montgomery CA Jr, Butel JS, Bradley A (1992) Mice deficient for p53 are developmentally normal but susceptible to spontaneous tumours. Nature 356(6366):215–221
56. Dudgeon C, Chan C, Kang W, Sun Y, Emerson R, Robins H, Levine AJ (2014) The evolution of thymic lymphomas in p53 knockout mice. Genes Dev 28(23):2613–2620. https://doi.org/10.1101/gad.252148.114
57. Dudley DD, Chaudhuri J, Bassing CH, Alt FW (2005) Mechanism and Control of V(D)J Recombination versus Class Switch Recombination: Similarities and Differences. Adv Immunol 86:43–112
58. Dujka ME, Puebla-Osorio N, Tavana O, Sang M, Zhu C (2010) ATM and p53 are essential in the cell-cycle containment of DNA breaks during V(D)J recombination in vivo. Oncogene 29(7):957–965. doi:onc2009394 [pii]. https://doi.org/10.1038/onc.2009.394
59. Duquette ML, Pham P, Goodman MF, Maizels N (2005) AID binds to transcription-induced structures in c-MYC that map to regions associated with translocation and hypermutation. Oncogene 24(38):5791–5798. https://doi.org/10.1038/sj.onc.1208746
60. Ehrlich LA, Yang-Iott K, Bassing CH (2014) Tcrdelta translocations that delete the Bcl11b haploinsufficient tumor suppressor gene promote atm-deficient T cell acute lymphoblastic leukemia. Cell Cycle 13(19):3076–3082. https://doi.org/10.4161/15384101.2014.949144
61. Escribano-Diaz C, Orthwein A, Fradet-Turcotte A, Xing M, Young JT, Tkac J, Cook MA, Rosebrock AP, Munro M, Canny MD, Xu D, Durocher D (2013) A cell cycle-dependent regulatory circuit composed of 53BP1-RIF1 and BRCA1-CtIP controls DNA repair pathway choice. Mol Cell. https://doi.org/10.1016/j.molcel.2013.01.001
62. Fang EF, Scheibye-Knudsen M, Chua KF, Mattson MP, Croteau DL, Bohr VA (2016) Nuclear DNA damage signalling to mitochondria in ageing. Nat Rev Mol Cell Biol 17(5):308–321. https://doi.org/10.1038/nrm.2016.14
63. Franco S, Alt FW, Manis JP (2006a) Pathways that suppress programmed DNA breaks from progressing to chromosomal breaks and translocations. DNA Repair (Amst) 5(9-10):1030–1041. doi:S1568-7864(06)00163-7 [pii]. https://doi.org/10.1016/j.dnarep.2006.05.024
64. Franco S, Gostissa M, Zha S, Lombard DB, Murphy MM, Zarrin AA, Yan C, Tepsuporn S, Morales JC, Adams MM, Lou Z, Bassing CH, Manis JP, Chen J, Carpenter PB, Alt FW (2006b) H2AX prevents DNA breaks from progressing to chromosome breaks and translocations. Mol Cell 21(2):201–214
65. Ginno PA, Lott PL, Christensen HC, Korf I, Chedin F (2012) R-loop formation is a distinctive characteristic of unmethylated human CpG island promoters. Mol Cell 45(6):814–825. https://doi.org/10.1016/j.molcel.2012.01.017

66. Gorgoulis VG, Vassiliou LV, Karakaidos P, Zacharatos P, Kotsinas A, Liloglou T, Venere M, Ditullio RA Jr, Kastrinakis NG, Levy B, Kletsas D, Yoneta A, Herlyn M, Kittas C, Halazonetis TD (2005) Activation of the DNA damage checkpoint and genomic instability in human precancerous lesions. Nature 434(7035):907–913
67. Gostissa M, Schwer B, Chang A, Dong J, Meyers RM, Marecki GT, Choi VW, Chiarle R, Zarrin AA, Alt FW (2014) IgH class switching exploits a general property of two DNA breaks to be joined in cis over long chromosomal distances. Proc Natl Acad Sci U S A 111(7):2644–2649. https://doi.org/10.1073/pnas.1324176111
68. Graham TG, Walter JC, Loparo JJ (2016) Two-stage synapsis of DNA ends during non-homologous end joining. Mol Cell 61(6):850–858. https://doi.org/10.1016/j.molcel.2016.02.010
69. Guo G, Sun X, Chen C, Wu S, Huang P, Li Z, Dean M, Huang Y, Jia W, Zhou Q, Tang A, Yang Z, Li X, Song P, Zhao X, Ye R, Zhang S, Lin Z, Qi M, Wan S, Xie L, Fan F, Nickerson ML, Zou X, Hu X, Xing L, Lv Z, Mei H, Gao S, Liang C, Gao Z, Lu J, Yu Y, Liu C, Li L, Fang X, Jiang Z, Yang J, Li C, Zhao X, Chen J, Zhang F, Lai Y, Lin Z, Zhou F, Chen H, Chan HC, Tsang S, Theodorescu D, Li Y, Zhang X, Wang J, Yang H, Gui Y, Wang J, Cai Z (2013) Whole-genome and whole-exome sequencing of bladder cancer identifies frequent alterations in genes involved in sister chromatid cohesion and segregation. Nat Genet 45(12):1459–1463. https://doi.org/10.1038/ng.2798
70. Guo Z, Kozlov S, Lavin MF, Person MD, Paull TT (2010) ATM activation by oxidative stress. Science 330(6003):517–521. doi:330/6003/517 [pii] https://doi.org/10.1126/science.1192912
71. Hartlova A, Erttmann SF, Raffi FA, Schmalz AM, Resch U, Anugula S, Lienenklaus S, Nilsson LM, Kroger A, Nilsson JA, Ek T, Weiss S, Gekara NO (2015) DNA damage primes the type I interferon system via the cytosolic DNA sensor STING to promote anti-microbial innate immunity. Immunity 42(2):332–343. https://doi.org/10.1016/j.immuni.2015.01.012
72. Harvey M, McArthur MJ, Montgomery CA Jr, Bradley A, Donehower LA (1993) Genetic background alters the spectrum of tumors that develop in p53-deficient mice. FASEB J 7(10):938–943
73. Hatchi E, Skourti-Stathaki K, Ventz S, Pinello L, Yen A, Kamieniarz-Gdula K, Dimitrov S, Pathania S, McKinney KM, Eaton ML, Kellis M, Hill SJ, Parmigiani G, Proudfoot NJ, Livingston DM (2015) BRCA1 recruitment to transcriptional pause sites is required for R-loop-driven DNA damage repair. Mol Cell 57(4):636–647. https://doi.org/10.1016/j.molcel.2015.01.011
74. Helmink BA, Bredemeyer AL, Lee BS, Huang CY, Sharma GG, Walker LM, Bednarski JJ, Lee WL, Pandita TK, Bassing CH, Sleckman BP (2009) MRN complex function in the repair of chromosomal Rag-mediated DNA double-strand breaks. J Exp Med 206(3):669–679. https://doi.org/10.1084/jem.20081326
75. Helmink BA, Tubbs AT, Dorsett Y, Bednarski JJ, Walker LM, Feng Z, Sharma GG, McKinnon PJ, Zhang J, Bassing CH, Sleckman BP (2011) H2AX prevents CtIP-mediated DNA end resection and aberrant repair in G1-phase lymphocytes. Nature 469(7329):245–249. doi:nature09585 [pii]. https://doi.org/10.1038/nature09585
76. Helmrich A, Ballarino M, Tora L (2011) Collisions between replication and transcription complexes cause common fragile site instability at the longest human genes. Mol Cell 44(6):966–977. https://doi.org/10.1016/j.molcel.2011.10.013
77. Heppell A, Butterworth SV, Hollis RJ, Kennaugh AA, Beatty DW, Taylor AM (1988) Breakage of the T cell receptor alpha chain locus in non malignant clones from patients with ataxia telangiectasia. Hum Genet 79(4):360–364
78. Hopfner KP, Craig L, Moncalian G, Zinkel RA, Usui T, Owen BA, Karcher A, Henderson B, Bodmer JL, McMurray CT, Carney JP, Petrini JH, Tainer JA (2002) The Rad50 zinc-hook is a structure joining Mre11 complexes in DNA recombination and repair. Nature 418(6897):562–566. doi:https://doi.org/10.1038/nature00922 nature00922 [pii]
79. Hu J, Tepsuporn S, Meyers RM, Gostissa M, Alt FW (2014) Developmental propagation of V(D)J recombination-associated DNA breaks and translocations in mature B cells via dicentric chromosomes. Proc Natl Acad Sci U S A 111(28):10269–10274. https://doi.org/10.1073/pnas.1410112111
80. Huang CY, Sharma GG, Walker LM, Bassing CH, Pandita TK, Sleckman BP (2007) Defects in coding joint formation in vivo in developing ATM-deficient B and T lymphocytes. J Exp Med 204(6):1371–1381. doi:jem.20061460 [pii]. https://doi.org/10.1084/jem.20061460
81. Huen MS, Grant R, Manke I, Minn K, Yu X, Yaffe MB, Chen J (2007) RNF8 transduces the DNA-damage signal via histone ubiquitylation and checkpoint protein assembly. Cell 131(5):901–914. doi:S0092-8674(07)01270-6 [pii]. https://doi.org/10.1016/j.cell.2007.09.041
82. Huertas P, Aguilera A (2003) Cotranscriptionally formed DNA:RNA hybrids mediate transcription elongation impairment and transcription-associated recombination. Mol Cell 12(3):711–721
83. Hustedt N, Durocher D (2016) The control of DNA repair by the cell cycle. Nat Cell Biol 19(1):1–9. https://doi.org/10.1038/ncb3452
84. Jacks T, Remington L, Williams BO, Schmitt EM, Halachmi S, Bronson RT, Weinberg RA (1994) Tumor spectrum analysis in p53-mutant mice. Curr Biol 4(1):1–7
85. Jackson SP, Durocher D (2013) Regulation of DNA damage responses by ubiquitin and SUMO. Mol Cell 49(5):795–807. https://doi.org/10.1016/j.molcel.2013.01.017

86. Jares P, Colomer D, Campo E (2012) Molecular pathogenesis of mantle cell lymphoma. J Clin Invest 122(10):3416–3423. https://doi.org/10.1172/JCI61272
87. Jasin M, Rothstein R (2013) Repair of strand breaks by homologous recombination. Cold Spring Harb Perspect Biol 5(11):a012740. https://doi.org/10.1101/cshperspect.a012740
88. Jiang H, Chang FC, Ross AE, Lee J, Nakayama K, Desiderio S (2005) Ubiquitylation of RAG-2 by Skp2-SCF links destruction of the V(D)J recombinase to the cell cycle. Mol Cell 18(6):699–709. doi:S1097-2765(05)01317-1 [pii]. https://doi.org/10.1016/j.molcel.2005.05.011
89. Jiang W, Lee BJ, Li C, Dubois RL, Gostissa M, Alt FW, Zha S (2015) Aberrant TCRdelta rearrangement underlies the T-cell lymphocytopenia and t(12;14) translocation associated with ATM deficiency. Blood 125(17):2665–2668. https://doi.org/10.1182/blood-2015-01-622621
90. Kamimura K, Ohi H, Kubota T, Okazuka K, Yoshikai Y, Wakabayashi Y, Aoyagi Y, Mishima Y, Kominami R (2007) Haploinsufficiency of Bcl11b for suppression of lymphomagenesis and thymocyte development. Biochem Biophys Res Commun 355(2):538–542. https://doi.org/10.1016/j.bbrc.2007.02.003
91. Kastan MB, Zhan Q, el-Deiry WS, Carrier F, Jacks T, Walsh WV, Plunkett BS, Vogelstein B, Fornace AJ Jr (1992) A mammalian cell cycle checkpoint pathway utilizing p53 and GADD45 is defective in ataxia-telangiectasia. Cell 71(4):587–597
92. Kennaugh AA, Butterworth SV, Hollis R, Baer R, Rabbitts TH, Taylor AM (1986) The chromosome breakpoint at 14q32 in an ataxia telangiectasia t(14;14) T cell clone is different from the 14q32 breakpoint in Burkitts and an inv(14) T cell lymphoma. Hum Genet 73(3):254–259
93. Klein IA, Resch W, Jankovic M, Oliveira T, Yamane A, Nakahashi H, Di Virgilio M, Bothmer A, Nussenzweig A, Robbiani DF, Casellas R, Nussenzweig MC (2011) Translocation-capture sequencing reveals the extent and nature of chromosomal rearrangements in B lymphocytes. Cell 147 (1):95–106. doi:S0092-8674(11)00999-8 [pii]10.1016/j.cell.2011.07.048
94. Kojis TL, Gatti RA, Sparkes RS (1991) The cytogenetics of ataxia telangiectasia. Cancer Genet Cytogenet 56(2):143–156
95. Kojis TL, Schreck RR, Gatti RA, Sparkes RS (1989) Tissue specificity of chromosomal rearrangements in ataxia-telangiectasia. Hum Genet 83(4):347–352
96. Kolas NK, Chapman JR, Nakada S, Ylanko J, Chahwan R, Sweeney FD, Panier S, Mendez M, Wildenhain J, Thomson TM, Pelletier L, Jackson SP, Durocher D (2007) Orchestration of the DNA-damage response by the RNF8 ubiquitin ligase. Science 318(5856):1637–1640. doi:1150034 [pii]. https://doi.org/10.1126/science.1150034
97. Korz C, Pscherer A, Benner A, Mertens D, Schaffner C, Leupolt E, Dohner H, Stilgenbauer S, Lichter P (2002) Evidence for distinct pathomechanisms in B-cell chronic lymphocytic leukemia and mantle cell lymphoma by quantitative expression analysis of cell cycle and apoptosis-associated genes. Blood 99(12):4554–4561
98. Krangel MS, Carabana J, Abbarategui I, Schlimgen R, Hawwari A (2004) Enforcing order within a complex locus: current perspectives on the control of V(D)J recombination at the murine T-cell receptor alpha/delta locus. Immunol Rev 200:224–232. https://doi.org/10.1111/j.0105-2896.2004.00155.x
99. Lavin MF (2013) The appropriateness of the mouse model for ataxia-telangiectasia: neurological defects but no neurodegeneration. DNA Repair (Amst) 12(8):612–619. https://doi.org/10.1016/j.dnarep.2013.04.014
100. Lens D, Matutes E, Catovsky D, Coignet LJ (2000) Frequent deletions at 11q23 and 13q14 in B cell prolymphocytic leukemia (B-PLL). Leukemia 14(3):427–430
101. Liao MJ, Zhang XX, Hill R, Gao J, Qumsiyeh MB, Nichols W, Van Dyke T (1998) No requirement for V(D)J recombination in p53-deficient thymic lymphoma. Mol Cell Biol 18(6):3495–3501
102. Lieber MR (2010) The mechanism of double-strand DNA break repair by the nonhomologous DNA end-joining pathway. Annu Rev Biochem 79:181–211. https://doi.org/10.1146/annurev.biochem.052308.093131
103. Lieber MR (2016) Mechanisms of human lymphoid chromosomal translocations. Nat Rev Cancer 16(6):387–398. https://doi.org/10.1038/nrc.2016.40
104. Lieber MR, Ma Y, Pannicke U, Schwarz K (2003) Mechanism and regulation of human non-homologous DNA end-joining. Nat Rev Mol Cell Biol 4(9):712–720. https://doi.org/10.1038/nrm1202. nrm1202 [pii]
105. Liyanage M, Weaver Z, Barlow C, Coleman A, Pankratz DG, Anderson S, Wynshaw-Boris A, Ried T (2000) Abnormal rearrangement within the alpha/delta T-cell receptor locus in lymphomas from Atm-deficient mice. Blood 96(5):1940–1946
106. Loffert D, Schaal S, Ehlich A, Hardy RR, Zou YR, Muller W, Rajewsky K (1994) Early B-cell development in the mouse: insights from mutations introduced by gene targeting. Immunol Rev 137:135–153
107. Lou Z, Minter-Dykhouse K, Franco S, Gostissa M, Rivera MA, Celeste A, Manis JP, van Deursen J, Nussenzweig A, Paull TT, Alt FW, Chen J (2006) MDC1 maintains genomic stability by participating in the amplification of ATM-dependent DNA damage signals. Mol Cell 21(2):187–200
108. Lovejoy CA, Cortez D (2009) Common mechanisms of PIKK regulation. DNA Repair (Amst) 8(9):1004–1008. https://doi.org/10.1016/j.dnarep.2009.04.006
109. Luo K, Yuan J, Lou Z (2011) Oligomerization of MDC1 protein is important for proper DNA dam-

age response. J Biol Chem 286(32):28192–28199. https://doi.org/10.1074/jbc.M111.258087
110. Mailand N, Bekker-Jensen S, Faustrup H, Melander F, Bartek J, Lukas C, Lukas J (2007) RNF8 ubiquitylates histones at DNA double-strand breaks and promotes assembly of repair proteins. Cell 131(5):887–900. doi:S0092-8674(07)01271-8 [pii]. https://doi.org/10.1016/j.cell.2007.09.040
111. Manis JP, Morales JC, Xia Z, Kutok JL, Alt FW, Carpenter PB (2004) 53BP1 links DNA damage-response pathways to immunoglobulin heavy chain class-switch recombination. Nat Immunol 5(5):481–487
112. Matsuoka S, Ballif BA, Smogorzewska A, McDonald ER, 3rd, Hurov KE, Luo J, Bakalarski CE, Zhao Z, Solimini N, Lerenthal Y, Shiloh Y, Gygi SP, Elledge SJ (2007) ATM and ATR substrate analysis reveals extensive protein networks responsive to DNA damage. Science 316(5828):1160–1166. doi:316/5828/1160 [pii] https://doi.org/10.1126/science.1140321
113. Matthews AJ, Zheng S, DiMenna LJ, Chaudhuri J (2014) Regulation of immunoglobulin class-switch recombination: choreography of noncoding transcription, targeted DNA deamination, and long-range DNA repair. Adv Immunol 122:1–57. https://doi.org/10.1016/B978-0-12-800267-4.00001-8
114. Metcalfe JA, Parkhill J, Campbell L, Stacey M, Biggs P, Byrd PJ, Taylor AM (1996) Accelerated telomere shortening in ataxia telangiectasia. Nat Genet 13(3):350–353. https://doi.org/10.1038/ng0796-350
115. Methot SP, Di Noia JM (2017) Molecular Mechanisms of Somatic Hypermutation and Class Switch Recombination. Adv Immunol 133:37–87. https://doi.org/10.1016/bs.ai.2016.11.002
116. Morales JC, Franco S, Murphy MM, Bassing CH, Mills KD, Adams MM, Walsh NC, Manis JP, Rassidakis GZ, Alt FW, Carpenter PB (2006) 53BP1 and p53 synergize to suppress genomic instability and lymphomagenesis. Proc Natl Acad Sci U S A 103(9):3310–3315
117. Morales JC, Xia Z, Lu T, Aldrich MB, Wang B, Rosales C, Kellems RE, Hittelman WN, Elledge SJ, Carpenter PB (2003) Role for the BRCA1 C-terminal repeats (BRCT) protein 53BP1 in maintaining genomic stability. J Biol Chem 278(17):14971–14977
118. Muramatsu M, Kinoshita K, Fagarasan S, Yamada S, Shinkai Y, Honjo T (2000) Class switch recombination and hypermutation require activation-induced cytidine deaminase (AID), a potential RNA editing enzyme. Cell 102(5):553–563
119. Myers JS, Cortez D (2006) Rapid activation of ATR by ionizing radiation requires ATM and Mre11. J Biol Chem 281(14):9346–9350. https://doi.org/10.1074/jbc.M513265200
120. Nissenkorn A, Levi YB, Vilozni D, Berkun Y, Efrati O, Frydman M, Yahav J, Waldman D, Somech R, Shenhod E, Menascu S, Ben-Zeev B (2011) Neurologic presentation in children with ataxia-telangiectasia: is small head circumference a hallmark of the disease? J Pediatr 159(3):466–471. e461. https://doi.org/10.1016/j.jpeds.2011.02.005
121. Nowak-Wegrzyn A, Crawford TO, Winkelstein JA, Carson KA, Lederman HM (2004) Immunodeficiency and infections in ataxia-telangiectasia. J Pediatr 144(4):505–511. https://doi.org/10.1016/j.jpeds.2003.12.046
122. Pan-Hammarstrom Q, Dai S, Zhao Y, van Dijk-Hard IF, Gatti RA, Borresen-Dale AL, Hammarstrom L (2003) ATM is not required in somatic hypermutation of VH, but is involved in the introduction of mutations in the switch mu region. J Immunol 170(7):3707–3716
123. Pan-Hammarstrom Q, Lahdesmaki A, Zhao Y, Du L, Zhao Z, Wen S, Ruiz-Perez VL, Dunn-Walters DK, Goodship JA, Hammarstrom L (2006) Disparate roles of ATR and ATM in immunoglobulin class switch recombination and somatic hypermutation. J Exp Med 203(1):99–110. https://doi.org/10.1084/jem.20050595
124. Pan-Hammarstrom Q, Zhao Y, Hammarstrom L (2007) Class switch recombination: a comparison between mouse and human. Adv Immunol 93:1–61. https://doi.org/10.1016/S0065-2776(06)93001-6
125. Pandita TK, Pathak S, Geard CR (1995) Chromosome end associations, telomeres and telomerase activity in ataxia telangiectasia cells. Cytogenet Cell Genet 71(1):86–93
126. Panier S, Durocher D (2009) Regulatory ubiquitylation in response to DNA double-strand breaks. DNA Repair (Amst) 8(4):436–443. doi:S1568-7864(09)00008-1 [pii]. https://doi.org/10.1016/j.dnarep.2009.01.013
127. Panier S, Durocher D (2013) Push back to respond better: regulatory inhibition of the DNA double-strand break response. Nat Rev Mol Cell Biol 14(10):661–672. https://doi.org/10.1038/nrm3659
128. Paula-Barbosa MM, Ruela C, Tavares MA, Pontes C, Saraiva A, Cruz C (1983) Cerebellar cortex ultrastructure in ataxia-telangiectasia. Ann Neurol 13(3):297–302. https://doi.org/10.1002/ana.410130312
129. Paull TT (2015) Mechanisms of ATM Activation. Annu Rev Biochem. https://doi.org/10.1146/annurev-biochem-060614-034335
130. Pekarsky Y, Hallas C, Isobe M, Russo G, Croce CM (1999) Abnormalities at 14q32.1 in T cell malignancies involve two oncogenes. Proc Natl Acad Sci U S A 96(6):2949–2951
131. Petersen S, Casellas R, Reina-San-Martin B, Chen HT, Difilippantonio MJ, Wilson PC, Hanitsch L, Celeste A, Muramatsu M, Pilch DR, Redon C, Ried T, Bonner WM, Honjo T, Nussenzweig MC, Nussenzweig A (2001) AID is required to initiate Nbs1/gamma-H2AX focus formation and mutations at sites of class switching. Nature 414(6864):660–665

132. Petiniot LK, Weaver Z, Barlow C, Shen R, Eckhaus M, Steinberg SM, Ried T, Wynshaw-Boris A, Hodes RJ (2000) Recombinase-activating gene (RAG) 2-mediated V(D)J recombination is not essential for tumorigenesis in Atm-deficient mice. Proc Natl Acad Sci U S A 97(12):6664–6669
133. Petiniot LK, Weaver Z, Vacchio M, Shen R, Wangsa D, Barlow C, Eckhaus M, Steinberg SM, Wynshaw-Boris A, Ried T, Hodes RJ (2002) RAG-mediated V(D)J recombination is not essential for tumorigenesis in Atm-deficient mice. Mol Cell Biol 22(9):3174–3177
134. Prakash R, Zhang Y, Feng W, Jasin M (2015) Homologous recombination and human health: the roles of BRCA1, BRCA2, and associated proteins. Cold Spring Harb Perspect Biol 7(4). https://doi.org/10.1101/cshperspect.a016600
135. Ramiro AR, Jankovic M, Callen E, Difilippantonio S, Chen HT, McBride KM, Eisenreich TR, Chen J, Dickins RA, Lowe SW, Nussenzweig A, Nussenzweig MC (2006) Role of genomic instability and p53 in AID-induced c-myc-Igh translocations. Nature 440(7080):105–109
136. Reiman A, Srinivasan V, Barone G, Last JI, Wootton LL, Davies EG, Verhagen MM, Willemsen MA, Weemaes CM, Byrd PJ, Izatt L, Easton DF, Thompson DJ, Taylor AM (2011) Lymphoid tumours and breast cancer in ataxia telangiectasia; substantial protective effect of residual ATM kinase activity against childhood tumours. Br J Cancer 105(4):586–591. https://doi.org/10.1038/bjc.2011.266
137. Reina-San-Martin B, Chen HT, Nussenzweig A, Nussenzweig MC (2004) ATM is required for efficient recombination between immunoglobulin switch regions. J Exp Med 200(9):1103–1110
138. Reina-San-Martin B, Chen J, Nussenzweig A, Nussenzweig MC (2007) Enhanced intra-switch region recombination during immunoglobulin class switch recombination in 53BP1-/- B cells. Eur J Immunol 37(1):235–239. https://doi.org/10.1002/eji.200636789
139. Reina-San-Martin B, Difilippantonio S, Hanitsch L, Masilamani RF, Nussenzweig A, Nussenzweig MC (2003) H2AX is required for recombination between immunoglobulin switch regions but not for intra-switch region recombination or somatic hypermutation. J Exp Med 197(12):1767–1778
140. Robbiani DF, Nussenzweig MC (2013) Chromosome translocation, B cell lymphoma, and activation-induced cytidine deaminase. Annu Rev Pathol 8:79–103. https://doi.org/10.1146/annurev-pathol-020712-164004
141. Roberts NJ, Jiao Y, Yu J, Kopelovich L, Petersen GM, Bondy ML, Gallinger S, Schwartz AG, Syngal S, Cote ML, Axilbund J, Schulick R, Ali SZ, Eshleman JR, Velculescu VE, Goggins M, Vogelstein B, Papadopoulos N, Hruban RH, Kinzler KW, Klein AP (2012) ATM mutations in patients with hereditary pancreatic cancer. Cancer Discov 2(1):41–46. https://doi.org/10.1158/2159-8290.CD-11-0194
142. Rogakou EP, Boon C, Redon C, Bonner WM (1999) Megabase chromatin domains involved in DNA double-strand breaks in vivo. J Cell Biol 146(5):905–916
143. Rothblum-Oviatt C, Wright J, Lefton-Greif MA, McGrath-Morrow SA, Crawford TO, Lederman HM (2016) Ataxia telangiectasia: a review. Orphanet J Rare Dis 11(1):159. https://doi.org/10.1186/s13023-016-0543-7
144. Ruiz JF, Gomez-Gonzalez B, Aguilera A (2011) AID induces double-strand breaks at immunoglobulin switch regions and c-MYC causing chromosomal translocations in yeast THO mutants. PLoS Genet 7(2):e1002009. https://doi.org/10.1371/journal.pgen.1002009
145. Russo G, Isobe M, Gatti R, Finan J, Batuman O, Huebner K, Nowell PC, Croce CM (1989) Molecular analysis of a t(14;14) translocation in leukemic T-cells of an ataxia telangiectasia patient. Proc Natl Acad Sci U S A 86(2):602–606
146. Rybanska-Spaeder I, Reynolds TL, Chou J, Prakash M, Jefferson T, Huso DL, Desiderio S, Franco S (2013) 53BP1 is limiting for NHEJ repair in ATM-deficient model systems that are subjected to oncogenic stress or radiation. Mol Cancer Res 11(10):1223–1234. https://doi.org/10.1158/1541-7786.MCR-13-0252-T
147. Santos-Pereira JM, Aguilera A (2015) R loops: new modulators of genome dynamics and function. Nat Rev Genet 16(10):583–597. https://doi.org/10.1038/nrg3961
148. Savic V, Yin B, Maas NL, Bredemeyer AL, Carpenter AC, Helmink BA, Yang-Iott KS, Sleckman BP, Bassing CH (2009) Formation of dynamic gamma-H2AX domains along broken DNA strands is distinctly regulated by ATM and MDC1 and dependent upon H2AX densities in chromatin. Mol Cell 34(3):298–310. doi:S1097-2765(09)00241-X [pii]. https://doi.org/10.1016/j.molcel.2009.04.012
149. Savitsky K, Bar-Shira A, Gilad S, Rotman G, Ziv Y, Vanagaite L, Tagle DA, Smith S, Uziel T, Sfez S, Ashkenazi M, Pecker I, Frydman M, Harnik R, Patanjali SR, Simmons A, Clines GA, Sartiel A, Gatti RA, Chessa L, Sanal O, Lavin MF, Jaspers NG, Taylor AM, Arlett CF, Miki T, Weissman SM, Lovett M, Collins FS, Shiloh Y (1995) A single ataxia telangiectasia gene with a product similar to PI-3 kinase. Science 268(5218):1749–1753
150. Schaffner C, Idler I, Stilgenbauer S, Dohner H, Lichter P (2000) Mantle cell lymphoma is characterized by inactivation of the ATM gene. Proc Natl Acad Sci U S A 97 (6):2773-2778. doi:https://doi.org/10.1073/pnas.050400997 050400997 [pii]
151. Schatz DG, Swanson PC (2011) V(D)J recombination: mechanisms of initiation. Annu Rev Genet 45:167–202. https://doi.org/10.1146/annurev-genet-110410-132552
152. Schlissel M, Constantinescu A, Morrow T, Baxter M, Peng A (1993) Double-strand signal sequence breaks in V(D)J recombination are blunt, 5′-phos-

phorylated, RAG-dependent, and cell cycle regulated. Genes Dev 7(12B):2520–2532
153. Shiloh Y, Ziv Y (2013) The ATM protein kinase: regulating the cellular response to genotoxic stress, and more. Nat Rev Mol Cell Biol 14(4):197–210
154. Sleckman BP, Bassing CH, Bardon CG, Okada A, Khor B, Bories JC, Monroe R, Alt FW (1998) Accessibility control of variable region gene assembly during T-cell development. Immunol Rev 165:121–130
155. Smilenov LB, Dhar S, Pandita TK (1999) Altered telomere nuclear matrix interactions and nucleosomal periodicity in ataxia telangiectasia cells before and after ionizing radiation treatment. Mol Cell Biol 19(10):6963–6971
156. Soulas-Sprauel P, Rivera-Munoz P, Malivert L, Le Guyader G, Abramowski V, Revy P, de Villartay JP (2007) V(D)J and immunoglobulin class switch recombinations: a paradigm to study the regulation of DNA end-joining. Oncogene 26(56):7780–7791. doi:1210875 [pii]. https://doi.org/10.1038/sj.onc.1210875
157. Spagnolo L, Rivera-Calzada A, Pearl LH, Llorca O (2006) Three-dimensional structure of the human DNA-PKcs/Ku70/Ku80 complex assembled on DNA and its implications for DNA DSB repair. Mol Cell 22(4):511–519
158. Stankovic T, Stewart GS, Byrd P, Fegan C, Moss PA, Taylor AM (2002) ATM mutations in sporadic lymphoid tumours. Leuk Lymphoma 43(8):1563–1571. https://doi.org/10.1080/1042819021000002884
159. Stiff T, O'Driscoll M, Rief N, Iwabuchi K, Lobrich M, Jeggo PA (2004) ATM and DNA-PK function redundantly to phosphorylate H2AX after exposure to ionizing radiation. Cancer Res 64(7):2390–2396
160. Stilgenbauer S, Liebisch P, James MR, Schroder M, Schlegelberger B, Fischer K, Bentz M, Lichter P, Dohner H (1996) Molecular cytogenetic delineation of a novel critical genomic region in chromosome bands 11q22.3-923.1 in lymphoproliferative disorders. Proc Natl Acad Sci U S A 93(21):11837–11841
161. Stilgenbauer S, Winkler D, Ott G, Schaffner C, Leupolt E, Bentz M, Moller P, Muller-Hermelink HK, James MR, Lichter P, Dohner H (1999) Molecular characterization of 11q deletions points to a pathogenic role of the ATM gene in mantle cell lymphoma. Blood 94(9):3262–3264
162. Stirling PC, Hieter P (2016) Canonical DNA repair pathways influence R-loop-driven genome instability. J Mol Biol. https://doi.org/10.1016/j.jmb.2016.07.014
163. Stoppa-Lyonnet D, Soulier J, Lauge A, Dastot H, Garand R, Sigaux F, Stern MH (1998) Inactivation of the ATM gene in T-cell prolymphocytic leukemias. Blood 91(10):3920–3926
164. Stracker TH, Petrini JH (2011) The MRE11 complex: starting from the ends. Nat Rev Mol Cell Biol 12(2):90–103. doi:nrm3047 [pii]. https://doi.org/10.1038/nrm3047
165. Taylor AM, Butterworth SV (1986) Clonal evolution of T-cell chronic lymphocytic leukaemia in a patient with ataxia telangiectasia. Int J Cancer 37(4):511–516
166. Taylor AM, Metcalfe JA, Thick J, Mak YF (1996) Leukemia and lymphoma in ataxia telangiectasia. Blood 87(2):423–438
167. Telatar M, Teraoka S, Wang Z, Chun HH, Liang T, Castellvi-Bel S, Udar N, Borresen-Dale AL, Chessa L, Bernatowska-Matuszkiewicz E, Porras O, Watanabe M, Junker A, Concannon P, Gatti RA (1998) Ataxia-telangiectasia: identification and detection of founder-effect mutations in the ATM gene in ethnic populations. Am J Hum Genet 62(1):86–97. https://doi.org/10.1086/301673
168. Tonegawa S (1983) Somatic generation of antibody diversity. Nature 302(5909):575–581
169. Tresini M, Warmerdam DO, Kolovos P, Snijder L, Vrouwe MG, Demmers JA, van IWF, Grosveld FG, Medema RH, Hoeijmakers JH, Mullenders LH, Vermeulen W, Marteijn JA (2015) The core spliceosome as target and effector of non-canonical ATM signalling. Nature 523(7558):53–58. https://doi.org/10.1038/nature14512
170. Vaziri H, West MD, Allsopp RC, Davison TS, Wu YS, Arrowsmith CH, Poirier GG, Benchimol S (1997) ATM-dependent telomere loss in aging human diploid fibroblasts and DNA damage lead to the post-translational activation of p53 protein involving poly(ADP-ribose) polymerase. EMBO J 16(19):6018–6033. https://doi.org/10.1093/emboj/16.19.6018
171. Verhagen MM, Abdo WF, Willemsen MA, Hogervorst FB, Smeets DF, Hiel JA, Brunt ER, van Rijn MA, Majoor Krakauer D, Oldenburg RA, Broeks A, Last JI, van't Veer LJ, Tijssen MA, Dubois AM, Kremer HP, Weemaes CM, Taylor AM, van Deuren M (2009) Clinical spectrum of ataxia-telangiectasia in adulthood. Neurology 73(6):430–437. https://doi.org/10.1212/WNL.0b013e3181af33bd
172. Verhagen MM, Last JI, Hogervorst FB, Smeets DF, Roeleveld N, Verheijen F, Catsman-Berrevoets CE, Wulffraat NM, Cobben JM, Hiel J, Brunt ER, Peeters EA, Gomez Garcia EB, van der Knaap MS, Lincke CR, Laan LA, Tijssen MA, van Rijn MA, Majoor-Krakauer D, Visser M, van't Veer LJ, Kleijer WJ, van de Warrenburg BP, Warris A, de Groot IJ, de Groot R, Broeks A, Preijers F, Kremer BH, Weemaes CM, Taylor MA, van Deuren M, Willemsen MA (2012a) Presence of ATM protein and residual kinase activity correlates with the phenotype in ataxia-telangiectasia: a genotype-phenotype study. Hum Mutat 33(3):561–571. https://doi.org/10.1002/humu.22016
173. Verhagen MM, Martin JJ, van Deuren M, Ceuterick-de Groote C, Weemaes CM, Kremer BH, Taylor MA, Willemsen MA, Lammens M (2012b) Neuropathology in classical and variant ataxia-telangiectasia. Neuropathology 32(3):234–244. https://doi.org/10.1111/j.1440-1789.2011.01263.x

174. Ward IM, Chen J (2001) Histone H2AX is phosphorylated in an ATR-dependent manner in response to replicational stress. J Biol Chem 276(51):47759–47762. https://doi.org/10.1074/jbc.C100569200
175. Ward IM, Difilippantonio S, Minn K, Mueller MD, Molina JR, Yu X, Frisk CS, Ried T, Nussenzweig A, Chen J (2005) 53BP1 cooperates with p53 and functions as a haploinsufficient tumor suppressor in mice. Mol Cell Biol 25(22):10079–10086
176. Ward IM, Minn K, van Deursen J, Chen J (2003) p53 Binding protein 53BP1 is required for DNA damage responses and tumor suppression in mice. Mol Cell Biol 23(7):2556–2563
177. Ward IM, Reina-San-Martin B, Olaru A, Minn K, Tamada K, Lau JS, Cascalho M, Chen L, Nussenzweig A, Livak F, Nussenzweig MC, Chen J (2004) 53BP1 is required for class switch recombination. J Cell Biol 165(4):459–464
178. Williams RS, Moncalian G, Williams JS, Yamada Y, Limbo O, Shin DS, Groocock LM, Cahill D, Hitomi C, Guenther G, Moiani D, Carney JP, Russell P, Tainer JA (2008) Mre11 dimers coordinate DNA end bridging and nuclease processing in double-strand-break repair. Cell 135(1):97–109. https://doi.org/10.1016/j.cell.2008.08.017
179. Wood LD, Halvorsen TL, Dhar S, Baur JA, Pandita RK, Wright WE, Hande MP, Calaf G, Hei TK, Levine F, Shay JW, Wang JJ, Pandita TK (2001) Characterization of ataxia telangiectasia fibroblasts with extended life-span through telomerase expression. Oncogene 20(3):278–288. https://doi.org/10.1038/sj.onc.1204072
180. Xie A, Kwok A, Scully R (2009) Role of mammalian Mre11 in classical and alternative nonhomologous end joining. Nat Struct Mol Biol 16(8):814–818. https://doi.org/10.1038/nsmb.1640
181. Xu Y, Ashley T, Brainerd EE, Bronson RT, Meyn MS, Baltimore D (1996) Targeted disruption of ATM leads to growth retardation, chromosomal fragmentation during meiosis, immune defects, and thymic lymphoma. Genes Dev 10(19):2411–2422
182. Xu Y, Price BD (2011) Chromatin dynamics and the repair of DNA double strand breaks. Cell Cycle 10(2):261–267. https://doi.org/10.4161/cc.10.2.14543
183. Yang Y, McBride KM, Hensley S, Lu Y, Chedin F, Bedford MT (2014) Arginine methylation facilitates the recruitment of TOP3B to chromatin to prevent R loop accumulation. Mol Cell 53(3):484–497. https://doi.org/10.1016/j.molcel.2014.01.011
184. Yeo AJ, Becherel OJ, Luff JE, Cullen JK, Wongsurawat T, Jenjaroenpun P, Kuznetsov VA, McKinnon PJ, Lavin MF (2014) R-loops in proliferating cells but not in the brain: implications for AOA2 and other autosomal recessive ataxias. PLoS One 9(3):e90219. https://doi.org/10.1371/journal.pone.0090219
185. Yin B, Lee BS, Yang-Iott KS, Sleckman BP, Bassing CH (2012) Redundant and nonredundant functions of ATM and H2AX in alphabeta T-lineage lymphocytes. J Immunol 189(3):1372–1379. https://doi.org/10.4049/jimmunol.1200829
186. Yin B, Savic V, Juntilla MM, Bredemeyer AL, Yang-Iott KS, Helmink BA, Koretzky GA, Sleckman BP, Bassing CH (2009) Histone H2AX stabilizes broken DNA strands to suppress chromosome breaks and translocations during V(D)J recombination. J Exp Med 206(12):2625–2639. https://doi.org/10.1084/jem.20091320
187. Yin B, Yang-Iott KS, Chao LH, Bassing CH (2011) Cellular context-dependent effects of H2ax and p53 deletion on the development of thymic lymphoma. Blood 117(1):175–185. https://doi.org/10.1182/blood-2010-03-273045
188. Yu K, Chedin F, Hsieh CL, Wilson TE, Lieber MR (2003) R-loops at immunoglobulin class switch regions in the chromosomes of stimulated B cells. Nat Immunol 4(5):442–451. https://doi.org/10.1038/ni919
189. Zha S, Bassing CH, Sanda T, Brush JW, Patel H, Goff PH, Murphy MM, Tepsuporn S, Gatti RA, Look AT, Alt FW (2010) ATM-deficient thymic lymphoma is associated with aberrant tcrd rearrangement and gene amplification. J Exp Med 207(7):1369–1380. doi:jem.20100285 [pii]. https://doi.org/10.1084/jem.20100285
190. Zha S, Sekiguchi J, Brush JW, Bassing CH, Alt FW (2008) Complementary functions of ATM and H2AX in development and suppression of genomic instability. Proc Natl Acad Sci U S A 105(27):9302–9306. doi:0803520105 [pii]. https://doi.org/10.1073/pnas.0803520105
191. Zhang Y, Gostissa M, Hildebrand DG, Becker MS, Boboila C, Chiarle R, Lewis S, Alt FW (2010) The role of mechanistic factors in promoting chromosomal translocations found in lymphoid and other cancers. Adv Immunol 106:93–133. doi:S0065-2776(10)06004-9 [pii]. https://doi.org/10.1016/S0065-2776(10)06004-9
192. Zhao Y, Brickner JR, Majid MC, Mosammaparast N (2014) Crosstalk between ubiquitin and other post-translational modifications on chromatin during double-strand break repair. Trends Cell Biol 24(7):426–434. https://doi.org/10.1016/j.tcb.2014.01.005
193. Zhou Y, Lee JH, Jiang W, Crowe JL, Zha S, Paull TT (2017) Regulation of the DNA damage response by DNA-PKcs inhibitory phosphorylation of ATM. Mol Cell 65(1):91–104. https://doi.org/10.1016/j.molcel.2016.11.004

Telomeres and Chromosomal Translocations

7

There's a Ligase at the End of the Translocation

Duncan M. Baird and Eric A. Hendrickson

Abstract

Chromosomal translocations are now well understood to not only constitute signature molecular markers for certain human cancers but often also to be causative in the genesis of that tumor. Despite the obvious importance of such events, the molecular mechanism of chromosomal translocations in human cells remains poorly understood. Part of the explanation for this dearth of knowledge is due to the complexity of the reaction and the need to archaeologically work backwards from the final product (a translocation) to the original unrearranged chromosomes to infer mechanism. Although not definitive, these studies have indicated that the aberrant usage of endogenous DNA repair pathways likely lies at the heart of the problem. An equally obfuscating aspect of this field, however, has also originated from the unfortunate species-specific differences that appear to exist in the relevant model systems that have been utilized to investigate this process. Specifically, yeast and murine systems (which are often used by basic science investigators) rely on different DNA repair pathways to promote chromosomal translocations than human somatic cells. In this chapter, we will review some of the basic concepts of chromosomal translocations and the DNA repair systems thought to be responsible for their genesis with an emphasis on underscoring the differences between other species and human cells. In addition, we will focus on a specific subset of translocations that involve the very end of a chromosome (a telomere). A better understanding of the relationship between DNA repair pathways and chromosomal translocations is guaranteed to lead to improved therapeutic treatments for cancer.

Keywords

Chromosomal translocations · DNA DSB Repair · HDR · C-NHEJ · A-NHEJ · LIGIII · LIGIV

D. M. Baird
Division of Cancer and Genetics, School of Medicine, Cardiff University, Cardiff, UK

E. A. Hendrickson (✉)
Department of Biochemistry, Molecular Biology, and Biophysics, University of Minnesota Medical School, Minneapolis, MN, USA
e-mail: hendr064@umn.edu

Abbreviations

A-NHEJ	Alternative nonhomologous end joining
APOBEC3	Apolipoprotein B editing complex 3
BFB	breakage-fusion-bridging

Y. Zhang (ed.), *Chromosome Translocation*, Advances in Experimental Medicine and Biology 1044,
https://doi.org/10.1007/978-981-13-0593-1_7

BRCA1	Breast cancer allele 1
BRCA2	Breast cancer allele 2
BLM	Bloom syndrome gene
C-NHEJ	Classic nonhomologous end joining
CtIP	C-terminal interacting protein
DNA2	DNA exonuclease 2
DNA-PK$_{cs}$	DNA dependent protein kinase catalytic subunit
DNA-PK	DNA dependent protein kinase complex
DN-hTERT	Dominant-negative version of human telomerase
DSBs	DNA double-strand breaks
EME1	Essential meiotic endonuclease 1
ERCC1	Excision repair cross-complementing 1
EXO1	Exonuclease 1
FANCN	Fanconi anemia complementation group N
GEN1	General endonuclease homolog 1
HDR	Homology-dependent repair
indels	Insertions and/or deletions
Ku	Ku70:Ku86 heterodimer
LIGI	DNA ligase I
LIGIII	DNA ligase III
LIGIV	DNA ligase IV
MRE11	Meiotic recombination defective 11
MRN	MRE11/RAD50/NBS1
MUS81	Mutagen sensitive 81
NBS1	Nijmegen breakage syndrome 1
NRT	Non-reciprocal translocation
nt	Nucleotide
PARP1	Poly(ADP-ribose) polymerase 1
PAXX	Paralog of XRCC4 and XLF
POLQ	DNA polymerase *theta*
RAD50	Radiation sensitive 50
RAD51	Radiation sensitive 51
RAD54	Radiation sensitive 54
RMI1	RecQ-mediated genome instability homolog 1
RPA	Replication protein A
SLX1	Synthetically lethal with unknown function (X) 1
SLX4	Synthetically lethal with unknown function (X) 4
ssDNA	Single-stranded DNA
STELA	Single telomere length analysis
TALEN	Transcription activator-like effector nuclease
TRF2	Telomere recognition factor 2
TOPO3α	Topoisomerase 3α
XLF	XRCC-4-like factor/Cernunnos
XPF	Xeroderma pigmentosum gene F
XRCC4	X-ray cross complementing group 4

7.1 Introduction

The concept of chromosomal translocations — in which a portion of one chromosome breaks off and fuses inappropriately to another chromosome — has been part of the scientific consciousness for the better part of eight decades. Chromosomal translocations were first described by Karl Sax in 1938 [84] and then elaborated by Barbara McClintock in the construction of her seminal "breakage-fusion-bridging" [BFB] model during the 1940s [64]. Chromosomal translocations gained significant clinical relevance a couple of decades later when it was demonstrated that a single recurring chromosomal translocation (the Philadelphia chromosome) was often found in patients suffering from leukemia [71, 83]. Chromosomal translocations are now well understood to not only constitute signature molecular markers of human cancers (solid tumors in addition to blood cancers) but to be causative in their genesis as well [34, 56]. As such, these translocations become extremely important for clinical diagnostics as well as treatment-related options, respectively. Moreover, with the advent of comprehensive cancer genome sequencing, it is now appreciated that translocations, causative or not, are a common feature of human tumors [16, 94]. It is not surprisingly, therefore, that interest in identifying and quantitating chromosomal translocations has increased exponentially in the past decade. As a consequence of this interest — and the experimentation associated with it — tens of thousands of translocations in a veritable bevy of different human cancers have been catalogued and characterized [56, 94]. While incredibly rich in molecular information, most of these studies suffer (biologically speaking) in being retrospective; that is, the mechanism for how the translocation occurred is often (correctly or incorrectly) simply inferred after the fact from the junctional sequences present at the site of a chromosomal translocation.

To try and address this dearth of mechanistic knowledge, this chapter will focus on the relationship between DNA repair (specifically DNA double-strand break [DSB] repair) and chromosomal translocations. An understanding of DNA DSB repair is paramount to our discussion because it seems obvious, if only intuitively, that a chromosomal translocation is the result of aberrant DNA DSB repair [13, 42, 81]. However, DNA DSB repair is infrequently — and probably only rarely — aberrant because it is responsible for the stability of the genome. Thus, it needs to be appreciated and emphasized that chromosomal translocations are by far the exception to the rule of the normally helpful processes (predominately DNA DSB repair) that keep the genome stable.

It is a tautology and a fact appreciated by all cancer researchers that a stable genome is highly desirable and is inherently anti-oncogenic. While this perspective is basically sensible, it is also important to remember that complete stability is antithetical with evolution/life. That is, perfect immutability is contrary to the process of evolution and thus nature must maintain a balance between accurate DNA repair and the formation of mutations (*i.e.*, the lack of — or mis-repair of — DNA) upon which selection can act. Thus, all DNA repair processes, DNA DSB repair included, have a bit of "sloppiness" inherent in their mechanism. It is likely that chromosomal translocations are the result of one of these rare sloppy DNA DSB repair events. Trying to identify how, why and when such events occur; and perhaps most importantly — whether they can be abrogated — is the clinically relevant goal for this field.

7.2 DNA Damage

As elaborated above, the human genome needs to be nearly (but not completely) immutable in order to ensure the survival of the species. This turns out to be an extremely tall order as the human genome is constantly being chemically assaulted by both endogenous and exogenous factors. The endogenous sources of damage likely vary from cell type to cell type, but can include: lesions associated with aberrant lymphoid gene recombination, DNA replication errors, transcriptional errors, the formation of reactive oxygen species during oxidative phosphorylation, as well as the spontaneous depurination or deamination of nucleotides [nts] due to the proximity of DNA to adjacent water molecules [100] or the aberrant action of cellular deaminases [96]. Exogenous sources of DNA damage include, but are certainly not limited to, exposure to ultraviolet light, chemotherapeutic drugs, or ionizing radiation. Indeed, *in toto*, it is estimated that each human cell sustains an astronomical ~70,000 lesions per day [100]. Importantly, however approximately 69,975 of these lesions result in DNA damage on only one strand of the DNA duplex. Thus, although the type of damage can vary extensively and certain types of lesions require discrete DNA repair pathways (expanded upon in the next section) these processes are inherently high fidelity as a consequence of having an undamaged DNA strand upon which to template the repair events. Thus, in human cells there is a surprisingly high level of DNA damage occurring on a daily basis that is nearly completely neutralized by conservative DNA repair pathways that utilize undamaged DNA to restore genome integrity.

In contrast to all the other types of lesions combined, human cells suffer only about 25 DSB lesions per cell per day [100]. Again, the exact cause of the DSB can vary greatly depending upon the cell type. Some likely occur due to aberrant lymphoid recombination processes [56], whereas others may be due to reactive metabolic oxygen production, DNA replication errors [7] or the inappropriate action of cytidine deaminases [48]. Whatever their exact origin, DNA DSBs are uniquely toxic to cells because when both strands of the chromosome are damaged most of the time the only way to restore the chromosome to its original state is if an undamaged homologous chromosome (or sister chromatid if the DSB should occur during S phase of the cell cycle) is available to template the repair event. As a consequence, DSBs are inherently more mutagenic than most other types of lesions because of the difficultly in enacting their proper repair. A second parameter, which is relevant to this chapter, is that for all the other lesions, not only can the undamaged DNA strand help to enact error-free repair, but it also perforce holds the chromosome intact. In contrast, the formation of a DSB generates a window of opportunity, however small, for the two chromosomal fragments to move away from one another. If this happens, the chances of one of those fragments "repairing" itself onto another chromosome (*i.e.*, causing a translocation) rises astronomically.

In summary, the vast majority of the DNA lesions that a human cell experiences on a daily basis are generally rapidly and correctly repaired and are likely not relevant for the genesis of chromosomal translocations. Importantly, this is not to say that these types of lesions cannot cause chromosomal translocations. It is just likely that it is not a single single-stranded lesion *per se* that can trigger translocations, but the juxtaposition of two closely spaced single-strand lesions that give rise to a *de facto* DSB that are the culprit. Thus, DSBs and DNA DSB repair (or the lack thereof) have been firmly established as being mechanistically responsible for chromosomal translocations.

7.3 DNA Repair

7.3.1 DNA Repair Involving Only a Single Strand

Due to the broad spectrum of lesions that can occur to DNA it is not surprising that discrete DNA repair pathways have evolved to correct these life-threatening alterations. Of all the lesions that damage only a single-strand of DNA most result in the formation of only a singly modified nt or an abasic site. These lesions are readily repaired by the base excision repair pathway (Fig. 7.1a). This process involves the action of DNA gylcosylases, apurinic or apyrimidinic endonucleases and phosphodiesterases that ultimately convert the lesion into a single-stranded nick. This nick is then filled in by a DNA polymerase and sealed by a DNA ligase [104].

When the DNA lesion is bulkier than a single standard nt or when nts are fused together (*e.g.* via the formation of pyrimidine dimers) then a more complicated repair pathway, nucleotide excision repair (Fig. 7.1b), is utilized that is capable of restoring stretches of nts (up to 24 nts in humans) in one event. In nucleotide excision repair, the bulkier lesion is recognized by a multi-subunit protein complex that introduces nicks 5′ and 3′ of the lesion. The offending lesion is then removed as an oligonucleotide by the action of a helicase and the resulting ~20 nt gap is filled in by a polymerase and then sealed by a DNA ligase [92].

A third common type of lesion is the misincorporation of nucleotides and/or generation of small insertions or deletions [indels] during DNA replication. These types of lesions are repaired by mismatch repair (Fig. 7.1c). The mismatch repair machinery consists of large heterodimeric complexes that scan DNA and look for helical distortions due to the mispairing or indels. These complexes recruit additional factors including endonucleases that nick and exonucleases that degrade one of the strands resulting in the removal of the offending mispaired nucleotide and some flanking nucleotides. As before, the resulting gap is subsequently filled in by a DNA polymerase and sealed by a DNA ligase [53].

All three of the above processes (base excision repair, nucleotide excision repair and mismatch repair) are critical for cellular and organismal well-being. Mutation of any of the factors associated with these pathways is generally either lethal or oncogenic, (although, pertinently, in the latter scenario not usually associated with chromosomal translocations). The importance of single-stranded DNA repair for genome stability is further evidenced by the awarding of the 2015 Nobel Prize in Physiology or Medicine to the investigators responsible for the discovery and/or initial characterization of these DNA repair pathways [50].

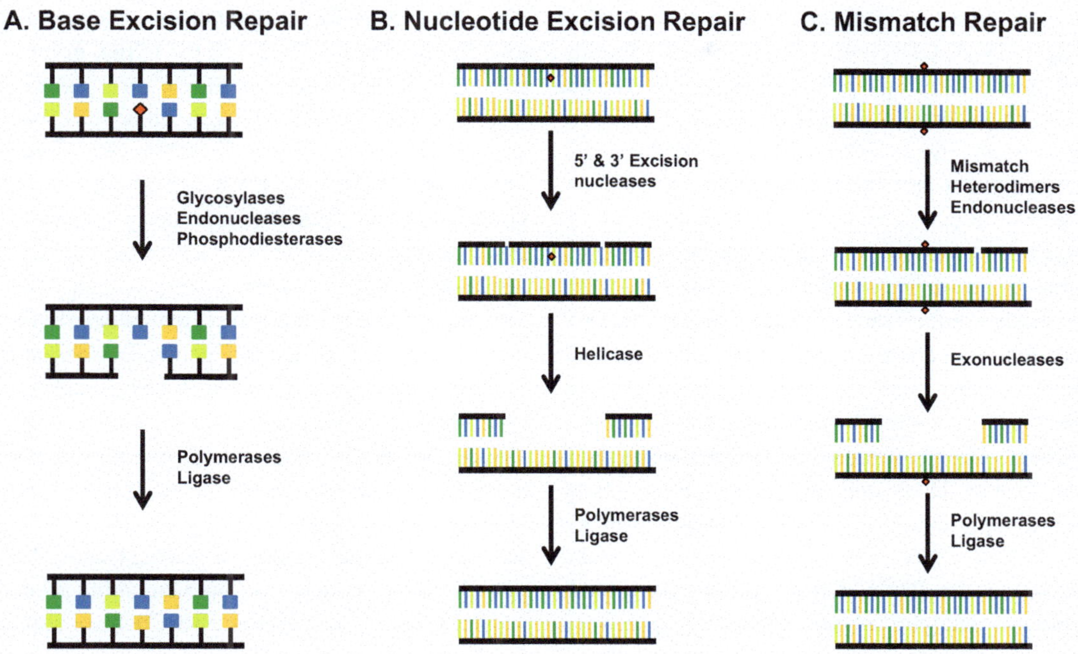

Fig. 7.1 Repair of lesions where one DNA strand is still intact. (**a**) Base Excision Repair. The schematic shows a small piece of double-stranded DNA (colored rectangles represent nucleotides) containing a singly modified nt (red diamond). This lesion is repaired by the action of DNA gylcosylases, endonucleases and phosphodiesterases that ultimately convert the lesion into a single-stranded nick. This nick is then filled in by a DNA polymerase and sealed by a DNA ligase. (**b**) Nucleotide Excision Repair. The schematic shows two nts (colored lines) that are fused together (red diamond). This lesion is recognized by a multi-subunit protein complex that introduces nicks 5′ and 3′ of the lesion. The offending lesion is then removed as an oligonucleotide by the action of a helicase and the resulting ~20 nt gap is filled in by a polymerase and then sealed by a DNA ligase. (**c**) Mismatch Repair. The schematic shows two nts (colored lines) where one base pair is mispaired (red diamonds). The mispaired nts are recognized by mismatch heterodimeric complexes that recruit endonucleases that nick the DNA. Exonucleases then degrade one of the strands resulting in the removal of the offending nt as well as some flanking nts. As before, the resulting gap is subsequently filled in by a DNA polymerase and sealed by a DNA ligase

7.3.2 DNA Double-Strand Break Repair

7.3.2.1 C-NHEJ

Although DNA DSB lesions occur proportionately much less frequently than single-stranded lesions, they are so toxic that cells have evolved multiple pathways that utilize hundreds of genes to make sure the DSBs are quickly and (semi)-accurately repaired. The two major pathways are nonhomologous end joining [NHEJ] [55, 110] and homology-dependent repair [HDR] [44].

In higher eukaryotes, DNA DSB repair proceeds most frequently by a process that does not require extended regions of homology. Specifically, mammalian cells — and humans in particular — have evolved a highly efficient ability to join nonhomologous DNA molecules together [80]. This pathway is referred to as classic NHEJ [C-NHEJ] and it is generally error-prone. The evolution of a repair pathway that is error-prone may seem paradoxical but is likely due to 1) the increased percentage of non-coding DNA in higher eukaryotes, a feature that more readily tolerates imprecise rejoining (a luxury that bacteria and lower eukaryotes do not have) and 2) the requirement for productive error-prone repair during lymphoid recombination processes to generate a large immune repertoire.

Many of the details of C-NHEJ have been worked out, and the process is well (albeit certainly not completely) understood. Following the introduction of a DSB into a chromosome, DNA Ligase IV [LIGIV] will often (if possible) attempt to immediately and precisely rejoin the broken ends to generate a perfect repair event. Exactly how frequently such "error-free" C-NHEJ repair occurs is not known [10], but it is now appreciated that it can occur much more frequently than had been believed [73, 107]. At least some fraction of the time, however, the ends cannot be properly rejoined (due, for example, to the loss of nts and/or to aberrant adducts at the break site). In these instances, the ends are bound by the Ku86:Ku70 heterodimer [Ku; reviewed by [37]], a highly abundant protein complex that binds to the broken DNA ends to prevent unnecessary DNA degradation (Fig. 7.2). The binding of Ku to the free DNA ends subsequently recruits and activates the DNA-dependent protein kinase complex catalytic subunit [DNA-PK$_{cs}$, [11, 45]]. DNA-PK$_{cs}$:DNA-PK$_{cs}$ homotypic interactions (one molecule on each end of the DSB), in turn, are the critical feature required for synapsis, which retains the two broken ends near one another [87, 91]. Once a Ku:DNA-PK$_{cs}$ dimer [also referred to as the DNA dependent protein kinase complex; DNA-PK] is properly assembled at the broken ends it, in turn, activates a tightly-associated nuclease, Artemis [69], to help trim any damaged DNA ends. The extent of deletion is usually only a few nts and generally does not extend much beyond 25 nts with few exceptions [32, 38, 55]. Subsequently, the X family polymerases *mu* and *lambda* fill in missing nucleotides [55]. The rejoining of the DNA DSB requires the recruitment [24] of LIGIV and accessory factors: Paralog of XRCC4 and XLF [PAXX, [72]], X-ray cross complementing group 4 [XRCC4, [23, 52]] and/or XRCC4-like factor/

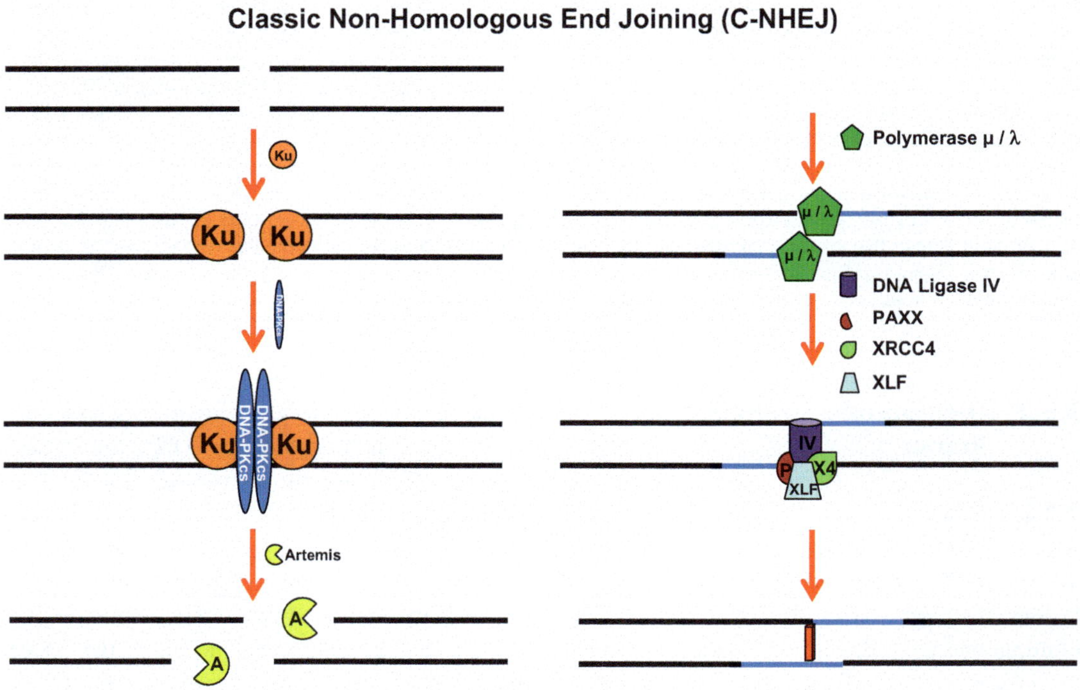

Fig. 7.2 A schematic depicting Classic Non-Homologous End Joining (C-NHEJ). The black lines represent strands of DNA. First, the Ku heterodimer (orange ball) binds onto the ends of the DNA. Ku then recruits DNA-PK$_{cs}$ (blue oval) and the homotypic interactions between two DNA-PK$_{cs}$ molecules tethers the ends together. The nuclease Artemis (yellow PacMan™), which is physically associated with DNA-PK$_{cs}$, can then remove any mispaired or damaged nucleotides from the ends. Most missing nts are then replaced by the DNA polymerases μ or λ (green pentagon). Finally, a ligase complex, consisting of DNA ligase IV (purple cylinder) and the accessory proteins PAXX (red half oval), XRCC4 (light green tear drop) and XLF (Carolina blue cropped pyramid) then religates the ends back together. This process often results in indels (denoted by the red rectangle) at the site of repair

Cernunnos [XLF, [2, 15]] (Fig. 7.2). Finally, it is relevant to note that most of the cells in a human being are either not cycling or in G_1 phase of the cell cycle. Because HDR (described below) is predominately restricted to cells in S phase (when a sister chromatid may be available for repair) C-NHEJ is perforce the preferred repair pathway in human cells and this accounts for its frequent usage. In summary, in humans the predominant pathway of DSB repair is C-NHEJ and it first utilizes LIGIV to try and simply re-ligate the ends of a DSB back together. Failing at that, C-NHEJ keeps the ends in proximity, polishes them up by limited resection and polymerization as needed and then uses LIGIV to religate the ends. Due to the nuclease and polymerase action on one or both of the ends, small indels are a classic and frequent hallmark of chromosomal junctions repaired by C-NHEJ.

7.3.2.2 A-NHEJ

It had long been appreciated that the kinetics of C-NHEJ were biphasic — most (~80%) of the ends were rejoined quickly (within 15′ to 30′ of the chromosome breaking), but some ends could take hours to finally be rejoined. For many years this was interpreted simply as some DSBs being "easier" to repair than others. It was the laboratory of George Iliakis that first suggested that the slow phase of DSB repair may in fact represent a completely separate repair pathway [105]. This hypothesis dovetailed nicely with earlier work done in yeast, which had genetically documented that in the absence of Ku, DSBs could be repaired by an alternative error-prone end-joining process that utilized microhomology [14]. Although still somewhat controversial [76] significant evidence has accumulated over the past 15 years to substantiate the Iliakis hypothesis. The hallmarks of this pathway (generally referred to as alternative nonhomologous end joining [A-NHEJ] to distinguish it from C-NHEJ) are that it is Ku-independent and utilizes small (3^+ nt) regions of homology [14, 30, 42] to facilitate end joining. The process of A-NHEJ is mechanistically simple and straightforward: both ends of the DSB are resected to generate 3′ overhangs that are intermediate in length from those generated during C-NHEJ (which are a few nts at most) and HDR (which are often hundreds or thousands of nts long). These resected ends can then base pair using now exposed stretches of "microhomology" (probably 3^+ nts). Nucleases are recruited to trim the flaps that are often generated and the nicks/ends are then sealed by a ligase [30, 42]. Thus, A-NHEJ is inherently an error-prone repair process as it always generates deletions including one of the two regions of microhomology and all the DNA in between the two patches of microhomology.

Although the intellectual concept of how A-NHEJ occurs is clear, the genetics and biochemistry of the synapsis, processing and ligation of an A-NHEJ DSB repair event are still quite obscure leading to heavy debate by investigators in the field as to the precise mechanism. Several studies have suggested that, like Ku for C-NHEJ, the protein poly(ADP-ribose) polymerase 1 [PARP1] may bind to the DNA ends (Fig. 7.3). Indeed, there is evidence that PARP1 may even compete with Ku for access to the ends thereby determining the choice of the NHEJ pathways used for the repair of specific DSBs [20]. Alternatively, the repair complex meiotic recombination 11/radiation sensitive 50/Nijmegen breakage syndrome 1 [MRE11/RAD50/NBS1; MRN] has also been implicated as the A-NHEJ end tethering activity [28, 114]. Regardless of whether recognition or tethering of the ends is carried out by either PARP1 or MRN, resection is required to reveal the microhomology that will subsequently be used to mediate the repair event. The initial resection is thought to be carried out by MRN and an associated nuclease, C-terminal interacting protein [CtIP]. The short resection mediated by MRN/CtIP is then elongated by DNA exonuclease 2 [DNA2] and/or exonuclease 1 [EXO1] [9, 68]. Once sufficient 3′-single-stranded DNA [ssDNA] overhangs have been generated the strands can anneal through the exposed microhomology (Fig. 7.3). Moreover, the presence of microhomology modulates further resection activity and stabilizes the junction to facilitate ligation [77]. Finally, there are likely an additional number of enzymatic activities required for A-NHEJ including DNA

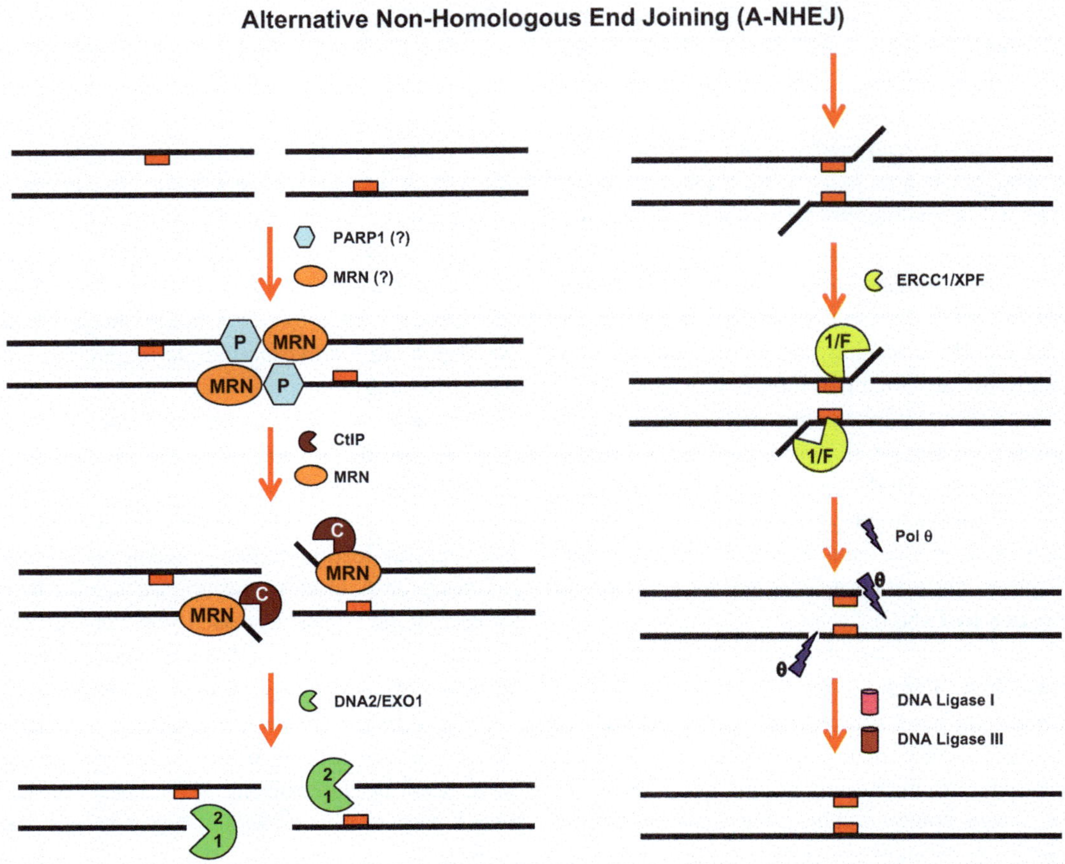

Fig. 7.3 A schematic depicting Alternative Non-Homologous End Joining (A-NHEJ). The black lines represent strands of DNA and the red rectangles blocks of microhomology. The broken ends may be held together either by PARP1 (Carolina blue hexagon) or by the MRN complex (orange oval). The initial resection is carried out by MRN and CtIP (crimson PacMan™). This short resection is then elongated by DNA2 and/or EXO1 (light green PacMan™). Once sufficient 3′-single-stranded DNA [ssDNA] overhangs have been generated the strands can anneal through the exposed microhomology. The presence of microhomology stabilizes the junction. The resulting flaps are likely cleaved off by the ERCC1/XPF heterodimer (yellow PacMan™). Before ligation occurs it is also likely that POLQ (purple lightning bolt) may act on the DNA ends. Ultimately, the repaired DSB junction is religated using either LIGIII (red cylinder) or LIGI (pink cylinder)

polymerases and helicases, but most of these have only been inferred and not rigorously identified. Perhaps the only two enzymatic activities which seem clearly required are DNA polymerase *theta* [POLQ] and a flap endonuclease activity needed to clip off mispaired strands. In the case of POLQ, the evidence is strong that much of the microhomology introduced at DSB repair junctions is dependent upon this inherently error-prone enzyme [19, 62]. The flap endonuclease activity is very likely supplied by the structure-specific nuclease complex excision repair cross-complementing 1/xeroderma pigmentosum gene F [ERCC1/XPF] [1]. Ultimately, the repaired DSB junction needs to be religated and DNA ligase III [LIGIII] appears to be the principal ligase used [5, 27, 106] although it is now clear that DNA ligase I [LIGI] can functionally substitute for LIGIII as well [3, 59, 73] (Fig. 7.3).

In summary, in order for A-NHEJ to occur, the broken DNA ends must somehow bypass being repaired by C-NHEJ (how this occurs is poorly understood). If the ends are then subjected to significant, but nonetheless limited resection, they can utilize exposed microhomology to facilitate the repair event in a fashion that always generates deletions. Although most [albeit certainly not all, [76]] investigators now accept that A-NHEJ is a discrete DNA repair pathway, it is confounded by two serious shortcomings. First, the most prominent feature of this pathway is the residual microhomology left at a repaired DNA DSB junction. However, the definition of microhomology is often investigator-arbitrary and may include microhomologies as short as 1 or 2 nts. Thus, there are a myriad of published studies where A-NHEJ is the inferred DNA repair mechanism because short microhomologies were observed at the repair junctions. This is unfortunate because C-NHEJ can also use and generate microhomologies of 1 or 2 nts during repair [76]. Thus, to be rigorous, at least 3 nt of homology is probably required before an assignment of A-NHEJ can confidently be given. When this criterion is utilized, the vast majority of DNA repair events suggested to be caused by A-NHEJ is dramatically reduced [see, for example [21]]. The second failing of A-NHEJ is that there is no specific factor required for the repair event. LIGIII was one of the best candidates for such a factor, but recent work has shown that even this enzyme is dispensable for A-NHEJ [3, 73]. Thus, until this situation is clarified it seems as if the best operational definition of A-NHEJ is: 1) a DNA DSB process that is Ku- and LIGIV-independent but POLQ-dependent, 2) relies upon LIGIII or LIGI and 3) generates repair junctions with 3^{+} nts of microhomology.

7.3.2.3 HDR

Whereas C-NHEJ is the major DNA DSB repair pathway in human cells, DNA DSBs that occur in S phase of the cell cycle can instead be, and often are, repaired by HDR (Fig. 7.4). In HDR [reviewed by [44]], the DNA ends of the incoming DNA are likely extensively resected to yield 3′-single-stranded DNA overhangs. As described above for A-NHEJ, the nuclease(s) responsible for this resection are the MRN:CtIP complex (which generates the initial resection) followed by the action of DNA2 and EXOI [9, 68]. The resulting overhangs are then coated by replication protein A [RPA], a heterotrimeric single-stranded DNA binding protein, which removes the secondary structures from the overhangs [reviewed by [41]]. The breast cancer allele 1 and 2 [BRCA1 and BRCA2, respectively] proteins and the Fanconi anemia complementation group N protein [FANCN] then help to recruit radiation sensitive 51 [RAD51] to the overhangs [103]. RAD51 is the key strand exchange protein in HDR [reviewed by [98]]. In humans, there are at least seven Rad51 family members and almost all of them have been implicated in some aspect of HDR and also in human disease. Strand invasion into the homologous chromosomal sequence requires RAD54 [radiation sensitive 54] and DNA replication. Rad54 is a double-stranded DNA-dependent ATPase that can remodel chromatin and it probably plays critical roles at several steps in the recombination process [reviewed by [39]]. In particular, Rad54 is critical for stabilizing the Rad51-dependent joint molecule formation as well as for promoting the disassembly of Rad51 following exchange [90]. Strand exchange generates an interdigitated set of strands that can be resolved into a complicated set of products. In mitotic cells most of the intermediates are resolved as non-crossover products by dissolving the interdigitated strands back into their original duplexes after sufficient DNA replication has occurred to restore the genetic information lost at the site of the DSB (Fig. 7.4a). The dissolution process requires the action of the Bloom syndrome gene, topoisomerase 3α and RecQ-mediated genome instability homolog 1 [BLM, TOPO 3α, RMI1, respectively] complex [113]. Less frequently the second end of DNA is captured and a covalently closed "Holliday junction" [40] is formed that can be resolved as either non-crossover products (which are functionally identical to dissolution) or crossover products (Fig. 7.4b). The resolution of Holliday junctions is complicated and in human cells appears to be carried out by at least three partially, redundant resolvases consisting of mutagen sensitive 81/essential meiotic endonuclease 1 [MUS81/EME1, respectively], synthetically lethal with genes of unknown function (X) 1 and 4 [SLX1 and SLX4, respectively], and general homolog of endonuclease 1 GEN1] [63]. Finally, LIGI is utilized to covalently seal any nicks left in the DNA.

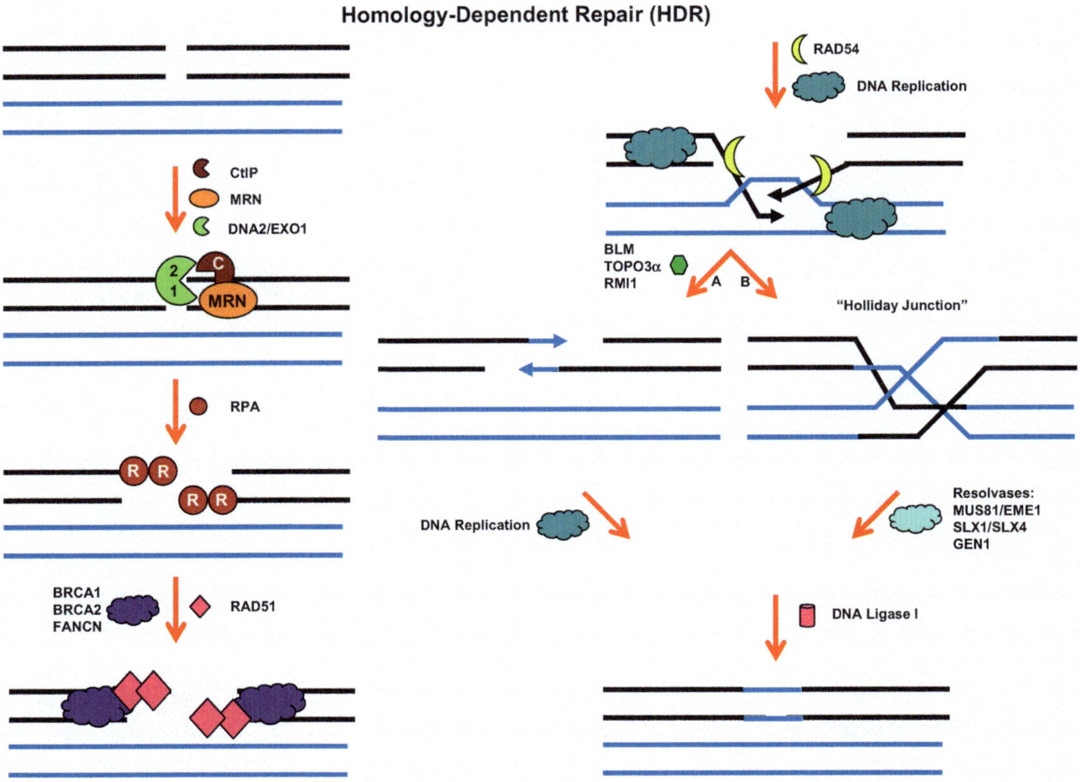

Fig. 7.4 A schematic depicting Homology-Dependent Repair (HDR). The black lines represent strands of DNA where a DSB has occurred and the blue lines represent an undamaged sister chromatid or a homologue. In HDR, the DNA ends of the DSB are extensively resected to yield 3′-single-stranded DNA overhangs. As described for A-NHEJ, the nuclease(s) responsible for this resection are the MRN:CtIP complex (which generates the initial resection; orange oval and crimson PacMan™, respectively) followed by the action of DNA2 and EXOI (light green PacMan™). The resulting overhangs are then coated by RPA (red circles), which removes the secondary structures from the overhangs. A complex of proteins including BRCA1, BRCA2, and FANCN (purple cloud) then help to recruit RAD51 (pink diamond) to the overhangs. Strand invasion into the homologous chromosomal sequence requires RAD54 (yellow moon) and DNA replication (blue-green cloud). Strand exchange generates an interdigitated set of strands that can be resolved into a complicated set of products. (**a**) In mitotic cells most of the intermediates are resolved as non-crossover products by dissolving the interdigitated strands back into their original duplexes after sufficient DNA replication has occurred to restore the genetic information lost at the site of the DSB. The dissolution process requires the action of the BLM/TOPO 3α/RMI1 complex (green hexagon). (**b**) Less frequently the second end of DNA is captured and a covalently closed "Holliday junction" is formed that can be resolved as either non-crossover products (which are functionally identical to dissolution) or crossover products. The resolution of Holliday junctions is complicated and in human cells appears to be carried out by at least three partially redundant resolvases consisting of mutagen sensitive 81/essential meiotic endonuclease 1 [MUS81/EME1, respectively], synthetically lethal with genes of unknown function (X) 1 and 4 [SLX1 and SLX4, respectively], and general homolog of endonuclease 1 GEN1] (1) (light blue cloud). Finally, LIGI (pink cylinder) is utilized to covalently seal any nicks left in the DNA. Note that only the non-crossover product for Holliday Junction resolution is diagrammed

Although HDR is often referred to as error-free repair, that characterization is only partially true. Thus, in the case of non-crossover events in which the repair is templated from a sister chromatid the DSB is in fact repaired in an error-free fashion. However, when a homolog, rather than a sister chromatid, is utilized there is a risk of the loss of heterozygosity and uniparental disomy as observed in several developmental disorders and numerous tumor types [101]. In summary, human somatic cells express all of the gene products needed to carry out HDR. These events occur, however, only at very low frequency and usually only in S phase due to the preferred usage of NHEJ.

In summary, human cells can repair DNA DSBs by at least three discrete pathways: C-NHEJ, A-NHEJ and HDR. How pathway choice (which pathway is utilized in which cells during which phases in the cell cycle, *etc.*) is biochemically determined is the focus of much research. Regardless, from a logistical perspective, one thing that clearly differentiates these repair pathways is their reliance on different DNA ligases to complete the reaction. Thus, C-NHEJ utilizes exclusively LIGIV, whereas A-NHEJ prefers to use LIGIII (although it can utilize LIGI) and HDR uses exclusively LIGI. In conclusion, until better biochemical or genetic markers become available, ligation is one the most distinguishing features of these repair pathways.

7.4 Translocations

7.4.1 DSBs and Translocations

As enumerated above, DNA DSBs in human cells can occur either spontaneously or through exposure of the cells to environmental toxins. The vast majority of the time, the two ends of a DSB are rejoined back to one another either by C-NHEJ, A-NHEJ or HDR with a varying loss of genetic information, but in a fashion that almost always restores genome stability. Rarely, one or both ends of a DSB will be incorrectly rejoined to another DSB end resulting in a translocation. The biological consequences of this can be enormous as translocations can inactivate tumor suppressor genes, activate oncogenes or make new chimeric oncogenes [13, 16, 56]. All of these scenarios promote the formation of tumors.

Translocations can occur within a chromosome (an intrachromosomal translocation) which can result in inversions or to another chromosome (interchromosomal translocation). In the latter case, the simplest outcome is a reciprocal translocation where the proximal portion of one chromosome is joined to the distal portion of another chromosome and *vice versa*. Needless to say, since the occurrence of any one DSB is a relatively rare event (only 25 DSBs, per cell, per day) the likelihood of concomitant DSBs existing in the same cell at the same time is quite small and likely explains why translocations occur so much less frequently than other types of mutations. The only situation where two DSBs are not required is when one of the DSBs is the natural end of a chromosome; *i.e.*, a telomere. As we will discuss below, this is a specialized case of translocation. In all other cases, there is a requirement for 2 DSBs to exist simultaneously in order for a translocation to occur. The basic, but as yet still unanswered, question that drives virtually all research in this field is why are these DSBs simply not repaired normally? That is, two DSBs yield four DNA ends: 1 and 2 as well as 3 and 4. In normal repair reactions end 1 would get re-joined to 2 and end 3 would become re-joined to 4. In a translocation, however, 1 joins to 3 (or 4) and 2 joins to 4 (or 3). Why and how the ends of a DSB become available to join with an end other than the one they were normally connected with is key to understanding the genesis of translocations.

7.4.1.1 Of Men, Mice and Translocations

Before a discussion of the mechanistic aspects of this process can begin however, it is important to understand that the translocations that occur in human cells appear to arise by a different process than translocations that occur in the laboratory workhorse model organism, the mouse. This appears to be an exceptionally unfortunate biological difference as the mouse is used for a veritable plethora of cancer modeling studies and an enormous amount of pre-clinical cancer research is carried out with the mouse.

In the mouse, it is manifestly compelling that translocations are mechanistically dependent upon A-NHEJ. This conclusion rests upon at least three pieces of evidence. First, in the mouse, when genes involved in C-NHEJ are mutated the translocation frequency actually increases [12, 116]. This observation is consistent with the interpretation that in the absence of C-NHEJ that there is likely a greater cellular reliance on A-NHEJ. Second, when DNA sequence analyses are utilized to investigate the junctional diversity of translocations in the mouse, the frequency of microhomology — a quasi-hallmark of A-NHEJ — found at the repair site is quite high [21, 31]. As noted above, however, the appearance of microhomology in and of itself is not unequivocally proof of the use of A-NHEJ. Thus, in one very large study of translocations carried out by the Alt laboratory 75 to 90% of all translocations had microhomologies of 1 to 5 nts at the breakpoint junction. However, only 10% of those same junctions were 5 nts or longer [21]. In conclusion, while clearly not unequivocal, these data are consistent with the use of A-NHEJ. Third, and perhaps the strongest piece of data, comes from a demonstration that genetic ablation of nuclear LIGIII, reduced the occurrence of translocations in the mouse [88]. In summary, the increase in translocations when C-NHEJ is absent, the frequent use of microhomology at translocation breakpoints and the reduction of translocations when LIGIII is absent, compellingly indicate that translocations in the mouse are LIGIII-dependent and likely mediated by A-NHEJ.

In contrast (and certainly confusingly), by the same set of criteria it appears as if translocations in human cells are mediated by C-NHEJ. Thus, in contrast to the mouse, mutations in C-NHEJ genes LIGIV and XRCC4 greatly reduce the frequency of translocations in human somatic cells [33, 47, 54]. In addition, although microhomology can be found at translocations breakpoint junctions in human tumors [99], the frequency and amount of it is generally small [8, 93]. Finally, the functional inactivation of LIGIII has little to no impact on translocations in human somatic cells [33]. It should be noted, however, that inhibition of PARP1, an A-NHEJ gene, reduced translocations in some human cells [17, 111], but not in others [54]. This latter observation notwithstanding, the reduction in translocations when C-NHEJ is absent, the infrequent use of microhomology at translocation breakpoints and the lack of an impact on the frequency of translocations when LIGIII is absent, compellingly indicate that most translocations in human cells are LIGIV-dependent and likely mediated by C-NHEJ.

Needless to say, these observations raise the question of why a seemingly similar process should be mechanistically so different in these two organisms. To date, there is no clear answer. The most likely explanation has to do with species-specific differences in the factors that make up the DSB repair pathways. For example, DNA-PK_{cs} is the key C-NHEJ factor that tethers the two ends of a DSB together through homodimerization [87, 91]. Relevantly, DNA-PK_{cs} is more abundant (by ~ an order of magnitude) in human cells than it is in rodent cells [29]. Thus, the reduced quantities of DNA-PK_{cs} (and presumably therefore reduced numbers of tethered ends) may provide A-NHEJ in the mouse with additional windows of opportunity for the ends to dissociate and be conscripted by A-NHEJ factors, whereas in human cells, with a superabundance of DNA-PK_{cs} [there are estimated to be between one-half to one million molecules of DNA-PK_{cs} in every human cell; [66]] C-NHEJ is the dominant repair pathway. The obvious follow-up question of why human cells should contain so much more DNA-PK_{cs} than rodents is unfortunately not biochemically obvious, but the empirical fact that they do likely provides at least a partial answer for why the two organisms utilize the C-NHEJ and A-NHEJ pathways differentially. In addition, it is well known that chromatin organization and epigenetic modifications can affect the mutation rate across genomes [85, 100]. Specific chromatin features and epigenetic marks are unlikely to be highly conserved across species and these differences may also impact upon the process of translocations. Finally, it is now appreciated that at least some of the endogenous DSBs generated in vertebrate cells may be due to the aberrant action of apolipoprotein B editing complex 3 [APOBEC3], a cytidine deaminase capable of introducing closely spaced nicks into the DNA [96, 100]. Importantly, there is a single APOBEC3 gene in the mouse, whereas in humans that locus has been significantly expanded to eight functional isoforms. Thus, differences in APOBEC3 expression could certainly causes significant differences in either the frequency and/or location of DSBs in the genome. Whatever the correct answer(s) may be, it is important to appreciate that particular care must be taken in interpreting or extrapolating experimental results obtained in rodent model systems to humans since some of the basic biology appears to be different [discussed at length as well by [56]].

7.4.1.2 Spatial Karma and Translocations

Regardless of which pathway of end joining (C-NHEJ or A-NHEJ) is used for repair, why are these processes not always faithful? The correlation of translocations with aberrant A-NHEJ is easiest to reconcile. Thus, while PARP1 can bind tightly to DNA ends, it is not known to homodimerize. Moreover, while some studies have suggested that MRN, or subunits thereof, are capable of homodimerization [109] there is frankly no A-NHEJ factor comparable to DNA-PK_{cs}. As a consequence of this, it seems likely that the ends of a DSB that are being repaired by A-NHEJ may not be as synaptically as stable as ends being repaired by C-NHEJ and therefore simply stand a statistically higher chance of separating from one another before the repair event is completed. This model dovetails nicely with the reduced amounts of DNA-PK_{cs} observed in rodents and their correspondingly greater propensity to utilize A-NHEJ in the formation of translocations. The flip side of this rationalization is more complex. Thus, in humans, where C-NHEJ apparently predominates, why does the end of DSB ever become capable of joining to an end other than its cognate end? Indeed, it is well known (albeit mostly from mouse studies) that C-NHEJ is more likely to join DSBs intrachromosomally rather than interchromosomally [21, 60]. In essence then, when C-NHEJ is utilized it is simply less likely that a translocation will result. The most compelling explanation for the translocations that do result is that the DSBs may be spatially adjacent to one another. For example, even early experiments on the spatial organization of the human genome noted that translocations often involved regions that were physically closer to one another than to other regions of the genome [65]. These observations have been confirmed and extended over the past decade as technology has improved the characterization of the large-scale organization of chromosomes [13, 81, 82]. Thus, nuclear DSBs have a tendency not to move very much [43, 49] and this correlates well with the observation that more than 80% of DSBs translocate to regions that are physically located to within 2.5 μm of each other [81, 82]. In conclusion, the current best explanation for why translocations occur in human cells is "bad karma". That is, a translocation likely only occurs when two concomitant DSBs are also spatially close to one another in the nucleus such that a synaptic complex (likely a requirement for repair) can form — albeit in these rare instances between non-cognate ends.

7.4.1.3 Selection, Not the Translocation, Drives Cancer

It is well known that particular translocations are the hallmark of certain cancers [16, 56]. However, it is important to appreciate that the predominance of a translocation in a tumor is due solely to the subsequent selection that is imposed upon all the translocations that may have occurred during the genesis of that tumor. That is, if, and only if — and this is a stochastic probability — the translocation generates a novel chromosome that gives the cell a selective growth advantage, will these cells be subsequently amplified to generate the tumor. Indeed, translocations that are oncogenic have invariably inactivated a tumor suppressor gene, activated an oncogene and/or created a chimeric gene that is acting as an oncogene. This event, however, is independent from the mechanism of the translocation; that is, there is nothing inherently oncogenic about translocations. Both ends of a DSB have a similar propensity to translocate [21] and although there is a bias towards translocations happening near transcriptional start sites in the mouse [21], this bias is not observed in human cells [56] (yet another difference between mice and humans). Consequently, it is important to appreciate is that there is no evidence of directionality or specificity intrinsic to translocations themselves. Thus, both ends of a broken chromosome likely have the potential to translocate to an infinite number of chromosomal locations and this is likely limited only by the spatial parameters discussed above.

7.4.2 Considerations for When One DSB is a Telomeric End

Up until now, all of the translocations that have been discussed were canonical ones requiring the formation of two DSBs and the generation of four DNA ends. There is one biologically important scenario, however, where translocations can occur between a DSB and a "single-ended DSB" and hence only involve three DNA ends. This scenario occurs when the end of a chromosome, *i.e.*, a telomere, participates in the translocation reaction.

7.4.2.1 Telomeres Stabilize the Genome

There are 46 chromosomes in a normal diploid human cell and because each chromosome has 2 ends, there are in principle 92 natural DSBs constitutively present in a cell. Such a scenario, if it truly existed, would be lethal, so evolution has devised an answer in the form of telomeres. Telomeres are specialized nucleoprotein structures that are found at the extreme termini of linear eukaryotic chromosomes. Telomeres "cap" those ends and prevent the recognition of the chromosomal termini as DSBs by the cellular DNA damage response apparatus. Telomeres consist of a repetitive hexameric tract of DNA (TTAGGG) bound by an evolutionarily-conserved complex of proteins collectively called Shelterin [26]. Importantly, ongoing cell division (*i.e.*, aging) results in gradual telomere erosion [36], and ultimately, the loss of the end-capping function which, in the context of a functional DNA damage response, leads to the induction of a p53-dependent G_1/S cell cycle arrest, known as replicative senescence [25]. This cell-intrinsic limit on replicative lifespan provides a stringent tumor suppressive mechanism. However, in the absence of a fully functional DNA damage checkpoint response, older cells containing short dysfunctional telomeres (which are essentially one-ended DSBs) enter a state of crisis during which telomeres undergo fusion, either between sister chromatids (Fig. 7.5a), with interchromosomal telomeres (Fig. 7.5b) or with non-telomeric DSBs, creating dicentric chromosomes and initiating BFB cycles [22, 70]. This, in turn, leads to the creation of genomic rearrangements, including the translocations that are common in cells from many different tumor types [4, 86]. The development of single-molecule approaches to characterize the sequence of telomere fusion events, has revealed that short dysfunctional telomeres are capable of recombining with both telomeric and non-telomeric loci across the genome [51, 54]. Thus, whilst BFB cycles initiated because of telomere dysfunction can lead to chromosomal translocations [70], telomere fusions themselves can also lead directly to translocation events. Intra-chromosomal telomere fusion involving sister-chromatids predominates over inter-chromosomal telomere fusion, which in turn is more frequent than inter-chromosomal fusion between telomeres and non-telomeric loci [54]. The characteristics of the non-telomeric loci involved in telomere fusion have yet to be fully characterized, but thus far it is apparent that they

Fig. 7.5 (continued) Shelterin complex. Fusion between sister-chromatids results in the formation of a dicentric chromosome, that can form a bridge between daughter cells at anaphase, that is subjected to a breakage event. Depending on the position of the break, this can lead to a daughter cell that has lost terminal sequences, or has an addition copy of a gene — in this example, gene B. Further BFB cycles can lead to further amplification and deletion. This process can be stopped by the healing of a DSB via the acquisition of a *de novo* telomere, either by telomerase-mediated extension or by recombination with a pre-existing telomere. Centromeres are depicted as green ovals, telomeres by black and white rectangles and genes in colored squares. (**b**) Inter-chromosomal telomere fusion between short dysfunction telomeres (depicted), or with non-telomeric DSBs, can lead to the formation of dicentric chromosomes and the initiation of BFB cycles that can lead to the formation of non-reciprocal translocations (NRT) and deletions. This process can be prevented by chromosomal healing via the acquisition of new telomere

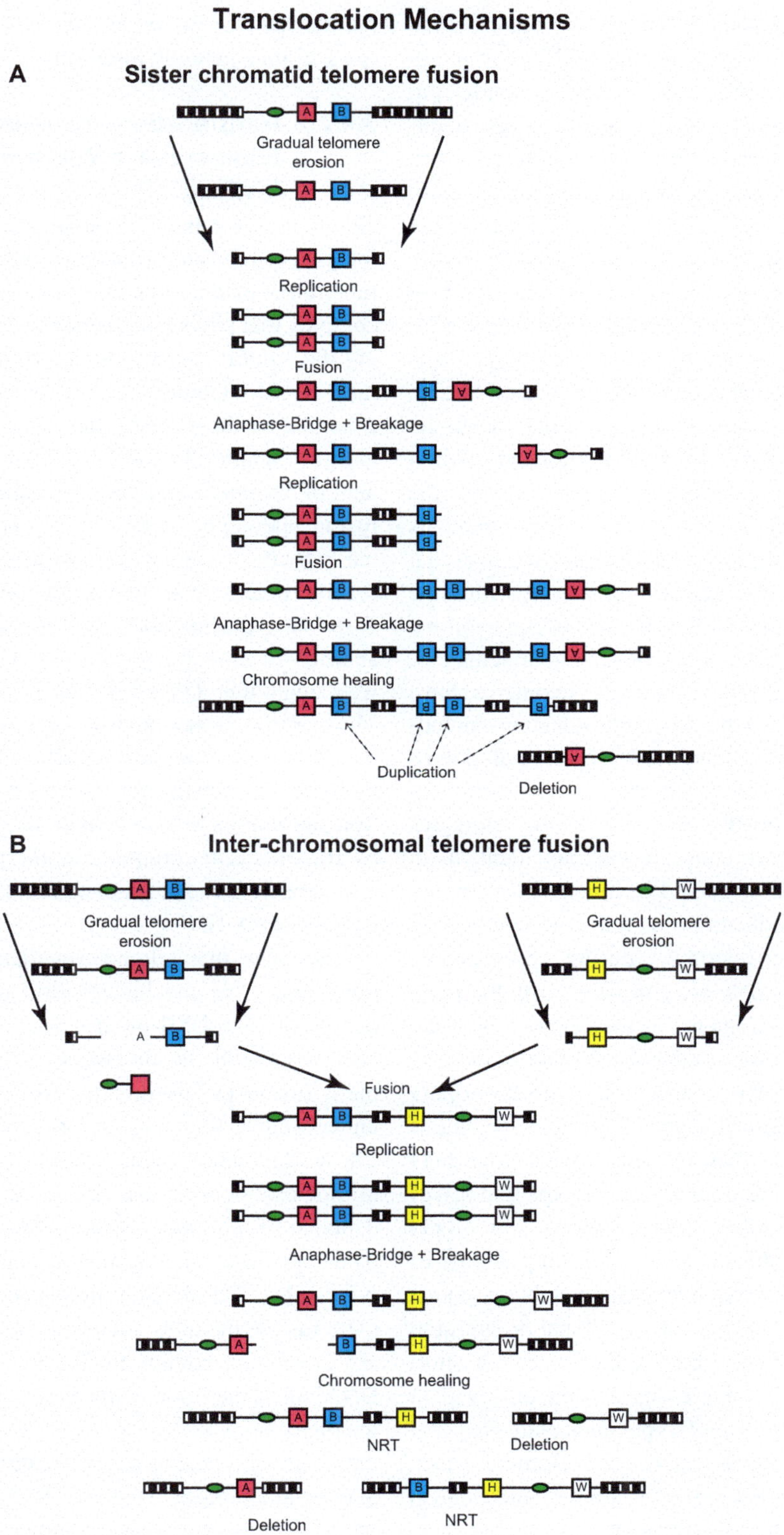

Fig. 7.5 Short dysfunctional telomeres can be subjected to sister-chromatid fusion, or inter-chromosomal fusion events to create amplifications, deletions and non-reciprocal translocations. (**a**) Gradual telomere erosion in the absence of functional DNA damage checkpoints, leads to short telomeres that are no longer protected by the

occur predominantly within coding regions of the genome; indicating a potential role for chromatin structure and replication timing in conferring sensitivity to fusion [54]. Larger datasets documenting the specific loci involved in telomere fusion are required before a definition of these fusagenic loci can be provided and potential hot spots identified.

Thus, telomere-dependent crisis is a key event in driving genomic instability and clonal evolution during the progression to malignancy; this is consistent with data and observations of telomere dynamics and fusion in a broad range of human tumor types in which extreme telomere erosion and fusion is observed that correlates with the presence of large-scale genomic rearrangements [57, 67, 79]. Moreover, patients with tumors that display short dysfunctional telomeres, within the length ranges at which fusion can occur, exhibit a poorer prognosis and response to treatments [58, 95, 108]. Short dysfunctional telomeres have been identified in the very earliest lesions, including very small adenomatous colorectal polyps [79] and in leukemias prior to clinical progression [57]. Importantly, the short telomeres observed in early stage lesions are identical in length to those observed in more advanced disease clones, indicating that telomere length does not vary considerably during progression. Together these data are consistent with the presence of short telomeres in the cell in which the initiating mutation occurred and that this dictates the telomere length distribution of the developing clone. In this model, if the initiating cell contains short telomeres then the subsequent clone may have a "telomere-mutator" phenotype that drives genomic instability, translocations and clonal progression, whereas a cell with long telomeres gives rise to a clone with a more stable genome, which exhibits slower rates of clonal progression (92). Finally, it is important to note that short dysfunctional telomeres have been observed in the majority of tumor types analyzed [46] and thus it appears that a period of telomere-driven genome instability may be a common mechanism underlying the progression to malignancy. Therefore, there is a requirement to understand the mechanisms by which telomere dysfunction can facilitate genome instability.

7.4.2.2 Translocations Involving Telomeres are Mechanistically Distinct

Telomere fusion is clearly an important and physiologically relevant mutational event. Key to the function of mammalian telomeres is the Shelterin complex that plays a fundamental role in protecting the natural chromosomal termini from aberrant NHEJ-mediated joining events [26]. For example, in the mouse, the abrogation of telomere recognition factor 2 [TRF2], a core component of Shelterin, confers a widespread telomere fusion phenotype [102] that is dependent upon the activity of LIGIV. In contrast, fusions were readily detected in telomerase-deficient mice, with short dysfunctional telomeres, despite the absence of core components of C-NHEJ pathway, including DNA-PK$_{cs}$ or LIGIV [61, 78]. Thus, in the mouse and in the context of short dysfunctional telomeres, which is likely the most biologically relevant form of telomere dysfunction, telomeres are no longer fully recognized by the Shelterin complex and the processing of telomere fusion appears to be mediated by either C-NHEJ or A-NHEJ.

The view that telomere-mediated translocations may be mechanistically distinct from canonical two DSB-mediated translocations is consistent with the molecular analysis of telomere fusion events directly from human cells undergoing a telomere-driven crisis in culture. These data show that fusion between short telomeres — ones that are almost completely denuded of telomere repeats — is accompanied by deletion and microhomology across the fusion points [18]. The deletion that accompanies telomere fusion, includes not just the telomere repeat array itself, but extends into the telomere-adjacent DNA, up to the limit of the assays used (6.1 kb), the distribution of fusion points from the start of the telomere repeat arrays, indicates that deletion may be much more extensive. This characteristic profile is also observed at telomere fusion junctions isolated from some human malignancies, including early-stage and pre-malignant lesions [57, 79], as well normal human cells, in which rare stochastic telomeric deletion results in fusion [18, 57, 79]. Finally, molecular analysis of fusion events following replicative telomere erosion in human cells carrying hypomorphic MRE11 alleles revealed a change in the mutational spectrum with an increase in insertions at the fusion point [97]. The reliance on MRE11, the extensive deletion and the high degree of microhomology that accompanied these human telomere fusions was indicative of error-prone processing of short dysfunctional telomeres via the A-NHEJ pathway and suggested that telomere-mediated fusions in human cells may be mechanistically fundamentally different than canonical two DSB-mediated translocations, which, as detailed above, appear to be predominately mediated by C-NHEJ.

7.4.2.3 Translocations Involving Human Telomeres can be Mediated by LIGIII or LIGIV

To experimentally test this idea, a study was undertaken utilizing human cell lines in which either nuclear LIGIII [73] or LIGIV [74] (and presumably A-NHEJ or C-NHEJ, respectively) had been inactivated by gene targeting. A dominant-negative version of human telomerase [DN-hTERT; [35]] was then expressed in these cells to cause gradual telomere shortening and the status of the telomere stability was assessed by a single telomere length analysis [STELA; [6]] and single-molecule telomere fusion analyses. These approaches allow one to either (1) quantitate the length of a single telomere, (2) detect and characterize the DNA sequence of translocations or (3) detect and quantitate sister chromatid:sister chromatid fusions/translocations. These experiments demonstrated that translocations involving telomeres occurred in either LIGIII- or LIGIV-null cells [47]. Thus, unlike canonical translocations, which are heavily dependent upon LIGIV, a high frequency of telomere-mediated translocations was still observed in LIGIV-null cells. There were, however, some parallels with canonical translocations. Thus, the majority of the translocations that occurred in LIGIII-null cells (*i.e.*, translocations perforce mediated by LIGIV) were biased 3:1 towards interchromosomal translocations, as is observed for canonical translocations. Similarly, in LIGIV-null cells (*i.e.*, translocations perforce mediated by LIGIII) while there were still interchromosomal translocations, telomere fusions were now biased 52:1 towards intrachromosomal sister chromatid fusion events [47]. These biases were so significant that they had a profound biological effect — cells that were LIGIII-null were not able to survive the DN-hTERT-induced crisis whereas those that were either wild type or LIGIV-null readily survived. A parsimonious interpretation of this data is that the LIGIV-mediated interchromosomal translocations were predominately toxic und ultimately lethal for cells whereas the LIGIII-mediated intrachromosomal fusions provided a growth advantage that could be selected for during crisis. This interpretation is consistent with the gene duplications and localized amplifications that are associated with sister:sister fusion events that are not observed with interchromosomal translocations [70].

These experiments beg the question as to why a telomere-mediated translocation (as compared to a interchromosomal DSB-mediated translocation) should be less reliant on C-NHEJ (and/or more reliant on A-NHEJ). The most obvious difference is simply that while a shortened telomere can bind a single DNA-PK complex, there is no corresponding end to bind a second DNA-PK complex and hence there is a greatly reduced chance of forming a synaptic complex. The lack of a synaptic complex presumably now permits the recruitment of A-NHEJ factors to the end and/or the displacement of the DNAPK complex from the end such that a higher frequency of A-NHEJ-mediated fusions can occur. Another factor that might influence the relative activities of A- and C-NHEJ at telomeres may be the nature of a short telomere, compared to a non-telomeric DSB. The telomeres terminate not with a blunted-ended DSB, but instead have a large (200 to 300 nt) overhang composed of TTAGGG repeats [112]. This unique structure has the potential to fold into G-quadruplex structures [115] and may represent a non-canonical substrate for DNA repair activity that may favor the slower kinetics of the A-NHEJ pathway over that of C-NHEJ.

7.4.2.4 Translocations Involving Human Telomeres can be Mediated by LIGI

The above data strongly suggested that the geometry of the DNA ends and the availability of a requisite DNA ligase controls the type of translocations that can occur in human cells. To extend these observations a follow-up study was carried out in which the frequency and kind of translocation was quantitated in cells that were genetically engineered to be deficient for both LIGIII and LIGIV, where, presumably, both C-NHEJ and A-NHEJ would be ablated. In this experimental set-up the telomere was not gradually exposed by the expression of DN-hTERT as before, but was rapidly deleted by the use of a transcription activator-like effector nuclease [TALEN; [75]]. A TALEN pair was designed to introduce a DSB 14 base pairs from the start of the telomeric TTAGGG repeat on the petite arm of chromosome 17. Thus, this experimental system is somewhat of a hybrid between those measuring canonical fusions and the system to gradually uncover a telomere end by enforced DN-hTERT expression. Specifically, the TALEN should generate a DSB with two ends, however, one of those ends is only a couple of kilobases long and consists solely of the telomeric TTAGGG hexameric repeat. It is unclear whether this end can function in a fashion similar to a canonical chromosomal end. With this caveat in mind, it was reassuringly observed that in the absence of LIGIV a greatly decreased frequency of interchromosomal translocations was observed [54]. Very surprisingly, however, in the combined absence of LIGIII and LIGIV significant amounts of both inter-and intrachromosomal translocations were observed although inter-chromosomal translocations were detected at a reduced frequency [54]. Interestingly, whilst the frequency of inter-chromosomal fusion events was decreased in the absence of LIGIV, intra-chromosomal sister chromatid fusion events appeared to be largely unchanged in the different genetic backgrounds tested. Moreover, there were differences in the utilization of microhomology, with significantly greater microhomology observed at intra-chromosomal events compared to inter-chromosomal events. Taken to together these data are consistent with a role for LIGIV-dependent C-NHEJ in driving interchromosomal telomere fusion and A-NHEJ being predominant for intrachromosomal sister chromatid telomere fusion. These data were also important because they provided the first demonstration in human cells that LIGI can facilitate chromosomal translocations — both inter-chromosomal and intra-chromosomal sister chromatid translocations. Moreover, these data revealed considerable redundancy in the utilization of the specific ligases for end-joining, with LIGI being able to facilitate intra-chromosomal fusion as well as inter-chromosomal fusion, albeit less efficiently. This may be discouraging from the clinical perspective, as these data indicate that attempts to inhibit human translocations using small molecule inhibitors to LIGIII and LIGIV [89] are destined to fail due to the robust ability of LIGI to compensate for their absence. That said, any intervention that can skew the fusion spectrum towards inter-chromosomal events, creating a larger mutational burden on cells and influencing their ability to escape a telomere-driven crisis, may have clinical utility. A deeper understanding of the key proteins involved in A-NHEJ and telomere fusion may identify additional therapeutic targets that could allow for more selective interventions into these pathways.

7.5 Summary and Future Considerations

In summary, DSBs are normally repaired with high fidelity in the sense that the pieces of DNA that were contiguous before the DSB are contiguous after DNA repair, regardless of the "sloppiness" of the actual join. In order for a canonical chromosomal translocation to occur there needs to be two contemporaneous DSBs within a cell (which is a low frequency event) and the ends that were contiguous with one another before the

DSBs, need to be rejoined aberrantly. This (mis) rejoining of the ends is likely mediated, at least in part, by their spatial proximity within the nucleus with the closer that the DSB ends are to each other the greater the likelihood of a translocation occurring. In the mouse, these events are predominately mediated by LIGIII/A-NHEJ whereas in human cells they are mediated by LIGIV/C--NHEJ. When a telomere shortens or when it loses its protective proteinaceous cap, the Shelterin complex, it is treated by the cell as a one-ended DSB and can engage in the formation of translocations as well. In this instance, however, both A-NHEJ and C-NHEJ seem to play an active role in mediating the resulting translocations. Layered over all of this is an additional layer of complexity provided by the recent demonstration that LIGI can fully compensate for translocations that were previously exclusively or predominately ascribed to LIGIII/A-NHEJ or LIGIV/C-NHEJ.

As is often the case in biology, reality is often much more complex than first envisioned. In the beginning, most models of chromosomal translocations invoked the aberrant use of either LIGIII or LIGIV. It is now clear that the situation is significantly more complicated with all three DNA ligases capable of generating translocations in a fashion that likely depends upon the state of the cell cycle, the level of expression of the various ligases within a given cell type and whether one of the translocating ends is telomeric or not. As a consequence, simplistic approaches of inhibiting a single ligase [and such specific inhibitors are not even currently available; [89]] are likely destined to fail. Nonetheless, it is clear that in a human cell where all three ligases are expressed that inhibiting LIGIV will significantly decrease interchromosomal translocations, which could potentially be used to therapeutic benefit. What is clinically perhaps more relevant however, is trying to inhibit the intrachromosomal sister chromatid:sister chromatid fusions as these appear to be critical for cells to escape crisis and thus become oncogenic [47]. In this scenario, inhibition of both LIGI and LIGIII will likely be necessary to achieve a therapeutic outcome. Given that LIGI also has important functions in DNA replication (an essential cellular process) it is likely that such approaches will have significant toxic side effects. Nonetheless, as more is learned about all three DNA ligases, and especially about how pathway choice for DSB repair is regulated there is still significant cause for optimism that windows of opportunity for therapeutic intervention will be uncovered.

Acknowledgements Work in the Hendrickson laboratory was supported in part by grants from the NIH (GM088351) and the NCI (CA154461 and CA190492). Work in the Baird laboratory was supported by the Cancer Research UK (C17199/A18246). Both authors thank previous and current members of their respective laboratories for their contributions to this work. EAH thanks Dr. Anja-Katrin Bielinsky (University of Minnesota) for her comments on the manuscript.

References

1. Ahmad A, Robinson AR, Duensing A, van Drunen E, Beverloo HB, Weisberg DB, Hasty P, Hoeijmakers JH, Niedernhofer LJ (2008) ERCC1-XPF endonuclease facilitates DNA double-strand break repair. Mol Cell Biol 28(16):5082–5092. https://doi.org/10.1128/MCB.00293-08
2. Ahnesorg P, Smith P, Jackson SP (2006) XLF interacts with the XRCC4-DNA ligase IV complex to promote DNA nonhomologous end-joining. Cell 124(2):301–313. https://doi.org/10.1016/j.cell.2005.12.031
3. Arakawa H, Bednar T, Wang M, Paul K, Mladenov E, Bencsik-Theilen AA, Iliakis G (2012) Functional redundancy between DNA ligases I and III in DNA replication in vertebrate cells. Nucleic Acids Res 40(6):2599–2610. https://doi.org/10.1093/nar/gkr1024
4. Artandi SE, Chang S, Lee SL, Alson S, Gottlieb GJ, Chin L, DePinho RA (2000) Telomere dysfunction promotes non-reciprocal translocations and epithelial cancers in mice. Nature 406(6796):641–645. https://doi.org/10.1038/35020592
5. Audebert M, Salles B, Calsou P (2004) Involvement of poly(ADP-ribose) polymerase-1 and XRCC1/DNA ligase III in an alternative route for DNA double-strand breaks rejoining. J Biol Chem 279(53):55117–55126. https://doi.org/10.1074/jbc.M404524200
6. Baird DM, Rowson J, Wynford-Thomas D, Kipling D (2003) Extensive allelic variation and ultrashort telomeres in senescent human cells. Nat Genet 33(2):203–207. https://doi.org/10.1038/ng1084
7. Barnes R, Eckert K (2017) Maintenance of genome integrity: how mammalian cells orchestrate genome duplication by coordinating replicative and special-

ized DNA polymerases. Genes (Basel) 8(1). https://doi.org/10.3390/genes8010019
8. Berger MF, Hodis E, Heffernan TP, Deribe YL, Lawrence MS, Protopopov A, Ivanova E, Watson IR, Nickerson E, Ghosh P, Zhang H, Zeid R, Ren X, Cibulskis K, Sivachenko AY, Wagle N, Sucker A, Sougnez C, Onofrio R, Ambrogio L, Auclair D, Fennell T, Carter SL, Drier Y, Stojanov P, Singer MA, Voet D, Jing R, Saksena G, Barretina J, Ramos AH, Pugh TJ, Stransky N, Parkin M, Winckler W, Mahan S, Ardlie K, Baldwin J, Wargo J, Schadendorf D, Meyerson M, Gabriel SB, Golub TR, Wagner SN, Lander ES, Getz G, Chin L, Garraway LA (2012) Melanoma genome sequencing reveals frequent PREX2 mutations. Nature 485(7399):502–506. https://doi.org/10.1038/nature11071
9. Bernstein KA, Rothstein R (2009) At loose ends: resecting a double-strand break. Cell 137(5):807–810. https://doi.org/10.1016/j.cell.2009.05.007
10. Betermier M, Bertrand P, Lopez BS (2014) Is non-homologous end-joining really an inherently error-prone process? PLoS Genet 10(1):e1004086. https://doi.org/10.1371/journal.pgen.1004086
11. Blackford AN, Jackson SP (2017) ATM, ATR, and DNA-PK: The Trinity at the Heart of the DNA Damage Response. Mol Cell 66(6):801–817. https://doi.org/10.1016/j.molcel.2017.05.015
12. Boboila C, Jankovic M, Yan CT, Wang JH, Wesemann DR, Zhang T, Fazeli A, Feldman L, Nussenzweig A, Nussenzweig M, Alt FW (2010) Alternative end-joining catalyzes robust IgH locus deletions and translocations in the combined absence of ligase 4 and Ku70. Proc Natl Acad Sci U S A 107(7):3034–3039. https://doi.org/10.1073/pnas.0915067107
13. Bohlander SK, Kakadia PM (2015) DNA repair and chromosomal translocations. Recent Results Cancer Res 200:1–37. https://doi.org/10.1007/978-3-319-20291-4_1
14. Boulton SJ, Jackson SP (1996) Saccharomyces cerevisiae Ku70 potentiates illegitimate DNA double-strand break repair and serves as a barrier to error-prone DNA repair pathways. EMBO J 15(18):5093–5103
15. Buck D, Malivert L, de Chasseval R, Barraud A, Fondaneche MC, Sanal O, Plebani A, Stephan JL, Hufnagel M, le Deist F, Fischer A, Durandy A, de Villartay JP, Revy P (2006) Cernunnos, a novel nonhomologous end-joining factor, is mutated in human immunodeficiency with microcephaly. Cell 124(2):287–299. https://doi.org/10.1016/j.cell.2005.12.030
16. Bunting SF, Nussenzweig A (2013) End-joining, translocations and cancer. Nat Rev Cancer 13(7):443–454. https://doi.org/10.1038/nrc3537
17. Byrne M, Wray J, Reinert B, Wu Y, Nickoloff J, Lee SH, Hromas R, Williamson E (2014) Mechanisms of oncogenic chromosomal translocations. Ann N Y Acad Sci 1310:89–97. https://doi.org/10.1111/nyas.12370
18. Capper R, Britt-Compton B, Tankimanova M, Rowson J, Letsolo B, Man S, Haughton M, Baird DM (2007) The nature of telomere fusion and a definition of the critical telomere length in human cells. Genes Dev 21(19):2495–2508. https://doi.org/10.1101/gad.439107
19. Ceccaldi R, Liu JC, Amunugama R, Hajdu I, Primack B, Petalcorin MI, O'Connor KW, Konstantinopoulos PA, Elledge SJ, Boulton SJ, Yusufzai T, D'Andrea AD (2015) Homologous-recombination-deficient tumours are dependent on Poltheta-mediated repair. Nature 518(7538):258–262. https://doi.org/10.1038/nature14184
20. Cheng Q, Barboule N, Frit P, Gomez D, Bombarde O, Couderc B, Ren GS, Salles B, Calsou P (2011) Ku counteracts mobilization of PARP1 and MRN in chromatin damaged with DNA double-strand breaks. Nucleic Acids Res 39(22):9605–9619. https://doi.org/10.1093/nar/gkr656
21. Chiarle R, Zhang Y, Frock RL, Lewis SM, Molinie B, Ho YJ, Myers DR, Choi VW, Compagno M, Malkin DJ, Neuberg D, Monti S, Giallourakis CC, Gostissa M, Alt FW (2011) Genome-wide translocation sequencing reveals mechanisms of chromosome breaks and rearrangements in B cells. Cell 147(1):107–119. https://doi.org/10.1016/j.cell.2011.07.049
22. Counter CM, Avilion AA, LeFeuvre CE, Stewart NG, Greider CW, Harley CB, Bacchetti S (1992) Telomere shortening associated with chromosome instability is arrested in immortal cells which express telomerase activity. EMBO J 11(5):1921–1929
23. Critchlow SE, Bowater RP, Jackson SP (1997a) Mammalian DNA double-strand break repair protein XRCC4 interacts with DNA ligase IV. Curr Biol 7(8):588–598
24. Critchlow SE, Bowater RP, Jackson SP (1997b) Mammalian DNA double-strand break repair protein XRCC4 interacts with DNA ligase IV. Curr Biol 7:588–598
25. d'Adda di Fagagna F, Reaper PM, Clay-Farrace L, Fiegler H, Carr P, Von Zglinicki T, Saretzki G, Carter NP, Jackson SP (2003) A DNA damage checkpoint response in telomere-initiated senescence. Nature 426(6963):194–198. https://doi.org/10.1038/nature02118
26. de Lange T (2005) Shelterin: the protein complex that shapes and safeguards human telomeres. Genes Dev 19(18):2100–2110. https://doi.org/10.1101/gad.1346005
27. Della-Maria J, Zhou Y, Tsai MS, Kuhnlein J, Carney JP, Paull TT, Tomkinson AE (2011) Human Mre11/human Rad50/Nbs1 and DNA ligase IIIalpha/XRCC1 protein complexes act together in an alternative nonhomologous end joining pathway. J Biol Chem 286(39):33845–33853. https://doi.org/10.1074/jbc.M111.274159
28. Dinkelmann M, Spehalski E, Stoneham T, Buis J, Wu Y, Sekiguchi JM, Ferguson DO (2009) Multiple functions of MRN in end-joining pathways dur-

ing isotype class switching. Nat Struct Mol Biol 16(8):808–813. https://doi.org/10.1038/nsmb.1639
29. Finnie NJ, Gottlieb TM, Blunt T, Jeggo PA, Jackson SP (1995) DNA-dependent protein kinase activity is absent in xrs-6 cells: implications for site-specific recombination and DNA double-strand break repair. Proc Natl Acad Sci U S A 92(1):320–324
30. Frit P, Barboule N, Yuan Y, Gomez D, Calsou P (2014) Alternative end-joining pathway(s): bricolage at DNA breaks. DNA Repair (Amst) 17:81–97. https://doi.org/10.1016/j.dnarep.2014.02.007
31. Frock RL, Hu J, Meyers RM, Ho YJ, Kii E, Alt FW (2015) Genome-wide detection of DNA double-stranded breaks induced by engineered nucleases. Nat Biotechnol 33(2):179–186. https://doi.org/10.1038/nbt.3101
32. Gauss GH, Lieber MR (1996) Mechanistic constraints on diversity in human V(D)J recombination. Mol Cell Biol 16(1):258–269
33. Ghezraoui H, Piganeau M, Renouf B, Renaud JB, Sallmyr A, Ruis B, Oh S, Tomkinson AE, Hendrickson EA, Giovannangeli C, Jasin M, Brunet E (2014) Chromosomal translocations in human cells are generated by canonical nonhomologous end-joining. Mol Cell 55(6):829–842. https://doi.org/10.1016/j.molcel.2014.08.002
34. Grade M, Difilippantonio MJ, Camps J (2015) Patterns of chromosomal aberrations in solid tumors. Recent Results Cancer Res 200:115–142. https://doi.org/10.1007/978-3-319-20291-4_6
35. Hahn WC, Stewart SA, Brooks MW, York SG, Eaton E, Kurachi A, Beijersbergen RL, Knoll JH, Meyerson M, Weinberg RA (1999) Inhibition of telomerase limits the growth of human cancer cells. Nat Med 5(10):1164–1170. https://doi.org/10.1038/13495
36. Harley CB, Futcher AB, Greider CW (1990) Telomeres shorten during ageing of human fibroblasts. Nature 345(6274):458–460. https://doi.org/10.1038/345458a0
37. Hendrickson EA, Huffman JL, Tainer JA (2006) Structural aspects of Ku and the DNA-dependent protein kinase complex. In: Seide W, Kow YW, Doetsch P (eds) DNA Damage Recognition. Taylor and Francis Group, New York, pp 629–684
38. Hendrickson EA, Schlissel MS, Weaver DT (1990) Wild-type V(D)J recombination in scid pre-B cells. Mol Cell Biol 10(10):5397–5407
39. Heyer WD, Li X, Rolfsmeier M, Zhang XP (2006) Rad54: the Swiss Army knife of homologous recombination? Nucl Acids Res 34:4115–4125
40. Holliday R (1964) The induction of mitotic recombination by mitomycin C in Ustilago and Saccharomyces. Genetics 50:323–335
41. Iftode C, Daniely Y, Borowiec JA (1999) Replication protein A (RPA): the eukaryotic SSB. Crit Rev Biochem Mol Biol 34(3):141–180. https://doi.org/10.1080/10409239991209255
42. Iliakis G, Murmann T, Soni A (2015) Alternative end-joining repair pathways are the ultimate backup for abrogated classical non-homologous end-joining and homologous recombination repair: Implications for the formation of chromosome translocations. Mutat Res Genet Toxicol Environ Mutagen 793:166–175. https://doi.org/10.1016/j.mrgentox.2015.07.001
43. Jakob B, Splinter J, Durante M, Taucher-Scholz G (2009) Live cell microscopy analysis of radiation-induced DNA double-strand break motion. Proc Natl Acad Sci U S A 106(9):3172–3177. https://doi.org/10.1073/pnas.0810987106
44. Jasin M, Rothstein R (2013) Repair of strand breaks by homologous recombination. Cold Spring Harb Perspect Biol 5(11):a012740. https://doi.org/10.1101/cshperspect.a012740
45. Jette N, Lees-Miller SP (2015) The DNA-dependent protein kinase: A multifunctional protein kinase with roles in DNA double strand break repair and mitosis. Prog Biophys Mol Biol 117(2-3):194–205. https://doi.org/10.1016/j.pbiomolbio.2014.12.003
46. Jones CH, Pepper C, Baird DM (2012) Telomere dysfunction and its role in haematological cancer. Br J Haematol 156(5):573–587. https://doi.org/10.1111/j.1365-2141.2011.09022.x
47. Jones RE, Oh S, Grimstead JW, Zimbric J, Roger L, Heppel NH, Ashelford KE, Liddiard K, Hendrickson EA, Baird DM (2014) Escape from telomere-driven crisis is DNA ligase III dependent. Cell Rep 8(4):1063–1076. https://doi.org/10.1016/j.celrep.2014.07.007
48. Knisbacher BA, Gerber D, Levanon EY (2016) DNA editing by APOBECs: a genomic preserver and transformer. Trends Genet 32(1):16–28. https://doi.org/10.1016/j.tig.2015.10.005
49. Kruhlak MJ, Celeste A, Dellaire G, Fernandez-Capetillo O, Muller WG, McNally JG, Bazett-Jones DP, Nussenzweig A (2006) Changes in chromatin structure and mobility in living cells at sites of DNA double-strand breaks. J Cell Biol 172(6):823–834. https://doi.org/10.1083/jcb.200510015
50. Kunkel TA (2015) Celebrating DNA's Repair Crew. Cell 163(6):1301–1303. https://doi.org/10.1016/j.cell.2015.11.028
51. Letsolo BT, Rowson J, Baird DM (2010) Fusion of short telomeres in human cells is characterized by extensive deletion and microhomology, and can result in complex rearrangements. Nucleic Acids Res 38(6):1841–1852. https://doi.org/10.1093/nar/gkp1183
52. Li Z, Otevrel T, Gao Y, Cheng HL, Seed B, Stamato TD, Taccioli GE, Alt FW (1995) The XRCC4 gene encodes a novel protein involved in DNA double-strand break repair and V(D)J recombination. Cell 83(7):1079–1089
53. Li Z, Pearlman AH, Hsieh P (2016) DNA mismatch repair and the DNA damage response. DNA

Repair (Amst) 38:94–101. https://doi.org/10.1016/j.dnarep.2015.11.019
54. Liddiard K, Ruis B, Takasugi T, Harvey A, Ashelford KE, Hendrickson EA, Baird DM (2016) Sister chromatid telomere fusions, but not NHEJ-mediated interchromosomal telomere fusions, occur independently of DNA ligases 3 and 4. Genome Res 26(5):588–600. https://doi.org/10.1101/gr.200840.115
55. Lieber MR (2010) The mechanism of double-strand DNA break repair by the nonhomologous DNA end-joining pathway. Annu Rev Biochem 79:181–211. https://doi.org/10.1146/annurev.biochem.052308.093131
56. Lieber MR (2016) Mechanisms of human lymphoid chromosomal translocations. Nat Rev Cancer 16(6):387–398. https://doi.org/10.1038/nrc.2016.40
57. Lin TT, Letsolo BT, Jones RE, Rowson J, Pratt G, Hewamana S, Fegan C, Pepper C, Baird DM (2010) Telomere dysfunction and fusion during the progression of chronic lymphocytic leukemia: evidence for a telomere crisis. Blood 116(11):1899–1907. https://doi.org/10.1182/blood-2010-02-272104
58. Lin TT, Norris K, Heppel NH, Pratt G, Allan JM, Allsup DJ, Bailey J, Cawkwell L, Hills R, Grimstead JW, Jones RE, Britt-Compton B, Fegan C, Baird DM, Pepper C (2014) Telomere dysfunction accurately predicts clinical outcome in chronic lymphocytic leukaemia, even in patients with early stage disease. Br J Haematol 167(2):214–223. https://doi.org/10.1111/bjh.13023
59. Lu G, Duan J, Shu S, Wang X, Gao L, Guo J, Zhang Y (2016) Ligase I and ligase III mediate the DNA double-strand break ligation in alternative end-joining. Proc Natl Acad Sci U S A 113(5):1256–1260. https://doi.org/10.1073/pnas.1521597113
60. Mahowald GK, Baron JM, Mahowald MA, Kulkarni S, Bredemeyer AL, Bassing CH, Sleckman BP (2009) Aberrantly resolved RAG-mediated DNA breaks in Atm-deficient lymphocytes target chromosomal breakpoints in cis. Proc Natl Acad Sci U S A 106(43):18339–18344. https://doi.org/10.1073/pnas.0902545106
61. Maser RS, Wong KK, Sahin E, Xia H, Naylor M, Hedberg HM, Artandi SE, DePinho RA (2007) DNA-dependent protein kinase catalytic subunit is not required for dysfunctional telomere fusion and checkpoint response in the telomerase-deficient mouse. Mol Cell Biol 27(6):2253–2265. https://doi.org/10.1128/MCB.01354-06
62. Mateos-Gomez PA, Gong F, Nair N, Miller KM, Lazzerini-Denchi E, Sfeir A (2015) Mammalian polymerase theta promotes alternative NHEJ and suppresses recombination. Nature 518(7538):254–257. https://doi.org/10.1038/nature14157
63. Matos J, West SC (2014) Holliday junction resolution: regulation in space and time. DNA Repair (Amst) 19:176–181. https://doi.org/10.1016/j.dnarep.2014.03.013
64. McClintock B (1941) The stability of broken ends of chromosomes in Zea Mays. Genetics 26(2):234–282
65. Meaburn KJ, Misteli T, Soutoglou E (2007) Spatial genome organization in the formation of chromosomal translocations. Semin Cancer Biol 17(1):80–90. https://doi.org/10.1016/j.semcancer.2006.10.008
66. Meek K, Dang V, Lees-Miller SP (2008) DNA-PK: the means to justify the ends? Adv Immunol 99:33–58. https://doi.org/10.1016/S0065-2776(08)00602-0
67. Meeker AK, Hicks JL, Iacobuzio-Donahue CA, Montgomery EA, Westra WH, Chan TY, Ronnett BM, De Marzo AM (2004) Telomere length abnormalities occur early in the initiation of epithelial carcinogenesis. Clin Cancer Res 10(10):3317–3326. https://doi.org/10.1158/1078-0432.CCR-0984-03
68. Mimitou EP, Symington LS (2008) Sae2, Exo1 and Sgs1 collaborate in DNA double-strand break processing. Nature 455(7214):770–774. https://doi.org/10.1038/nature07312
69. Moshous D, Callebaut I, de Chasseval R, Poinsignon C, Villey I, Fischer A, de Villartay JP (2003) The V(D)J recombination/DNA repair factor artemis belongs to the metallo-beta-lactamase family and constitutes a critical developmental checkpoint of the lymphoid system. Ann N Y Acad Sci 987:150–157
70. Murnane JP (2012) Telomere dysfunction and chromosome instability. Mutat Res 730(1-2):28–36. https://doi.org/10.1016/j.mrfmmm.2011.04.008
71. Nowell PC (1962) The minute chromosome (Phl) in chronic granulocytic leukemia. Blut 8:65–66
72. Ochi T, Blackford AN, Coates J, Jhujh S, Mehmood S, Tamura N, Travers J, Wu Q, Draviam VM, Robinson CV, Blundell TL, Jackson SP (2015) PAXX, a paralog of XRCC4 and XLF, interacts with Ku to promote DNA double-strand break repair. Science 347(6218):185–188. https://doi.org/10.1126/science.1261971
73. Oh S, Harvey A, Zimbric J, Wang Y, Nguyen T, Jackson PJ, Hendrickson EA (2014) DNA ligase III and DNA ligase IV carry out genetically distinct forms of end joining in human somatic cells. DNA Repair (Amst) 21:97–110. https://doi.org/10.1016/j.dnarep.2014.04.015
74. Oh S, Wang Y, Zimbric J, Hendrickson EA (2013) Human LIGIV is synthetically lethal with the loss of Rad54B-dependent recombination and is required for certain chromosome fusion events induced by telomere dysfunction. Nucleic Acids Res 41(3):1734–1749. https://doi.org/10.1093/nar/gks1326
75. Ousterout DG, Gersbach CA (2016) The development of TALE nucleases for biotechnology. Methods Mol Biol 1338:27–42. https://doi.org/10.1007/978-1-4939-2932-0_3
76. Pannunzio NR, Li S, Watanabe G, Lieber MR (2014) Non-homologous end joining often uses microhomology: implications for alternative end joining. DNA Repair (Amst) 17:74–80. https://doi.org/10.1016/j.dnarep.2014.02.006
77. Paull TT, Gellert M (2000) A mechanistic basis for Mre11-directed DNA joining at microhomologies. Proc Natl Acad Sci U S A 97(12):6409–6414. https://doi.org/10.1073/pnas.110144297

78. Rai R, Zheng H, He H, Luo Y, Multani A, Carpenter PB, Chang S (2010) The function of classical and alternative non-homologous end-joining pathways in the fusion of dysfunctional telomeres. EMBO J 29(15):2598–2610. https://doi.org/10.1038/emboj.2010.142
79. Roger L, Jones RE, Heppel NH, Williams GT, Sampson JR, Baird DM (2013) Extensive telomere erosion in the initiation of colorectal adenomas and its association with chromosomal instability. J Natl Cancer Inst 105(16):1202–1211. https://doi.org/10.1093/jnci/djt191
80. Roth DB, Wilson JH (1985) Relative rates of homologous and nonhomologous recombination in transfected DNA. Proc Natl Acad Sci U S A 82(10):3355–3359
81. Roukos V, Misteli T (2014) The biogenesis of chromosome translocations. Nat Cell Biol 16(4):293–300. https://doi.org/10.1038/ncb2941
82. Roukos V, Voss TC, Schmidt CK, Lee S, Wangsa D, Misteli T (2013) Spatial dynamics of chromosome translocations in living cells. Science 341(6146):660–664. https://doi.org/10.1126/science.1237150
83. Rowley JD (1973) A new consistent chromosomal abnormality in chronic myelogenous leukaemia identified by quinacrine fluorescence and Giemsa staining. Nature 243(5405):290–293
84. Sax K (1938) Chromosome aberrations induced by X-rays. Genetics 23(5):494–516
85. Schuster-Bockler B, Lehner B (2012) Chromatin organization is a major influence on regional mutation rates in human cancer cells. Nature 488(7412):504–507. https://doi.org/10.1038/nature11273
86. Shih IM, Zhou W, Goodman SN, Lengauer C, Kinzler KW, Vogelstein B (2001) Evidence that genetic instability occurs at an early stage of colorectal tumorigenesis. Cancer Res 61(3):818–822
87. Sibanda BL, Chirgadze DY, Ascher DB, Blundell TL (2017) DNA-PKcs structure suggests an allosteric mechanism modulating DNA double-strand break repair. Science 355(6324):520–524. https://doi.org/10.1126/science.aak9654
88. Simsek D, Brunet E, Wong SY, Katyal S, Gao Y, McKinnon PJ, Lou J, Zhang L, Li J, Rebar EJ, Gregory PD, Holmes MC, Jasin M (2011) DNA ligase III promotes alternative nonhomologous end-joining during chromosomal translocation formation. PLoS Genet 7(6):e1002080. https://doi.org/10.1371/journal.pgen.1002080
89. Singh DK, Krishna S, Chandra S, Shameem M, Deshmukh AL, Banerjee D (2014) Human DNA ligases: a comprehensive new look for cancer therapy. Med Res Rev 34(3):567–595. https://doi.org/10.1002/med.21298
90. Solinger JA, Kiianitsa K, Heyer WD (2002) Rad54, a Swi2/Snf2-like recombinational repair protein, disassembles Rad51:dsDNA filaments. Mol Cell 10(5):1175–1188
91. Spagnolo L, Rivera-Calzada A, Pearl LH, Llorca O (2006) Three-dimensional structure of the human DNA-PKcs/Ku70/Ku80 complex assembled on DNA and its implications for DNA DSB repair. Mol Cell 22(4):511–519. https://doi.org/10.1016/j.molcel.2006.04.013
92. Spivak G (2015) Nucleotide excision repair in humans. DNA Repair (Amst) 36:13–18. https://doi.org/10.1016/j.dnarep.2015.09.003
93. Stephens PJ, McBride DJ, Lin ML, Varela I, Pleasance ED, Simpson JT, Stebbings LA, Leroy C, Edkins S, Mudie LJ, Greenman CD, Jia M, Latimer C, Teague JW, Lau KW, Burton J, Quail MA, Swerdlow H, Churcher C, Natrajan R, Sieuwerts AM, Martens JW, Silver DP, Langerod A, Russnes HE, Foekens JA, Reis-Filho JS, van't Veer L, Richardson AL, Borresen-Dale AL, Campbell PJ, Futreal PA, Stratton MR (2009) Complex landscapes of somatic rearrangement in human breast cancer genomes. Nature 462(7276):1005–1010. https://doi.org/10.1038/nature08645
94. Stratton MR, Campbell PJ, Futreal PA (2009) The cancer genome. Nature 458(7239):719–724. https://doi.org/10.1038/nature07943
95. Strefford JC, Kadalayil L, Forster J, Rose-Zerilli MJ, Parker A, Lin TT, Heppel N, Norris K, Gardiner A, Davies Z, Gonzalez de Castro D, Else M, Steele AJ, Parker H, Stankovic T, Pepper C, Fegan C, Baird D, Collins A, Catovsky D, Oscier DG (2015) Telomere length predicts progression and overall survival in chronic lymphocytic leukemia: data from the UK LRF CLL4 trial. Leukemia 29(12):2411–2414. https://doi.org/10.1038/leu.2015.217
96. Swanton C, McGranahan N, Starrett GJ, Harris RS (2015) APOBEC enzymes: mutagenic fuel for cancer evolution and heterogeneity. Cancer Discov 5(7):704–712. https://doi.org/10.1158/2159-8290.CD-15-0344
97. Tankimanova M, Capper R, Letsolo BT, Rowson J, Jones RE, Britt-Compton B, Taylor AM, Baird DM (2012) Mre11 modulates the fidelity of fusion between short telomeres in human cells. Nucleic Acids Res 40(6):2518–2526. https://doi.org/10.1093/nar/gkr1117
98. Thacker J (2005) The RAD51 gene family, genetic instability and cancer. Cancer Lett 219(2):125–135. https://doi.org/10.1016/j.canlet.2004.08.018
99. Tsai AG, Lu H, Raghavan SC, Muschen M, Hsieh CL, Lieber MR (2008) Human chromosomal translocations at CpG sites and a theoretical basis for their lineage and stage specificity. Cell 135(6):1130–1142. https://doi.org/10.1016/j.cell.2008.10.035
100. Tubbs A, Nussenzweig A (2017) Endogenous DNA damage as a source of genomic instability in cancer. Cell 168(4):644–656. https://doi.org/10.1016/j.cell.2017.01.002
101. Tuna M, Knuutila S, Mills GB (2009) Uniparental disomy in cancer. Trends Mol Med 15(3):120–128. https://doi.org/10.1016/j.molmed.2009.01.005

102. van Steensel B, Smogorzewska A, de Lange T (1998) TRF2 protects human telomeres from end-to-end fusions. Cell 92(3):401–413
103. Venkitaraman AR (2014) Cancer suppression by the chromosome custodians, BRCA1 and BRCA2. Science 343(6178):1470–1475. https://doi.org/10.1126/science.1252230
104. Wallace SS (2014) Base excision repair: a critical player in many games. DNA Repair (Amst) 19:14–26. https://doi.org/10.1016/j.dnarep.2014.03.030
105. Wang H, Perrault AR, Takeda Y, Qin W, Wang H, Iliakis G (2003) Biochemical evidence for Ku-independent backup pathways of NHEJ. Nucleic Acids Res 31(18):5377–5388
106. Wang H, Rosidi B, Perrault R, Wang M, Zhang L, Windhofer F, Iliakis G (2005) DNA ligase III as a candidate component of backup pathways of nonhomologous end joining. Cancer Res 65(10):4020–4030. https://doi.org/10.1158/0008-5472.CAN-04-3055
107. Waters CA, Strande NT, Pryor JM, Strom CN, Mieczkowski P, Burkhalter MD, Oh S, Qaqish BF, Moore DT, Hendrickson EA, Ramsden DA (2014) The fidelity of the ligation step determines how ends are resolved during nonhomologous end joining. Nat Commun 5:4286. https://doi.org/10.1038/ncomms5286
108. Williams J, Heppel NH, Britt-Compton B, Grimstead JW, Jones RE, Tauro S, Bowen DT, Knapper S, Groves M, Hills RK, Pepper C, Baird DM, Fegan C (2017) Telomere length is an independent prognostic marker in MDS but not in de novo AML. Br J Haematol 178(2):240–249. https://doi.org/10.1111/bjh.14666
109. Williams RS, Moncalian G, Williams JS, Yamada Y, Limbo O, Shin DS, Groocock LM, Cahill D, Hitomi C, Guenther G, Moiani D, Carney JP, Russell P, Tainer JA (2008) Mre11 dimers coordinate DNA end bridging and nuclease processing in double-strand-break repair. Cell 135(1):97–109. https://doi.org/10.1016/j.cell.2008.08.017
110. Woodbine L, Gennery AR, Jeggo PA (2014) The clinical impact of deficiency in DNA non-homologous end-joining. DNA Repair (Amst) 16:84–96. https://doi.org/10.1016/j.dnarep.2014.02.011
111. Wray J, Williamson EA, Singh SB, Wu Y, Cogle CR, Weinstock DM, Zhang Y, Lee SH, Zhou D, Shao L, Hauer-Jensen M, Pathak R, Klimek V, Nickoloff JA, Hromas R (2013) PARP1 is required for chromosomal translocations. Blood 121(21):4359–4365. https://doi.org/10.1182/blood-2012-10-460527
112. Wright WE, Tesmer VM, Huffman KE, Levene SD, Shay JW (1997) Normal human chromosomes have long G-rich telomeric overhangs at one end. Genes Dev 11(21):2801–2809
113. Wu L, Hickson ID (2003) The Bloom's syndrome helicase suppresses crossing over during homologous recombination. Nature 426(6968):870–874. https://doi.org/10.1038/nature02253
114. Xie A, Kwok A, Scully R (2009) Role of mammalian Mre11 in classical and alternative nonhomologous end joining. Nat Struct Mol Biol 16(8):814–818. https://doi.org/10.1038/nsmb.1640
115. Xu Y, Ishizuka T, Kurabayashi K, Komiyama M (2009) Consecutive formation of G-quadruplexes in human telomeric-overhang DNA: a protective capping structure for telomere ends. Angew Chem Int Ed Engl 48(42):7833–7836. https://doi.org/10.1002/anie.200903858
116. Zhang Y, Gostissa M, Hildebrand DG, Becker MS, Boboila C, Chiarle R, Lewis S, Alt FW (2010) The role of mechanistic factors in promoting chromosomal translocations found in lymphoid and other cancers. Adv Immunol 106:93–133. https://doi.org/10.1016/S0065-2776(10)06004-9

3D Genome Organization Influences the Chromosome Translocation Pattern

8

Rachel Patton McCord and Adayabalam Balajee

Abstract

Recent imaging, molecular, and computational modeling studies have greatly enhanced our knowledge of how eukaryotic chromosomes are folded in the nuclear space. This work has begun to reveal how 3D genome structure contributes to various DNA-mediated metabolic activities such as replication, transcription, recombination, and repair. Failure of proper DNA repair can lead to the chromosomal translocations observed in human cancers and other diseases. Questions about the role of 3D genome structure in translocation mechanisms have interested scientists for decades. Recent applications of imaging and Chromosome Conformation Capture approaches have clarified the influence of proximal positioning of chromosomal domains and gene loci on the formation of chromosomal translocations. These approaches have revealed the importance of 3D genome structure not only in translocation partner selection, but also in repair efficiency, likelihood of DNA damage, and the biological implications of translocations. This chapter focuses on our current understanding of the role of 3D genome structure in chromosome translocation formation and its potential implications in disease outcome.

Keywords

3D Genome · Chromosome organization · Translocation mechanisms · Hi-C · Chromatin structure · Radiation exposure

8.1 Introduction

The structural integrity of chromosomes is vital for proper genome function. Disruptions in gene linkage caused by structural chromosomal aberrations can profoundly affect gene expression patterns and result in serious health consequences [1]. Chromosomal translocations involve an illegitimate fusion of two or more broken chromosomes, and are often associated with cancer [2–4]. The biological processes leading to the formation of chromosome translocations *in vivo* are highly complex. At the most basic level, translocations form when DNA double strand breaks (DSBs) at two different genomic loci are joined by DNA repair/recombination activity. Numerous factors can influence which

R. P. McCord (✉)
Department of Biochemistry & Cellular and Molecular Biology, University of Tennessee, Knoxville, TN, USA
e-mail: rmccord@utk.edu

A. Balajee
Radiation Emergency Assistance Center and Training Site, Cytogenetics Biodosimetry Laboratory, Oak Ridge Institute for Science and Education, Oak Ridge Associated Universities, Oak Ridge, TN, USA

Y. Zhang (ed.), *Chromosome Translocation*, Advances in Experimental Medicine and Biology 1044,
https://doi.org/10.1007/978-981-13-0593-1_8

translocations form, including nuclear architecture, DSB frequencies at different loci, and details of recombination and DNA repair pathways. Two models are often proposed for how DSBs become juxtaposed in translocation formation: (I) breakage first and (II) contact first [5]. The breakage first model suggests that broken chromosome ends move through the nucleus in search of a pairing partner for rejoining. The contact first model assumes that the translocation partners are pre-determined by the proximal positioning of chromosome domains and gene loci.

Extensive work has been carried out over the past several decades to investigate the relative roles of genome organization and DSB movement in translocation formation. Fixed and live cell imaging methods have contributed both foundational principles and more recent breakthroughs in our understanding of 3D genome structure and translocations. A recent explosion of data from DNA sequencing-based chromosome contact mapping has revealed additional layers of 3D genome structure. These data have allowed larger-scale studies comparing this genome structure to the landscape of translocations. In this chapter, we will describe how imaging and chromosome conformation capture-based methods have revealed the influence of 3D genome structure on translocations. We will discuss the extensive evidence supporting the idea that genomic locus proximity influences translocation partner selection. But, we will also describe the complexities introduced by variations in DSB frequency, repair efficiency, and mobility. A synthesis of the literature reveals that chromatin organization can affect many important steps of translocation formation, from DSB generation to the downstream biological implications of the translocation. We will discuss remaining important questions in this area, and describe implications for clinical interpretation.

8.2 The Impact of Spatial Proximity on Partner Selection in DNA Break Repair

8.2.1 Evidence from Imaging Methods

An astounding observation in early cancer biology was that different individuals with the same cancer all had the same spontaneously occurring translocation [6]. For decades, scientists have looked for causative factors of such cancer-specific "recurrent translocations". One commonly tested hypothesis is that translocations tend to occur between genomic regions already near each other in 3D nuclear space. To test whether translocated loci are near each other in pre-cancerous cells, researchers have often employed the fluorescence in situ hybridization (FISH) technique. FISH uses fluorescently-labeled gene locus-specific probes to measure the distance between loci across a population of normal cells (Fig. 8.1a). For example, the BCR and ABL loci, despite their location on different chromosomes (chr9 and chr22), exhibit translocations in 90% of chronic myelogenous leukemia patients. Using FISH, it was demonstrated that these loci are already closer to each other than to their own homologs in normal bone marrow cells [7]. Similarly, a translocation event involving *c-myc* on chromosome 8 and *IgH* on chromosome 14 has been observed in 70% of Burkitt's lymphoma patients. This is consistent with their proximal positioning commonly observed in the genome of normal B-lymphocytes. Meanwhile, *c-myc* rearranges less frequently with *Igκ* (chr2) and *Igλ* (chr22) in Burkitt's lymphoma. This corresponds to the observation that these loci are less frequently proximal in the normal B-cell genome [8]. Similar results have been shown for the PML-RARA loci that form a translocation in

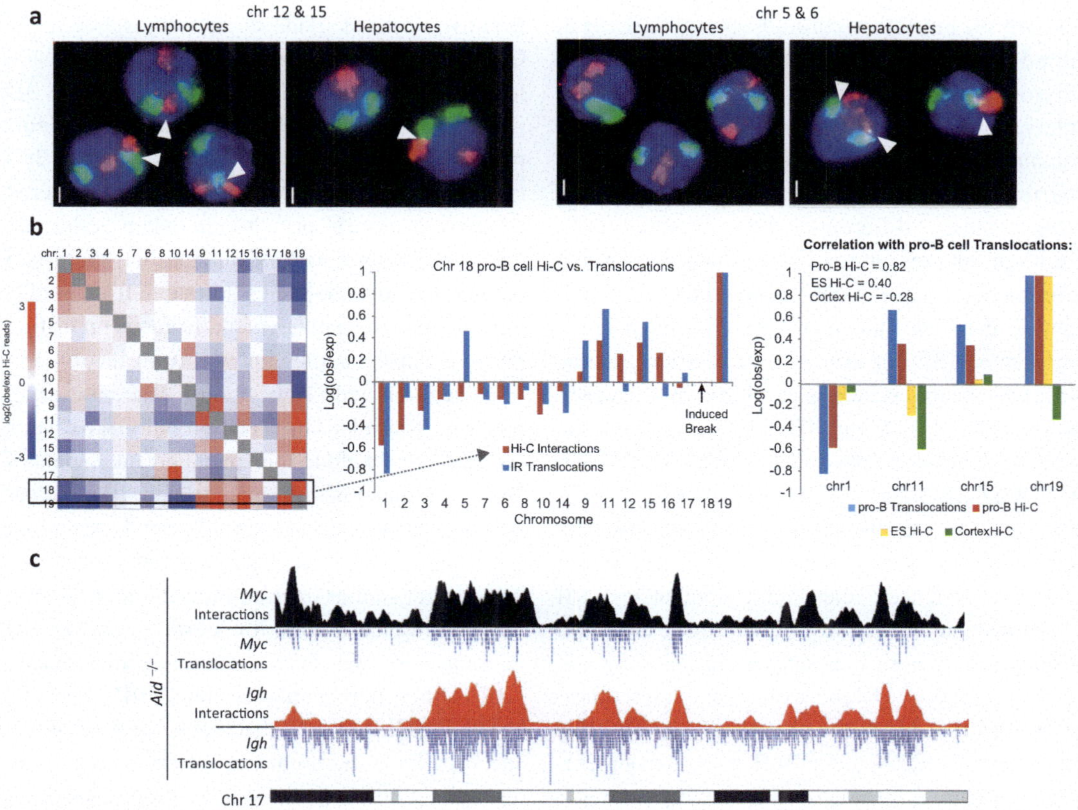

Fig. 8.1 3D Genome Structure Guides Translocations from Local to Global Scales. (**a**) FISH staining of chromosome territories. Frequent translocation partners in a given cell type (chr12 / 15 in lymphocytes or chr5/6 in hepatocytes) are more often spatially proximal in that cell type compared to a tissue in which translocations are not frequently observed. (Reprinted with permission under Open Access policy from Genome Biology: Ref. [10], Copyright Parada et al. 2004) (**b**) Hi-C data shows a correlation genome wide between interchromosomal interaction frequency and translocation frequency from an induced break on chr18. Pro-B cell translocations correlate much less well with Hi-C data from mismatched cell types. (Figure adapted with permission from Cell Press: Ref. [47], Copyright 2012) (**c**) Aid independent translocations in mouse are highly correlated with 4C-derived interaction frequency at a local scale. (Reprinted by permission from Macmillan Publishers Ltd: Ref. [46], Copyright 2012)

acute promyelocytic leukemia [9], and the proximity of IgH-BCL2 in B cell cancers [8]. Many of these studies only examine locus proximity in one cell type, often blood lymphocytes. But, as results of this type continue to accumulate, tissue specific translocations have been shown to correlate with tissue specific genome organization [10]. Certain chromosome pairs have distinctly different preferred neighbors in liver cells as compared to lymphocytes (Fig. 8.1a). Accordingly, hepatomas display a different spectrum of translocations than lymphomas [10].

Imaging studies have further revealed that translocations are influenced not only by the positioning of whole chromosomes, but also by folding structures within chromosomes. For example, the RET and H4 loci, which often form translocations in thyroid cancer, are separated by

30 Mb on the linear chromosome (chr10). But, these loci colocalize much more frequently in thyroid cells (35% of cells by FISH) than mammary epithelial cells (6%) [11]. In general, in mammalian cells, each chromosome exists in a territory, interacting much more within itself than with other different chromosomes [12]. Accordingly, researchers have found that broken chromosomes are much more likely to repair within their own territory than with other chromosomes [13]. For example, 11 rearranged gene loci have been reported in papillary thyroid carcinoma (PTC 1-11), but 90% of cases involve the two intrachromosomal translocations: PTC1 (H4, CCDC6)-RET and PTC3 (NCOA4)-RET [14]. Chromosome territories may also explain why plutonium workers exposed to high LET radiation experience elevated frequencies of chromosome inversions (inter and intra-arm exchanges) in their lymphocytes [3, 15, 16].

In yeast, no enrichment of intra- vs. interchromosomal translocations was observed. But, this apparent discrepancy is in fact even stronger evidence for the influence of 3D structure on translocations. Yeast do not have the strongly evident chromosome territories that higher eukaryotes exhibit, and thus experience much more promiscuous translocations [17]. Further supporting this claim, there is evidence that through evolutionary history the unique Rabl conformation of the yeast genome has contributed to break joining frequency. In the Rabl structure, chromosome arms run parallel to each other away from the centromere, interacting at the telomeres. There is a concordant pattern of genome rearrangements that favors joining regions from opposite chromosome arms that are equidistant from the centromere [18]. So, even in an organism with a fundamentally different 3D chromosome organization, the spatial arrangement influences translocations.

Further evidence for the influence of chromosome spatial proximity on translocations has been obtained recently in multiple myeloma. Spatial distances were determined for four pairs of chromosomes that often exhibit translocations in myeloma. These chromosomes were found to be proximal to each other more often than one negative control pair of chromosomes [19]. Certainly, each new example of this phenomenon lends additional credibility to idea that 3D genome proximity influences translocation formation. But, it is unclear exactly when such a collection of isolated examples can be considered to prove a general principle or when additional examples become redundant. Further, there is a danger that such isolated studies can from confirmation bias. Researchers tend to focus on testing pairs of chromosomes known to be involved in translocations with only one or two contrasting negative controls. Only occasionally have such comparisons extended to spatial arrangements of these chromosomes in other tissues. This selection bias can be circumvented by newly emerging methods for high content/high throughput imaging. Recent studies have employed either spectral karyotyping or multicolor FISH to image the positions of all chromosomes simultaneously [20]. Other new methods, like HIP-Map and HiBA-FISH, measure distances between many specific loci in numerous cells using high throughput FISH [21]. Similar high-throughput imaging approaches are being used to detect translocations in interphase cells [21]. But, for most FISH-based approaches, analysis of a large number of cells is still often time consuming and labor intensive. Further, inter-cellular variations related to cell cycle stage, cell size, cell shape, and imaging artifacts must be carefully monitored for proper interpretation of the results. Therefore, complementary genome-wide techniques with higher resolution are invaluable for analyzing the role of genome organization in translocation formation.

8.2.2 Implications of DNA DSB Movement

The formation of DSBs is pivotal for chromosome translocation. At minimum, two DSBs are required, one on each of the participating chromosomes or gene loci in translocation. Consistent with observations that proximally positioned chromosomes have a high probability of forming translocations, imaging studies in mammalian cells have shown that DSBs remain fairly close to their original positions. These DSBs show little movement in search of repair partners [22]. Live cell imaging using fluorescently tagged repair factors suggested a limited local DSB motion of approximately 1$\mu M^2 h^{-1}$. Further damaged chromatin in human fibroblasts was found to be largely static following exposure to ultrasoft X-rays [23]. In yeast, by contrast, DSB tracking has revealed substantial movements of broken DNA loci, as well as increased mobility of intact loci after DNA damage [24, 25]. This difference may in part be reconciled by considering the significantly smaller size of the yeast—the same amount of DSB motion would naturally cover a much larger fraction of the yeast nucleus as compared to a mammalian nucleus. Another possible explanation is that the yeast genome is only equipped with the homologous recombination (HR) repair pathway. As discussed below, HR may require DSBs to diffuse as they search for homology [26]. Evidence of substantial DSB movement has also been suggested in *Drosophila*, where heterochromatic regions were found to move out of their pre-existing domain for repair [27]. While such dramatic motion of heterochromatic DSBs could be specific to *Drosophila*, other work has shown a 30-40% decondensation of damaged chromatin within mammalian nuclei [28]. Building a more complete picture, Misteli and colleagues have shown the importance of both condensation and decondensation in the DNA repair process [29]. By imaging a photoactivated spot on chromatin after DNA damage, they demonstrated that chromatin undergoes an initial rapid decondensation followed by hypercondensation and finally relaxation in response to DNA damage induction. The extensive condensation process signals for the activation of the DNA damage response pathway. Inhibiting any of these phases: decondensation, hypercondensation, or recovery from condensation, interfered with DNA damage response signaling and DNA repair [29]. However, the chromatin did not undergo any large scale nuclear movements in these experiments despite the occurrence of relaxation and condensation events in succession.

Some discrepancies between observations of DSB motion and DSB stability could stem from the source of DNA damage and the type of repair that is necessary. For example, while only limited DSB movement was observed after X-ray irradiation, α-particle irradiation led to large scale (several micrometer) chromatin motion and clustering of damaged chromatin domains [20, 30]. Breaks induced by a topoisomerase II (TopoII) inhibitor also appear to be more mobile than irradiation-induced DSBs [31]. DSBs caused by this topological strain likely occur at different locations than those induced by irradiation, leading some to hypothesize that the location of breaks within the genome influences their mobility [32]. Further supporting this idea, DSBs that occur in different nuclear compartments (e.g. at the nuclear membrane, near the nucleolus, in the nuclear interior) display different mobility and tend to choose different repair pathways (non-homologous end joining, NHEJ vs. homologous recombination, HR) [33]. DSB mobility may be needed to facilitate a productive homology search for HR. Thus, HR translocation junctions may be more influenced by sequence similarity and less constrained by initial partner proximity [34], while proximity exerts a greater effect on NHEJ translocations.

Observations of increased chromatin movement and changes in chromatin condensation after DNA damage are important to consider. But, this does not negate the importance of 3D genome structure in translocation outcome. On the contrary, the evidence that DSB mobility is influenced by chromatin compaction and nuclear positioning suggests that DSB motion is yet be another step in the DNA damage repair process that is influenced by the 3D genome structure.

8.3 Conceptual Contributions from Chromosome Conformation Capture

Many studies relating 3D genome structure and translocation frequency by imaging methods suffer from two technical limitations: low resolution and relatively small sample sizes. Observing changes in the positions of whole chromosomes can be informative, but it is difficult for imaging approaches to measure the distance between regions separated by only 100 kb. And, despite the recent emergence of recent high throughput imaging techniques, most imaging studies comparing spatial positioning to translocations are confined to a few gene loci. The analysis of a few chromosomes or gene loci of interest introduces some bias in interpreting the significance of such interactions in the context of whole genome organization. In recent years, the suite of techniques based on chromosome conformation capture [35] have provided researchers with a complementary tool for examining 3D genome contact frequency. These approaches can measure the interaction frequency between many loci and even agnostic detection of common translocations genome-wide across a population of cells.

The chromosome conformation capture family of techniques (which includes 3C, 4C, 5C, Hi-C, ChIA-PET, Capture-C, and others) are all based on converting spatially proximal genomic regions into chimeric ligation products. 3D genome interactions can then be identified by probing for specific ligated pairs or high throughput sequencing of all ligation products [36–39]. Over the last decade, these techniques have allowed researchers to characterize genome structures across a range of length scales. The data reveal small-scale looping interactions between enhancers and promoters as well as large-scale segregation of chromosomes into individual territories [40–42]. At intermediate length scales, Hi-C and other methods have demonstrated that active and inactive chromatin is spatially segregated into compartments. Interaction maps have also revealed new building blocks of genome structure- most significantly, topologically associating domains (TADs). TADs, which are hundreds of kilobases in size, have been shown to structurally organize gene expression regulation [43–45]. 3C-based methods report interaction frequencies in a cell population between genomic loci at a resolution that can be as high as 1 kb, and thus provide a powerful tool for comparing genome structure to translocation patterns.

8.3.1 Characterizing Pre-existing 3D Genome Structure Before Translocations to Link Translocation Frequency to 3D Structure

Several 3C-based methods have been used to relate pre-existing 3D genome organization to translocation frequency (Fig. 8.1b, c). In B lymphocytes, 4C (circular chromosome conformation capture) experiments were used to identify all interacting partners of the *IgH* and c-*myc* loci genome-wide. These loci commonly form translocations in cancer. The pre-existing interaction patterns of these loci were then compared with patterns of translocations resulting from an induced break at either the *IgH* or the c-*myc* locus. Unlike previous imaging experiments, which compared distances between *c-myc* and just a few loci, this experiment revealed the full interaction profile of the *c-myc* locus. The translocation patterns were strikingly correlated to the interaction profiles of these loci. Notable exceptions to this correlation occurred where localized DNA breaks were highly frequent due to processes such as V(D)J recombination and the activity of the protein AID (Fig. 8.1c) [46]. To address the concern that the commonly-studied IgH and c-myc loci might not fairly represent the rest of the genome, a related study induced breaks at several random locations in mouse pro-B cells [47] (Fig. 8.1b). The translocation partners of these breaks were then compared with pre-existing genome structure measured by Hi-C (genome wide chromosome conformation capture), which measures interaction frequencies between all loci genome-wide. As in the 4C study, translocation and interaction frequencies were found to be correlated along the chromosome. The correlation was strongest when DSB frequency was equalized across the genome by irradiating the cells and disrupting their normal DNA damage response. Hi-C can provide information about the positioning of whole chromosome territories by measuring their overall frequencies of interaction (Fig. 8.1b). This is similar to the experiments performed by imaging labeled chromosomes, except that Hi-C collects information across millions of nuclei simultaneously, and reports contacts rather than distance. This measure showed that chromosomes 2 and 18, which have entirely different nuclear localization and chromosomal neighbors, also have correspondingly different translocation partner distributions across chromosomes [47]. Since the original study comparing Hi-C to translocation data, Hi-C has been used to characterize 3D genome structure in additional mouse cell types. Interchromosomal interaction frequencies differ dramatically for some chromosomes between pro-B, cortex, and embryonic stem cell types. Reanalysis of the published data shows that pro-B cell translocations follow the specific interactions of pro-B cell chromosomes but are discordant with the interaction patterns of other cell types (Fig. 8.1b). At a very local scale, the correlation between translocation partners and Hi-C interaction frequency was weak, perhaps due to the local dynamic motion of DSBs described earlier [48]. Both Hi-C and 4C comparisons of 3D genome structure to translocations show that both pre-existing locus positioning and DSB frequency have a large impact on translocation frequency.

Hi-C data from non-cancerous cell lines has also been used to evaluate the interaction frequency between regions that are associated with disease-related translocations [49]. This approach, like most FISH studies of proximity between cancer-associated translocation partners, is somewhat indirect. It cannot connect translocations to proximity as directly as the studies in which translocations were identified immediately after DNA damage in the same population of cells used for Hi-C or 4C. But, this approach does alleviate some of the technical issues of small sample size and biased gene locus selection frequently encountered in FISH-based imaging studies. With Hi-C data, the proximity of candidate loci can easily be compared to all surrounding loci as a control. Disease translocated loci are found to be more frequently proximal in a related non-cancerous cell type than would be expected at random. This shows that the effect of proximity is still visible even after accounting for other factors like selection that will have occurred in the development of the cancer cell line.

8.3.2 3C-Based Methods for Agnostic Translocation Identification

The utility of 3C-based methods in studying translocations is not limited to measuring pre-translocation interaction frequencies. Techniques such as 4C and Hi-C can also be used to identify translocations that occur frequently in a population of cells (Fig. 8.2a). Long-range PCR coupled with DNA sequencing is routinely performed to identify translocation breakpoints at the DNA sequence level but this technique involves designing of several overlapping primer sets for DNA amplification. Further, primer design requires the prior knowledge of gene loci involved in the translocation [50]. If regions involved in the translocation are not known, finding DNA sequences that happen to cross a translocation junction can be like looking for a needle in the haystack of the entire genome sequence. Even with high-throughput sequencing methods, and increasingly high throughput long-read sequencing, identifying unknown translocations remains challenging [51]. But, with chromosome conformation capture technologies, evidence for translocations comes not only from the sequence directly at the breakpoint, but also from dramatically increased interaction frequencies between translocated chromosomes for up to several megabases beyond the breakpoint (Fig. 8.2a). The average interaction frequency between unconnected chromosomes is normally at least 10 fold lower than regions separated by 1 Mb on the same chromosome. So, when two chromosomes are newly connected in a translocation, their interaction frequencies show a sharp increase for several megabases on each side of the breakpoint. Identified translocations match with those identified using imaging approaches such as SKY-FISH [52]. With even the very first low resolution Hi-C dataset, it was possible to identify the known BCR-ABL translocation in the K562 cancer cell line. The increased interac-

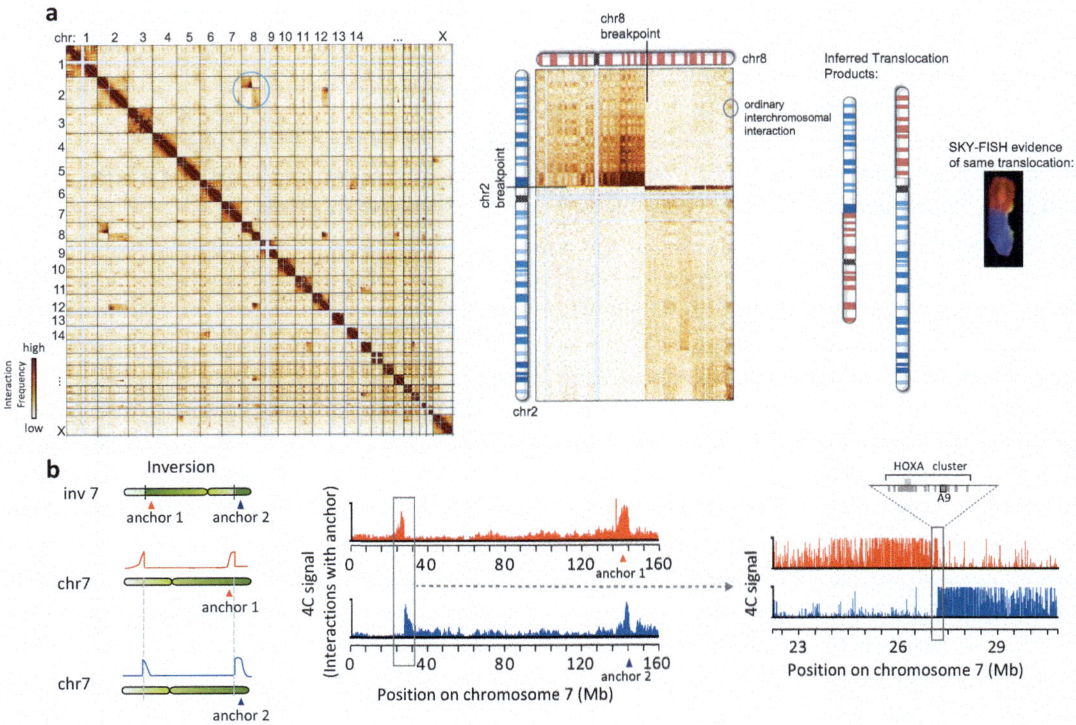

Fig. 8.2 Identifying Translocations with Hi-C and 4C. (**a**) (Left) Hi-C heatmap of genome-wide interaction pattern of breast cancer cell line MDA-MB-231. Unusually strong interactions between chromosomes indicate translocations *(blue circle)*. (Center) Translocation breakpoints are evident as sharp transitions between high and low interaction frequency. Level of true (not translocation) interchromosomal interactions shown for comparison *(grey circle)*. (Right) New hybrid chromosomes resulting from translocation can be inferred from Hi-C interactions. This matches SKY-FISH observation (Ref. [52]). (**b**) 4C measures interactions of anchors on either side of an inversion breakpoint *(red and blue triangles)* with the rest of chromosome 7. Each anchor shows an unexpectedly large peak of interactions with a region on the opposite end of the chromosome *(grey box)* because of an inversion. Zooming in on these unexpected peaks shows that anchor 1 interacts with regions upstream of the break while anchor 2 interacts with regions downstream of the break. The break occurs at the HOXA cluster as shown. (Adapted by permission from Nature Publishing Group: Ref. [59] Copyright 2009)

tions defining the translocation were clearly not present in the GM06990 lymphoblast Hi-C data [49]. In another example, most of the translocations identified in the MCF7 breast cancer cell line by spectral karyotyping and mFISH techniques were detected with Hi-C data [53]. When SNPs were used to generate maternal and paternal chromosome-specific Hi-C maps in diploid cells, translocations specific to the paternal homologs were identified. These translocations could be detected even though they were only present in about 1-5% of the cell population [54]. Hi-C has been used to characterize novel rearrangements in repetitive regions of highly rearranged HeLa cells [55]. These rearrangements were not detected by linear sequencing methods [56]. Beyond cancer cell lines, recent work has shown that Hi-C can characterize translocations and copy number variations in clinical samples from patients with cancer or constitutional translocations [57]. Occasionally, it may be hard to distinguish an infrequent translocation from an increased interaction between separate chromosomes in a cancer cell [58]. But, careful analysis of whether the Hi-C interactions are asymmetric on one side of a potential breakpoint (Fig. 8.2a, center) can help to resolve these ambiguities.

The 4C technique is particularly well suited to screen for translocations of a candidate locus since it measures detailed interactions of one locus genome-wide (Fig. 8.2b). 4C can identify translocations in small sub-populations of cells, complex rearrangements, and even breakpoints in repetitive DNA sequences, which are usually very hard to identify definitively by sequencing [59]. This translocation-detection approach has been shown to work in non-mammalian species as well. A variant of the Hi-C protocol called tethered chromosome conformation capture (TCC) was recently used to characterize translocations, deletions, and inversions in *C. elegans* [60].

One technical drawback for the Hi-C technique is that it effectively averages interactions across a large population of cells and thus would miss very infrequent interactions or translocations. Such problems can be in part mitigated by the development of the single cell Hi-C technique [61].

8.3.3 Interpreting the Biological/Medical Impact of Translocations

When translocations are identified in research or the clinic, the 3D genome structures identified by 3C-based methods can also be used to help interpret the biological impact of these translocations. Some well-studied cancer-causing translocations have obvious mechanisms: they may create new oncogenic fusion proteins by linking parts of two genes together, they may result in genomic region duplications that increase the copy number of oncogenes, or they bring strong regulatory elements close to an oncogene [62]. In many B and T cell malignancies, translocations have been shown to juxtapose proto-oncogenes with new cis regulatory elements, leading to their overexpression. Transposition of genes by translocation can also lead to inactivation of tumor suppressor genes such as P53 and PTEN either by genetic or epigenetic mechanisms. Despite such examples of enhancer hijacking after translocations [63, 64], until recently, no systematic theory was available to predict which regulatory regions would affect genes newly positioned by a rearrangement. Recent work has revealed that some these cancer-causing genomic aberrations are easily interpretable when placed in the context of TADs (Fig. 8.3).

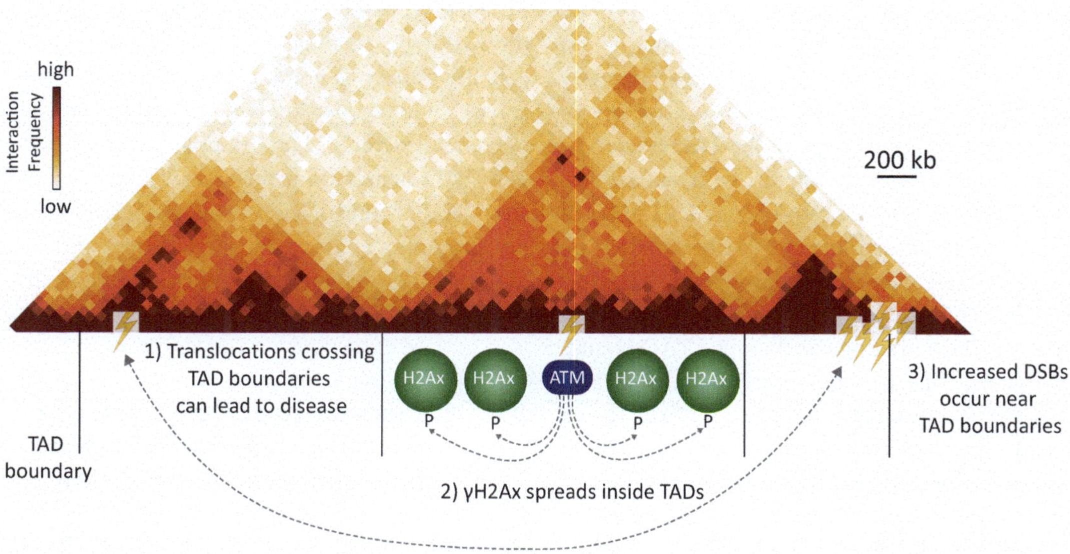

Fig. 8.3 TADs and translocations. The Hi-C heatmap (40 kb resolution) shows triangles of enriched interactions (TADs) separated by boundaries. Several ways that TADs influence translocations are depicted: (*1*) Translocations that cross TAD boundaries can lead to gene misregulation and diseases like cancer. (*2*) Intra-TAD interactions promote γH2Ax spread within TADs but restricts spread beyond TAD boundaries, favoring intra-TAD repair. (*3*) The enrichment of TopoII, topological strain, and actively expressed genes increases the likelihood of DNA DSBs near TAD boundaries

The detailed mechanisms by which TADs form are still being elucidated. But, a model called "loop extrusion" explains many features of TADs and their boundaries. In this model, chromatin loops are extruded by a factor such as cohesin, and extrusion is blocked by CTCF proteins binding in a certain orientation [65, 66]. If the DNA sequence defining a TAD boundary is deleted or the CTCF sites at that boundary are inverted, extrusion can proceed through the boundary, increasing contacts across the former boundary [44, 66]. Similarly, removing the CTCF protein drastically reduces TAD boundary contact insulation [67]. Correspondingly, clear 4C and Hi-C evidence has shown that congenital copy number variations of chromosomal regions can disrupt TAD structure [45, 68]. If the duplicated or deleted region includes a TAD boundary, there can be a loss or gain of enhancer-promoter interactions, resulting in deleterious gene expression.

Following the same principles, chromosomal aberrations in cancer can change the position of TAD boundaries by moving the corresponding CTCF sites. Recurrent microdeletions in T cell acute lymphoblastic leukemia (T-ALL) patients span TAD boundary sites as defined by 5C and ChIA-PET analysis of the local genomic region [69]. Experimentally deleting just this boundary region in normal cells is enough to drive proto-oncogene activation. Without the TAD boundary, nearby super enhancers interact with and activate the prominent T-ALL proto-oncogenes LMO2 or TAL1. Similar mutations of TAD boundaries are found in many cancers that have been characterized by The Cancer Genome Atlas. Even just altering CTCF binding can disrupt a TAD boundary in a way that activates oncogenes and leads to cancer [70]. Similar principles of TAD fusion would occur any time translocations cross TAD boundaries or juxtapose parts of TADs from different chromosomes. Further suggesting the danger of disrupting TADs, it appears that evolution has selected against genomic rearrangements within TADs. Retrotransposons have tended to integrate at TAD boundaries rather than inside TADs [43], and synteny breaks between organisms also tend not to disrupt TADs [71]. Overall, knowing the 3D genome structure of normal cells aids in the interpretation of cancer genome abnormalities.

A deletion commonly found in prostate cancer cells was found to result in a depletion of Hi-C interactions reminiscent of a new TAD boundary [72]. However, given the described model of TAD formation, deletions would usually lead to the loss of insulation rather than the formation of a new boundary. This discrepancy points to the need for continued work in this area and emphasizes that Hi-C data of highly rearranged cancer cells must be interpreted with caution. First, TAD detection can vary depending on algorithm, data quality, and resolution [73]. Indeed, the study reporting a new TAD boundary after a deletion reported an average TAD size of 7.8 Mb, even though this is typically the size associated with active/inactive compartments rather than TADs. In contrast, the study reporting the loss of TAD boundaries after a deletion described an average TAD of about 300 kb. So, such studies may be observing phenomena at different scales. Further, if cancer genome interaction data is mapped to the wildtype genome sequence, features that actually reflect genomic rearrangements themselves could be mistakenly interpreted as new 3D genome structures. For example, an apparently new TAD boundary at the site of a CNV could simply reflect that deleted regions inherently cannot form interactions with surrounding sites and amplified regions inherently increase linear distances between genomic regions.

8.4 Importance of Heterogeneity in Interpreting the Influence of 3D Genome Organization

If translocated loci are not found to be physically proximal on average in normal cells, it is tempting to conclude that DSBs must have moved a long distance to find their partners. However, it is also possible that these translocations occurred in the few cells out of the population in which the genomic loci were proximal. There are several potential explanations for the observation of translocations between loci that are only rarely proximal. First, the translocation may activate an oncogene that increases proliferation and confers a selective advantage to a given cell, leading eventually to a clonal population of cells. Or, DSBs at these loci may be much more frequent than at other places in the genome. Then, even on the rare occasions that these loci interact, they form translocations, while other much more frequently interacting loci are never broken and thus rarely form translocations. Further, proximity measurements during G1 may not reflect the state of all cells in the population. 3D genome structure exhibits large-scale stability within individual cells during G1 [74], but subtle changes in structure occur as the cell progresses through S and G2 phases [75]. Additionally, with each round of cell division, chromosome territory positioning and lamin association can change dramatically [74, 76]. Dynamic motion of chromatin also occurs within individual cells at smaller scales [77]. These factors lead to intercellular variability in chromatin domain organization. Indeed, single cell Hi-C experiments as well as single cell FISH imaging experiments have shown that individual cells of the same type and stage can exhibit large variations around the average trends of whole chromosome positioning [61, 78]. So, it is plausible that average distances between loci are not the best measure of whether those loci interact in any cells in the population. The heterogeneity of 3D genome structure may also contribute to heterogeneous DNA damage responses of cells within a population, leading to the death of some cells and translocations in others. It is generally accepted that the G2/M phase is the most radiation sensitive phase, and it would be interesting to determine whether this radiosensitivity is due to 3D genome organizational differences between G1 and G2/M cells.

8.5 Widespread DSBs Exacerbate 3D Genome Proximity Effects on Translocations: Implications for Radiation-Induced Translocations

As noted above, genome-wide studies have shown that translocation events are more closely correlated with 3D genome organization when cells in culture are exposed to ionizing radiation [47]. When DSBs are very rare, then endogenous sites of DNA breaks, such as V(D)J recombination loci in lymphocytes, are much more likely to participate in translocations, even if these sites are not the most frequently proximal in the cell population [79, 80]. But, if DSBs are more widely available, as is true after radiation, then spatial proximity predicts translocation partners well. The assumption of this principle has even led researchers to use chromosomal aberrations after LET radiation as a functional measure of chromosome spatial clustering [81]. And, indeed, the frequency of translocations from this study was later explicitly shown to correlate strongly with the pre-existing average intermingling volume of chromosome pairs in human lymphocytes [82]. The observation that IR-induced DSBs are less mobile than breaks from other causes further increases the effect of pre-existing genome structure on translocations that arise from radiation exposure [31]. Future work with 3C-based methods will be needed to clarify the effects of ionizing radiation-induced DNA damage on 3D genome structure.

The increased role of spatial proximity in governing translocations after radiation has also been suggested in clinical settings. Thyroid papillary carcinoma patients with a history of radiation exposure have a higher incidence of a RET/PTC translocation than sporadic cases. These loci are naturally proximal in normal thyroid cells, and irradiation also increases the frequency of RET/PTC translocation cultured cells in a dose dependent manner [83]. Similarly, as noted earlier, plutonium workers who were exposed to ionizing radiation show translocations exhibiting signatures of locus proximity within chromosome territories [3, 16]. These results suggest that patients who develop cancer after radiation incidents could have fundamentally different oncogenic translocations than patients with cancer of the same tissue that developed for other reasons. Now that Hi-C 3D genome structure profiles are becoming available across many human tissues and cell types [84], translocations discovered in cancer patients could be classified according to the likely initial spatial proximity of the exchanged chromosomal regions. Such analysis could help assess whether widespread DNA damage (e.g. irradiation) vs. localized recombination-mediated chromatin breaks likely played a larger role in the development of an individual cancer. By determining whether TADs are disrupted, pathologists could also use existing Hi-C data to evaluate the likelihood that a clinically observed translocation is an oncogenic driver of cancer vs. a side effect of genome instability.

8.6 3D Genome Contributions to DSB Incidence and Control of DNA Repair

Given that DSB frequency strongly influences the spectrum of translocations that can occur, it is important to note that 3D genome structure also influences the likelihood of DSB formation by DNA damaging agents. Biophysical alterations of chromatin state by salt-induced nuclear swelling or compaction has shown that condensed chromatin is less likely to be damaged by ionizing radiation than decondensed chromatin [85]. Similarly, in cancer treatment, histone deacetylase inhibitors, which tend to decondense chromatin, can render tumor cells more susceptible to damage by radiation treatment [86].

The physical stress caused by 3D chromatin packing can also increase the incidence of spontaneous DSBs at certain loci. If TopoII is inhibited, the topological stress of chromatin structure leads to increased DSBs [87, 88]. Related to this phenomenon, DSBs are likely to be more common at TAD boundaries (Fig. 8.3). A combination of factors likely lead to this enrichment of DSBs at TAD boundaries: 1) Gene promoters, which are more likely to experience breaks due to the activity of transcription machinery, are enriched at TAD boundaries [80, 89] and 2) Higher topological stress, demonstrated by the enrichment of TopoII, is likely to be found at TAD boundaries [90].

Recent work has also implicated 3D chromatin structure as a modulator of DNA repair. DSB repair is generally inhibited by condensed chromatin, which appears to both inhibit the DNA damage response pathway and decrease repair efficiency [91, 92]. With our new understanding of chromatin organization in TADs, the relation between this 3D genome folding and DNA repair has become even stronger. The ataxia telangiectasia mutated protein (ATM) aids DNA repair by locally decondensing chromatin through the phosphorylation of histone H2Ax and KAP-1 [92, 93]. It is now known that H2Ax phosphorylation occurs throughout a single TAD containing a DSB, but is restricted by TAD boundaries [93]. Specifically, chromatin immunoprecipitation experiments suggest that cohesin is a substrate of ATM, and that cohesin binding restricts the spread of H2Ax marking after DNA damage [94]. As noted earlier, cohesin is a protein complex implicated in extruding DNA loops as a part of TAD formation [65, 95]. So, it is plausible that loop extrusion would bring ATM at a DSB into contact with other DNA throughout the same TAD, spreading DNA damage marks. But, as cohesin loop extrusion is blocked by CTCF, the spread of DNA damage marks would be restricted by TAD boundary, largely confining repair within TADs (Fig. 8.3). Overall, we see that the 3D genome structure can inform not only likely translocation partners, but also the likelihood and fidelity of DNA damage and repair.

8.7 Key Persistent Questions

8.7.1 How Can We Reconcile 3D DSB Movement with 3D Genome Stability and Influence on Translocations?

Throughout this chapter, we have seen evidence of both DSB movement and spatial stability, and the degree to which each of these contributes to translocations. Some experimental factors may complicate observations of DSB movement and stability and lead to apparent conflicts. Although live cell imaging studies are valuable for monitoring the DSB movement analysis, such studies often rely on cancer cells for ease of transfection and exogenously expressed proteins. Cancer cells may have inherently different chromosome organization, epigenetics, and chromatin mobility, and exogenously expressed proteins may have different effects than their endogenous counterparts. So, other methods are needed to directly evaluate the relative roles of DSB movement and stability in translocation formation. These more direct methods would complement existing approaches evaluating DSB movement and the large number of studies correlating 3D genome proximity with translocations.

The best approaches to answer this question allow researchers to watch the same chromosomal locations from their initial spatial position through the process of DNA damage and repair. This shows more decisively whether broken chromosome ends do or do not move large distances to form translocations. A major advance in this direction was made by the development of a cellular system to induce and visualize chromosome damage at specific sites in living mammalian cells. This approach couples an I-SceI endonuclease cut site to a LacO array that can be fluorescently imaged with labeled LacI in live cells. This I-SceI LacO/LacI was integrated into the genome on one chromosome and a similar I-SceI TetO/TetR was integrated on another chromosome [96]. Thus, the motion and proximity of the two cut sites could be visualized with high throughput imaging before and after DSBs in living cells. In this work, colocalization of cut sites increased from 2% to 12% over the 36 hours after endonuclease activity, indicating DSB pairing. Occasionally, labeled DSBs traveled over large (4 um) distances during a 24 h imaging timecourse. But, in the majority of cases, DSBs that formed translocations were already within 2 μm of each other before DNA damage [97]. Future work can extend this idea by examining the hypothesis that even more translocations would occur with proximal regions in an individual cell if more DNA breaks were present. It will also be interesting to observe how these results vary when labeling pairs of chromosomes that are normally further from or closer to each other. A remaining challenge is the inability to visualize loci in live cells without bulky insertions of repetitive arrays that may already change the chromatin state of the locus before the experiment proceeds.

Hi-C generally reports on populations of cells, and thus cannot show movement of individual breaks. But, Hi-C data on cell populations after induction of widespread DNA damage could help to evaluate the effective extent of DSB movement within the 3D genome structure. It is possible that chromatin contact frequency and topology change very little on average even when average motion or condensation state changes after DNA damage. Alternatively, a blurring of the structure observed by Hi-C could indicate a substantial impact of DSB movement on 3D genome interactions.

8.7.2 How Much Do DSBs and Translocations Affect Existing 3D Genome Structure?

Over the past 15 years, researchers have found several examples in which typical chromosome territory positioning in the nucleus is apparently changed by translocations [98, 99]. For example, translocations which attach chr18 (normally peripheral) to chr19 (normally internal) can cause chr18 to adopt the internal positioning of chr19 in the resulting cancer cells [99]. This suggests that regions of the genome can have their normal spatial positions reprogrammed by translocations. Alternately, following the idea that translocations usually occur when chromosomes are proximal in an individual cell, it may also be that chr18 was already localized internally in some cells before the translocation. Perhaps some premalignant signal (changes in histone modifications, aberrant gene activation etc.) brought chr18 near chr19 in some cells, increasing the likelihood of the 18-19 translocation.

Hi-C evidence has demonstrated that cancer cells with chromosomal aberrations exhibit different spatial compartmentalization and looping contacts than normal cells of the same lineage [53, 72]. These changes could be a direct result of the translocations themselves, but may also reflect broader cancer-related gene expression changes. As noted earlier, it appears that translocations may primarily move or duplicate existing TAD boundaries rather than creating fundamentally new TAD boundaries at sequences that were not boundaries to begin with [69]. But, more work tracing the 3D genome structure of the same cell before and after translocations is needed to thoroughly address how translocations may change the broader genome structure.

8.7.3 Do Pre-cancerous 3D Genome Organization Changes Affect Which Translocations Occur and Thus Affect Cancer Phenotype or Outcome?

If 3D genome structure influences translocations, it is plausible that oncogenic translocations in an individual cancer patient would be affected by specific details of 3D genome organization of that individual and tissue type. Along these lines, there may be externally or internally induced changes in the 3D arrangement of chromosomes before the development of cancer that would influence which translocations are most likely to occur. Since different cancer genotypes have different aggressiveness and susceptibility to treatment [100–102], translocation frequency could be a mechanism by which the pre-cancerous genome structure could influence cancer prognosis. Indeed, some studies have shown spatial reorganization of specific loci in early tumorigenesis [103]. Even more intriguing, a study of chromosome positioning, not only before transformation, but during cancer treatment and relapse, has suggested the existence of "transitional nuclei." In these cells of the population, chromosomes that later form translocations in leukemia begin to approach each other, perhaps setting up the later initiation of leukemia [104]. Hi-C studies confirm that genes can change their spatial compartmentalization in cancer vs. non cancer cells [53] and as a result of oncogene overexpression [58], but it is unclear whether these might contribute to or be an effect of translocations.

It is also possible that changes in 3D genome structure after irradiation reposition certain loci in a manner that favors certain translocations. While many DSBs exhibit positional stability after DNA damage, commonly translocated loci like BCR and ABL have been observed to shift toward the center of the nucleus after gamma irradiation [105]. This would bring BCR and ABL closer together after gamma radiation exposure, and would make the CML-causing BCR/ABL translocation more likely. In this example, a pre-cancerous reorganization of chromatin structure in response to a stimulus (gamma radiation) would predispose a cell to cancer development. Some studies have suggested other genome-wide changes in 3D genome structure in response to radiation, such as a decrease in superhelical density of chromatin in radiosensitive cell lines [106]. Future work is needed to better characterize 3D genome organization changes that precede translocations or cancer development. Can such pre-cancerous changes in genome structure make oncogenic translocations more likely? Future work must distinguish whether cancer-specific genome folding changes are a downstream side-effect of initial translocations and mutations or an upstream driving force of cancerous translocations. Studies relating cancer genomics to environmental and underlying genetic factors, such as those encouraged by The Cancer Genome Atlas, may contribute to this future work as well [107]. Surveys of which cancer translocations correlate to which environmental exposures or underlying genotypes could be linked to the impact of those environmental conditions on the 3D genome.

8.8 Conclusion

The past decades of research have shown that the effect of 3D genome structure on chromosomal translocations is more complex than a simple choice between the "break first" or "contact first" model. Certainly, both imaging and molecular methods have shown that pre-existing proximity of loci in the genome influences their likelihood of forming a translocation. But, layered on top of this truth are important complexities. Variations in DSB frequency, DSB motion, and repair signaling pathways and efficiency, and heterogeneity in 3D structure across cell populations can all influence translocation patterns. Importantly, 3D genome structure influences many of these factors as well (Fig. 8.4). As we have described, TAD boundaries influence DSB frequency and constrain DNA damage signal spreading. Chromatin compaction and locations of genomic regions in the nuclear space influences DSB frequency, DNA repair efficiency and pathway choice, and DSB mobility. After translocations occur, the disruption or preservation of genomic structures like TADs influence the biological implications of translocations and their likelihood of driving diseases like cancer. So, while the translocation landscape is a product of many features, understanding the pre-existing 3D genome structure will shed light on aspects of all of them. The complexity of translocation mechanisms also makes it valuable to identify particular practical circumstances in which 3D genome structure is likely to play a large role. For example, in clinical cases where patients have been exposed to radiation, DSBs will be widespread and break mobility likely low. Therefore, the influence of pre-existing 3D genome structure on translocation outcomes is likely to be stronger in these patients.

Knowledge of different aspects chromatin remodeling, histone modifications, transcriptional programming and higher order chromatin organization is required to understand the biogenesis of chromosome translocations. With increasingly sensitive imaging and molecular

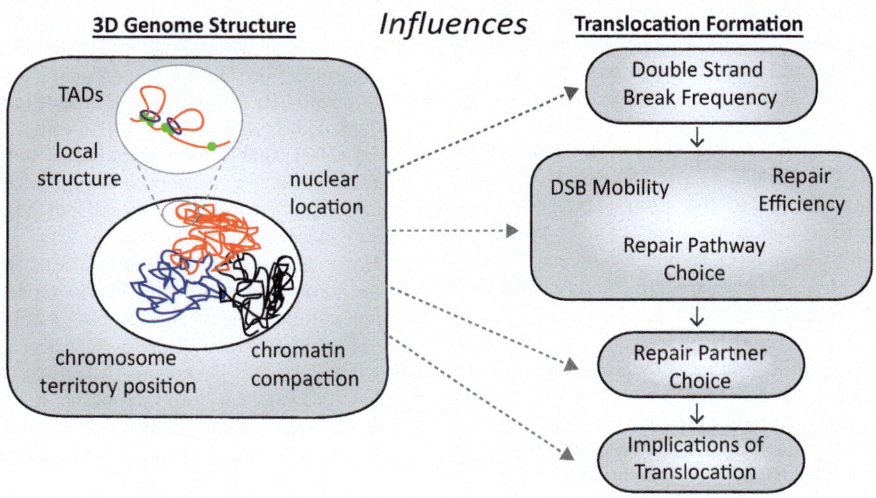

Fig. 8.4 Summary of 3D genome influence on translocations. All stages of translocation formation are influenced by various levels of 3D genome structure. TADs, chromosome territory positioning, nuclear compartments, and chromatin compaction can influence the initiation of DSBs, choice of repair partners, and downstream biological effects of translocations

methods for characterizing 3D genome structures and monitoring translocations, the field is poised for important advances in the next decade. Future work will transform a descriptive catalog of translocations and contributing causes to a predictive understanding of what conditions lead to different translocation profiles and the downstream implications for human health. Gaining a comprehensive knowledge of the driving forces for translocation formation may lead to effective therapeutic strategies for many of the human diseases triggered by translocation events that range from infertility to cancer.

Acknowledgements We thank Rosela Golloshi and Louis Becker for helpful advice and careful reading of the manuscript. The collaboration between A. Balajee and R.P. McCord was funded by an Oak Ridge Associated Universities Directed Research and Development Grant.

References

1. Kloosterman WP, Hochstenbach R (2014) Deciphering the pathogenic consequences of chromosomal aberrations in human genetic disease. Mol Cytogenet 7(1):100. https://doi.org/10.1186/s13039-014-0100-9
2. Aplan PD (2006) Causes of oncogenic chromosomal translocation. Trends Genet 22(1):46–55. https://doi.org/10.1016/j.tig.2005.10.002
3. Hande MP, Azizova TV, Burak LE, Khokhryakov VF, Geard CR, Brenner DJ (2005) Complex chromosome aberrations persist in individuals many years after occupational exposure to densely ionizing radiation: an mFISH study. Genes Chromosomes Cancer 44(1):1–9. https://doi.org/10.1002/gcc.20217
4. Yong LC, Sigurdson AJ, Ward EM, Waters MA, Whelan EA, Petersen MR, Bhatti P, Ramsey MJ, Ron E, Tucker JD (2009) Increased frequency of chromosome translocations in airline pilots with long-term flying experience. Occup Environ Med 66(1):56–62. https://doi.org/10.1136/oem.2008.038901
5. Misteli T, Soutoglou E (2009) The emerging role of nuclear architecture in DNA repair and genome maintenance. Nat Rev Mol Cell Biol 10(4):243–254. https://doi.org/10.1038/nrm2651
6. Nowell PC, Hungerford DA (1960) Chromosome studies on normal and leukemic human leukocytes. J Natl Cancer Inst 25:85–109
7. Lukasova E, Kozubek S, Kozubek M, Kjeronska J, Ryznar L, Horakova J, Krahulcova E, Horneck G (1997) Localisation and distance between ABL and BCR genes in interphase nuclei of bone marrow cells of control donors and patients with chronic myeloid leukaemia. Hum Genet 100(5-6):525–535
8. Roix JJ, McQueen PG, Munson PJ, Parada LA, Misteli T (2003) Spatial proximity of translocation-prone gene loci in human lymphomas. Nat Genet 34(3):287–291. https://doi.org/10.1038/ng1177
9. Neves H, Ramos C, da Silva MG, Parreira A, Parreira L (1999) The nuclear topography of ABL, BCR, PML, and RARalpha genes: evidence for gene proximity in specific phases of the cell cycle and stages of hematopoietic differentiation. Blood 93(4):1197–1207
10. Parada LA, McQueen PG, Misteli T (2004) Tissue-specific spatial organization of genomes. Genome Biol 5(7):R44. https://doi.org/10.1186/gb-2004-5-7-r44
11. Nikiforova MN, Stringer JR, Blough R, Medvedovic M, Fagin JA, Nikiforov YE (2000) Proximity of chromosomal loci that participate in radiation-induced rearrangements in human cells. Science 290(5489):138–141
12. Cremer T, Cremer M, Dietzel S, Muller S, Solovei I, Fakan S (2006) Chromosome territories--a functional nuclear landscape. Curr Opin Cell Biol 18(3):307–316. https://doi.org/10.1016/j.ceb.2006.04.007
13. Misteli T (2007) Beyond the sequence: cellular organization of genome function. Cell 128(4):787–800. https://doi.org/10.1016/j.cell.2007.01.028
14. Menicali E, Moretti S, Voce P, Romagnoli S, Avenia N, Puxeddu E (2012) Intracellular signal transduction and modification of the tumor microenvironment induced by RET/PTCs in papillary thyroid carcinoma. Front Endocrinol (Lausanne) 3:67. https://doi.org/10.3389/fendo.2012.00067
15. Hande MP, Azizova TV, Geard CR, Burak LE, Mitchell CR, Khokhryakov VF, Vasilenko EK, Brenner DJ (2003) Past exposure to densely ionizing radiation leaves a unique permanent signature in the genome. Am J Hum Genet 72(5):1162–1170. https://doi.org/10.1086/375041
16. Mitchell CR, Azizova TV, Hande MP, Burak LE, Tsakok JM, Khokhryakov VF, Geard CR, Brenner DJ (2004) Stable intrachromosomal biomarkers of past exposure to densely ionizing radiation in several chromosomes of exposed individuals. Radiat Res 162(3):257–263
17. Haber JE, Leung WY (1996) Lack of chromosome territoriality in yeast: promiscuous rejoining of broken chromosome ends. Proc Natl Acad Sci U S A 93(24):13949–13954
18. Khrameeva EE, Fudenberg G, Gelfand MS, Mirny LA (2016) History of chromosome rearrangements reflects the spatial organization of yeast chromosomes. J Bioinf Comput Biol 14(2):1641002. https://doi.org/10.1142/s021972001641002x
19. Sathitruangsak C, Righolt CH, Klewes L, Tung Chang D, Kotb R, Mai S (2017) Distinct and shared three-dimensional chromosome organization patterns in lymphocytes, monoclonal gammopathy of undetermined significance and multiple

myeloma. Intl J Cancer 140(2):400–410. https://doi.org/10.1002/ijc.30461
20. Anderson RM, Stevens DL, Goodhead DT (2002) M-FISH analysis shows that complex chromosome aberrations induced by alpha -particle tracks are cumulative products of localized rearrangements. Proc Natl Acad Sci U S A 99(19):12167–12172. https://doi.org/10.1073/pnas.182426799
21. Burman B, Misteli T, Pegoraro G (2015) Quantitative detection of rare interphase chromosome breaks and translocations by high-throughput imaging. Genome Biol 16(1):146. https://doi.org/10.1186/s13059-015-0718-x
22. Soutoglou E, Dorn JF, Sengupta K, Jasin M, Nussenzweig A, Ried T, Danuser G, Misteli T (2007) Positional stability of single double-strand breaks in mammalian cells. Nat Cell Biol 9(6):675–682. https://doi.org/10.1038/ncb1591
23. Nelms BE, Maser RS, MacKay JF, Lagally MG, Petrini JH (1998) In situ visualization of DNA double-strand break repair in human fibroblasts. Science 280(5363):590–592
24. Dion V, Kalck V, Horigome C, Towbin BD, Gasser SM (2012) Increased mobility of double-strand breaks requires Mec1, Rad9 and the homologous recombination machinery. Nat Cell Biol 14(5):502–509. https://doi.org/10.1038/ncb2465
25. Seeber A, Dion V, Gasser SM (2013) Checkpoint kinases and the INO80 nucleosome remodeling complex enhance global chromatin mobility in response to DNA damage. Genes Dev 27(18):1999–2008. https://doi.org/10.1101/gad.222992.113
26. Jasin M, Rothstein R (2013) Repair of strand breaks by homologous recombination. Cold Spring Harb Perspect Biol 5(11):a012740. https://doi.org/10.1101/cshperspect.a012740
27. Chiolo I, Minoda A, Colmenares SU, Polyzos A, Costes SV, Karpen GH (2011) Double-strand breaks in heterochromatin move outside of a dynamic HP1a domain to complete recombinational repair. Cell 144(5):732–744. https://doi.org/10.1016/j.cell.2011.02.012
28. Kruhlak MJ, Celeste A, Dellaire G, Fernandez-Capetillo O, Muller WG, McNally JG, Bazett-Jones DP, Nussenzweig A (2006) Changes in chromatin structure and mobility in living cells at sites of DNA double-strand breaks. J Cell Biol 172(6):823–834. https://doi.org/10.1083/jcb.200510015
29. Burgess RC, Burman B, Kruhlak MJ, Misteli T (2014) Activation of DNA damage response signaling by condensed chromatin. Cell Rep 9(5):1703–1717. https://doi.org/10.1016/j.celrep.2014.10.060
30. Aten JA, Stap J, Krawczyk PM, van Oven CH, Hoebe RA, Essers J, Kanaar R (2004) Dynamics of DNA double-strand breaks revealed by clustering of damaged chromosome domains. Science 303(5654):92–95. https://doi.org/10.1126/science.1088845
31. Krawczyk PM, Borovski T, Stap J, Cijsouw T, ten Cate R, Medema JP, Kanaar R, Franken NA, Aten JA (2012) Chromatin mobility is increased at sites of DNA double-strand breaks. J Cell Sci 125(Pt 9):2127–2133. https://doi.org/10.1242/jcs.089847
32. Marnef A, Legube G (2017) Organizing DNA repair in the nucleus: DSBs hit the road. Curr Opin Cell Biol 46:1–8. https://doi.org/10.1016/j.ceb.2016.12.003
33. Lemaitre C, Grabarz A, Tsouroula K, Andronov L, Furst A, Pankotai T, Heyer V, Rogier M, Attwood KM, Kessler P, Dellaire G, Klaholz B, Reina-San-Martin B, Soutoglou E (2014) Nuclear position dictates DNA repair pathway choice. Genes Dev 28(22):2450–2463. https://doi.org/10.1101/gad.248369.114
34. White RR, Sung P, Vestal CG, Benedetto G, Cornelio N, Richardson C (2013) Double-strand break repair by interchromosomal recombination: an in vivo repair mechanism utilized by multiple somatic tissues in mammals. PLoS One 8(12):e84379. https://doi.org/10.1371/journal.pone.0084379
35. Dekker J, Rippe K, Dekker M, Kleckner N (2002) Capturing chromosome conformation. Science 295(5558):1306–1311. https://doi.org/10.1126/science.1067799
36. Dekker J (2016) Mapping the 3D genome: Aiming for consilience. Nat Rev Mol Cell Biol 17(12):741–742. https://doi.org/10.1038/nrm.2016.151
37. Dekker J, Marti-Renom MA, Mirny LA (2013) Exploring the three-dimensional organization of genomes: interpreting chromatin interaction data. Nat Rev Genet 14(6):390–403. https://doi.org/10.1038/nrg3454
38. Dekker J, Mirny L (2016) The 3D genome as moderator of chromosomal communication. Cell 164(6):1110–1121. https://doi.org/10.1016/j.cell.2016.02.007
39. Fraser J, Williamson I, Bickmore WA, Dostie J (2015) An overview of genome organization and how we got there: from FISH to Hi-C. Microbiol Mol Biol Rev 79(3):347–372. https://doi.org/10.1128/MMBR.00006-15
40. Belton JM, McCord RP, Gibcus JH, Naumova N, Zhan Y, Dekker J (2012) Hi-C: a comprehensive technique to capture the conformation of genomes. Methods 58(3):268–276. https://doi.org/10.1016/j.ymeth.2012.05.001
41. Mifsud B, Tavares-Cadete F, Young AN, Sugar R, Schoenfelder S, Ferreira L, Wingett SW, Andrews S, Grey W, Ewels PA, Herman B, Happe S, Higgs A, LeProust E, Follows GA, Fraser P, Luscombe NM, Osborne CS (2015) Mapping long-range promoter contacts in human cells with high-resolution capture Hi-C. Nat Genet 47(6):598–606. https://doi.org/10.1038/ng.3286
42. Tolhuis B, Palstra RJ, Splinter E, Grosveld F, de Laat W (2002) Looping and interaction between hypersensitive sites in the active beta-globin locus. Mol Cell 10(6):1453–1465
43. Dixon JR, Selvaraj S, Yue F, Kim A, Li Y, Shen Y, Hu M, Liu JS, Ren B (2012) Topological domains in mammalian genomes identified by analysis of

chromatin interactions. Nature 485(7398):376–380. https://doi.org/10.1038/nature11082
44. Nora EP, Lajoie BR, Schulz EG, Giorgetti L, Okamoto I, Servant N, Piolot T, van Berkum NL, Meisig J, Sedat J, Gribnau J, Barillot E, Bluthgen N, Dekker J, Heard E (2012) Spatial partitioning of the regulatory landscape of the X-inactivation centre. Nature 485(7398):381–385. https://doi.org/10.1038/nature11049
45. Lupianez DG, Kraft K, Heinrich V, Krawitz P, Brancati F, Klopocki E, Horn D, Kayserili H, Opitz JM, Laxova R, Santos-Simarro F, Gilbert-Dussardier B, Wittler L, Borschiwer M, Haas SA, Osterwalder M, Franke M, Timmermann B, Hecht J, Spielmann M, Visel A, Mundlos S (2015) Disruptions of topological chromatin domains cause pathogenic rewiring of gene-enhancer interactions. Cell 161(5):1012–1025. https://doi.org/10.1016/j.cell.2015.04.004
46. Hakim O, Resch W, Yamane A, Klein I, Kieffer-Kwon K-R, Jankovic M, Oliveira T, Bothmer A, Voss TC, Ansarah-Sobrinho C, Mathe E, Liang G, Cobell J, Nakahashi H, Robbiani DF, Nussenzweig A, Hager GL, Nussenzweig MC, Casellas R (2012) DNA damage defines sites of recurrent chromosomal translocations in B lymphocytes. Nature 484(7392):69–74. https://doi.org/10.1038/nature10909
47. Zhang Y, McCord RP, Ho YJ, Lajoie BR, Hildebrand DG, Simon AC, Becker MS, Alt FW, Dekker J (2012) Spatial organization of the mouse genome and its role in recurrent chromosomal translocations. Cell 148(5):908–921. https://doi.org/10.1016/j.cell.2012.02.002
48. Seeber A, Gasser SM (2016) Chromatin organization and dynamics in double-strand break repair. Curr Opin Genet Dev 43:9–16. https://doi.org/10.1016/j.gde.2016.10.005
49. Engreitz JM, Agarwala V, Mirny LA (2012) Three-dimensional genome architecture influences partner selection for chromosomal translocations in human disease. PLoS One 7(9):e44196. https://doi.org/10.1371/journal.pone.0044196
50. Parker EP, Siebert R, Oo TH, Schneider D, Hayette S, Wang C (2013) Sequencing of t(2;7) translocations reveals a consistent breakpoint linking CDK6 to the IGK locus in indolent B-cell neoplasia. J Mol Diagn 15(1):101–109. https://doi.org/10.1016/j.jmoldx.2012.07.006
51. Rhoads A, Au KF (2015) PacBio Sequencing and Its Applications. Genomics Proteom Bioinform 13(5):278–289. https://doi.org/10.1016/j.gpb.2015.08.002
52. Singh B, Owens KM, Bajpai P, Desouki MM, Srinivasasainagendra V, Tiwari HK, Singh KK (2015) Mitochondrial DNA polymerase POLG1 disease mutations and germline variants promote tumorigenic properties. PLoS One 10(10):e0139846. https://doi.org/10.1371/journal.pone.0139846
53. Barutcu AR, Lajoie BR, McCord RP, Tye CE, Hong D, Messier TL, Browne G, van Wijnen AJ, Lian JB, Stein JL, Dekker J, Imbalzano AN, Stein GS (2015) Chromatin interaction analysis reveals changes in small chromosome and telomere clustering between epithelial and breast cancer cells. Genome Biol 16:214. https://doi.org/10.1186/s13059-015-0768-0
54. Rao SS, Huntley MH, Durand NC, Stamenova EK, Bochkov ID, Robinson JT, Sanborn AL, Machol I, Omer AD, Lander ES, Aiden EL (2014) A 3D map of the human genome at kilobase resolution reveals principles of chromatin looping. Cell 159(7):1665–1680. https://doi.org/10.1016/j.cell.2014.11.021
55. Burton JN, Adey A, Patwardhan RP, Qiu R, Kitzman JO, Shendure J (2013) Chromosome-scale scaffolding of de novo genome assemblies based on chromatin interactions. Nat Biotechnol 31(12):1119–1125. https://doi.org/10.1038/nbt.2727
56. Landry JJ, Pyl PT, Rausch T, Zichner T, Tekkedil MM, Stutz AM, Jauch A, Aiyar RS, Pau G, Delhomme N, Gagneur J, Korbel JO, Huber W, Steinmetz LM (2013) The genomic and transcriptomic landscape of a HeLa cell line. G3 (Bethesda) 3(8):1213–1224. https://doi.org/10.1534/g3.113.005777
57. Harewood L, Kishore K, Eldridge MD, Wingett S, Pearson D, Schoenfelder S, Collins VP, Fraser P (2017) Hi-C as a tool for precise detection and characterisation of chromosomal rearrangements and copy number variation in human tumours. Genome Biol 18(1):125. https://doi.org/10.1186/s13059-017-1253-8
58. Rickman DS, Soong TD, Moss B, Mosquera JM, Dlabal J, Terry S, MacDonald TY, Tripodi J, Bunting K, Najfeld V, Demichelis F, Melnick AM, Elemento O, Rubin MA (2012) Oncogene-mediated alterations in chromatin conformation. Proc Natl Acad Sci US A 109(23):9083–9088. https://doi.org/10.1073/pnas.1112570109
59. Simonis M, Klous P, Homminga I, Galjaard RJ, Rijkers EJ, Grosveld F, Meijerink JP, de Laat W (2009) High-resolution identification of balanced and complex chromosomal rearrangements by 4C technology. Nat Methods 6(11):837–842. https://doi.org/10.1038/nmeth.1391
60. Gabdank I, Ramakrishnan S, Villeneuve AM, Fire AZ (2016) A streamlined tethered chromosome conformation capture protocol. BMC Genomics 17(1):274. https://doi.org/10.1186/s12864-016-2596-3
61. Nagano T, Lubling Y, Stevens TJ, Schoenfelder S, Yaffe E, Dean W, Laue ED, Tanay A, Fraser P (2013) Single-cell Hi-C reveals cell-to-cell variability in chromosome structure. Nature 502(7469):59–64. https://doi.org/10.1038/nature12593
62. Vogelstein B, Kinzler KW (2004) Cancer genes and the pathways they control. Nat Med 10(8):789–799. https://doi.org/10.1038/nm1087
63. Groschel S, Sanders MA, Hoogenboezem R, de Wit E, Bouwman BA, Erpelinck C, van der Velden VH, Havermans M, Avellino R, van Lom K, Rombouts EJ, van Duin M, Dohner K, Beverloo HB, Bradner

JE, Dohner H, Lowenberg B, Valk PJ, Bindels EM, de Laat W, Delwel R (2014) A single oncogenic enhancer rearrangement causes concomitant EVI1 and GATA2 deregulation in leukemia. Cell 157(2):369–381. https://doi.org/10.1016/j.cell.2014.02.019

64. Northcott PA, Lee C, Zichner T, Stutz AM, Erkek S, Kawauchi D, Shih DJH, Hovestadt V, Zapatka M, Sturm D, Jones DTW, Kool M, Remke M, Cavalli FMG, Zuyderduyn S, Bader GD, VandenBerg S, Esparza LA, Ryzhova M, Wang W, Wittmann A, Stark S, Sieber L, Seker-Cin H, Linke L, Kratochwil F, Jager N, Buchhalter I, Imbusch CD, Zipprich G, Raeder B, Schmidt S, Diessl N, Wolf S, Wiemann S, Brors B, Lawerenz C, Eils J, Warnatz H-J, Risch T, Yaspo M-L, Weber UD, Bartholomae CC, von Kalle C, Turanyi E, Hauser P, Sanden E, Darabi A, Siesjo P, Sterba J, Zitterbart K, Sumerauer D, van Sluis P, Versteeg R, Volckmann R, Koster J, Schuhmann MU, Ebinger M, Grimes HL, Robinson GW, Gajjar A, Mynarek M, von Hoff K, Rutkowski S, Pietsch T, Scheurlen W, Felsberg J, Reifenberger G, Kulozik AE, von Deimling A, Witt O, Eils R, Gilbertson RJ, Korshunov A, Taylor MD, Lichter P, Korbel JO, Wechsler-Reya RJ, Pfister SM (2014) Enhancer hijacking activates GFI1 family oncogenes in medulloblastoma. Nature 511 (7510):428-434. doi:https://doi.org/10.1038/nature13379. http://www.nature.com/nature/journal/v511/n7510/abs/nature13379.html#supplementary-information
65. Fudenberg G, Imakaev M, Lu C, Goloborodko A, Abdennur N, Mirny LA (2016) Formation of Chromosomal Domains by Loop Extrusion. Cell Rep 15(9):2038–2049. https://doi.org/10.1016/j.celrep.2016.04.085
66. Sanborn AL, Rao SS, Huang SC, Durand NC, Huntley MH, Jewett AI, Bochkov ID, Chinnappan D, Cutkosky A, Li J, Geeting KP, Gnirke A, Melnikov A, McKenna D, Stamenova EK, Lander ES, Aiden EL (2015) Chromatin extrusion explains key features of loop and domain formation in wild-type and engineered genomes. Proc Natl Acad Sci U S A 112(47):E6456–E6465. https://doi.org/10.1073/pnas.1518552112
67. Nora EP, Goloborodko A, Valton AL, Gibcus JH, Uebersohn A, Abdennur N, Dekker J, Mirny LA, Bruneau BG (2017) Targeted degradation of CTCF decouples local insulation of chromosome domains from genomic compartmentalization. Cell 169(5):930–944. e922. https://doi.org/10.1016/j.cell.2017.05.004
68. Franke M, Ibrahim DM, Andrey G, Schwarzer W, Heinrich V, Schopflin R, Kraft K, Kempfer R, Jerkovic I, Chan WL, Spielmann M, Timmermann B, Wittler L, Kurth I, Cambiaso P, Zuffardi O, Houge G, Lambie L, Brancati F, Pombo A, Vingron M, Spitz F, Mundlos S (2016) Formation of new chromatin domains determines pathogenicity of genomic duplications. Nature 538(7624):265–269. https://doi.org/10.1038/nature19800
69. Hnisz D, Weintraub AS, Day DS, Valton AL, Bak RO, Li CH, Goldmann J, Lajoie BR, Fan ZP, Sigova AA, Reddy J, Borges-Rivera D, Lee TI, Jaenisch R, Porteus MH, Dekker J, Young RA (2016) Activation of proto-oncogenes by disruption of chromosome neighborhoods. Science 351(6280):1454–1458. https://doi.org/10.1126/science.aad9024
70. Flavahan WA, Drier Y, Liau BB, Gillespie SM, Venteicher AS, Stemmer-Rachamimov AO, Suva ML, Bernstein BE (2016) Insulator dysfunction and oncogene activation in IDH mutant gliomas. Nature 529(7584):110–114. https://doi.org/10.1038/nature16490
71. Nora EP, Dekker J, Heard E (2013) Segmental folding of chromosomes: a basis for structural and regulatory chromosomal neighborhoods? Bioessays 35(9):818–828. https://doi.org/10.1002/bies.201300040
72. Taberlay PC, Achinger-Kawecka J, Lun AT, Buske FA, Sabir K, Gould CM, Zotenko E, Bert SA, Giles KA, Bauer DC, Smyth GK, Stirzaker C, O'Donoghue SI, Clark SJ (2016) Three-dimensional disorganization of the cancer genome occurs coincident with long-range genetic and epigenetic alterations. Genome Res 26(6):719–731. https://doi.org/10.1101/gr.201517.115
73. Forcato M, Nicoletti C, Pal K, Livi CM, Ferrari F, Bicciato S (2017) Comparison of computational methods for Hi-C data analysis. Nat Methods 14(7):679–685. https://doi.org/10.1038/nmeth.4325
74. Strickfaden H, Zunhammer A, van Koningsbruggen S, Köhler D, Cremer T (2010) 4D chromatin dynamics in cycling cells: Theodor Boveri's hypotheses revisited. Nucleus 1(3):284–297. https://doi.org/10.4161/nucl.1.3.11969
75. Golloshi R, Sanders JT, McCord RP (2017) Genome organization during the cell cycle: unity in division. Wiley Interdiscip Rev Syst Biol Med. https://doi.org/10.1002/wsbm.1389
76. Kind J, Pagie L, Ortabozkoyun H, Boyle S, de Vries SS, Janssen H, Amendola M, Nolen LD, Bickmore WA, van Steensel B (2013) Single-cell dynamics of genome-nuclear lamina interactions. Cell 153(1):178–192. https://doi.org/10.1016/j.cell.2013.02.028
77. Gibcus JH, Dekker J (2013) The hierarchy of the 3D genome. Mol Cell 49(5):773–782. https://doi.org/10.1016/j.molcel.2013.02.011
78. Shachar S, Voss Ty C, Pegoraro G, Sciascia N, Misteli T (2015) Identification of Gene Positioning Factors Using High-Throughput Imaging Mapping. Cell 162(4):911–923. https://doi.org/10.1016/j.cell.2015.07.035
79. Burmeister T, Molkentin M, Schwartz S, Gokbuget N, Hoelzer D, Thiel E, Reinhardt R (2013) Erroneous class switching and false VDJ recombination: molecular dissection of t(8;14)/MYC-IGH translocations in Burkitt-type lymphoblastic leukemia/B--cell lymphoma. Mol Oncol 7(4):850–858. https://doi.org/10.1016/j.molonc.2013.04.006

80. McCord RP, Dekker J (2011) Translocation mapping exposes the risky lifestyle of B cells. Cell 147(1):20–22. https://doi.org/10.1016/j.cell.2011.09.005
81. Arsuaga J, Greulich-Bode KM, Vazquez M, Bruckner M, Hahnfeldt P, Brenner DJ, Sachs R, Hlatky L (2004) Chromosome spatial clustering inferred from radiogenic aberrations. Int J Radiat Biol 80(7):507–515
82. Branco MR, Pombo A (2006) Intermingling of chromosome territories in interphase suggests role in translocations and transcription-dependent associations. PLoS Biol 4(5):e138. https://doi.org/10.1371/journal.pbio.0040138
83. Caudill CM, Zhu Z, Ciampi R, Stringer JR, Nikiforov YE (2005) Dose-dependent generation of RET/PTC in human thyroid cells after in vitro exposure to gamma-radiation: a model of carcinogenic chromosomal rearrangement induced by ionizing radiation. J Clin Endocrinol Metab 90(4):2364–2369. https://doi.org/10.1210/jc.2004-1811
84. Schmitt AD, Hu M, Jung I, Xu Z, Qiu Y, Tan CL, Li Y, Lin S, Lin Y, Barr CL, Ren B (2016) A Compendium of Chromatin Contact Maps Reveals Spatially Active Regions in the Human Genome. Cell Rep 17(8):2042–2059. https://doi.org/10.1016/j.celrep.2016.10.061
85. Takata H, Hanafusa T, Mori T, Shimura M, Iida Y, Ishikawa K, Yoshikawa K, Yoshikawa Y, Maeshima K (2013) Chromatin compaction protects genomic DNA from radiation damage. PLoS One 8(10):e75622. https://doi.org/10.1371/journal.pone.0075622
86. Chinnaiyan P, Vallabhaneni G, Armstrong E, Huang SM, Harari PM (2005) Modulation of radiation response by histone deacetylase inhibition. Int J Radiat Oncol Biol Phys 62(1):223–229. https://doi.org/10.1016/j.ijrobp.2004.12.088
87. Cowell IG, Sondka Z, Smith K, Lee KC, Manville CM, Sidorczuk-Lesthuruge M, Rance HA, Padget K, Jackson GH, Adachi N, Austin CA (2012) Model for MLL translocations in therapy-related leukemia involving topoisomerase IIbeta-mediated DNA strand breaks and gene proximity. Proc Natl Acad Sci U S A 109(23):8989–8994. https://doi.org/10.1073/pnas.1204406109
88. Tammaro M, Barr P, Ricci B, Yan H (2013) Replication-dependent and transcription-dependent mechanisms of DNA double-strand break induction by the topoisomerase 2-targeting drug etoposide. PLoS One 8(11):e79202. https://doi.org/10.1371/journal.pone.0079202
89. Haffner MC, De Marzo AM, Meeker AK, Nelson WG, Yegnasubramanian S (2011) Transcription-induced DNA double strand breaks: both oncogenic force and potential therapeutic target? Clin Cancer Res 17(12):3858–3864. https://doi.org/10.1158/1078-0432.CCR-10-2044
90. Uuskula-Reimand L, Hou H, Samavarchi-Tehrani P, Rudan MV, Liang M, Medina-Rivera A, Mohammed H, Schmidt D, Schwalie P, Young EJ, Reimand J, Hadjur S, Gingras AC, Wilson MD (2016) Topoisomerase II beta interacts with cohesin and CTCF at topological domain borders. Genome Biol 17(1):182. https://doi.org/10.1186/s13059-016-1043-8
91. Cowell IG, Sunter NJ, Singh PB, Austin CA, Durkacz BW, Tilby MJ (2007) gammaH2AX foci form preferentially in euchromatin after ionising-radiation. PLoS One 2(10):e1057. https://doi.org/10.1371/journal.pone.0001057
92. Goodarzi AA, Noon AT, Deckbar D, Ziv Y, Shiloh Y, Lobrich M, Jeggo PA (2008) ATM signaling facilitates repair of DNA double-strand breaks associated with heterochromatin. Mol Cell 31(2):167–177. https://doi.org/10.1016/j.molcel.2008.05.017
93. Aymard F, Legube G (2016) A TAD closer to ATM. Mol Cell Oncol 3(3):e1134411. https://doi.org/10.1080/23723556.2015.1134411
94. Caron P, Choudjaye J, Clouaire T, Bugler B, Daburon V, Aguirrebengoa M, Mangeat T, Iacovoni JS, Alvarez-Quilon A, Cortes-Ledesma F, Legube G (2015) Non-redundant Functions of ATM and DNA-PKcs in Response to DNA Double-Strand Breaks. Cell Rep 13(8):1598–1609. https://doi.org/10.1016/j.celrep.2015.10.024
95. Alipour E, Marko JF (2012) Self-organization of domain structures by DNA-loop-extruding enzymes. Nucleic Acids Res 40(22):11202–11212. https://doi.org/10.1093/nar/gks925
96. Roukos V, Burgess RC, Misteli T (2014) Generation of cell-based systems to visualize chromosome damage and translocations in living cells. Nat Protoc 9(10):2476–2492. https://doi.org/10.1038/nprot.2014.167
97. Roukos V, Voss TC, Schmidt CK, Lee S, Wangsa D, Misteli T (2013) Spatial dynamics of chromosome translocations in living cells. Science 341(6146):660–664. https://doi.org/10.1126/science.1237150
98. Taslerová R, Kozubek S, Bártová E, Gajdušková P, Kodet R, Kozubek M (2006) Localization of genetic elements of intact and derivative chromosome 11 and 22 territories in nuclei of Ewing sarcoma cells. J Struct Biol 155(3):493–504. https://doi.org/10.1016/j.jsb.2006.05.005
99. Cremer M, Küpper K, Wagler B, Wizelman L, Jv H, Weiland Y, Kreja L, Diebold J, Speicher MR, Cremer T (2003) Inheritance of gene density–related higher order chromatin arrangements in normal and tumor cell nuclei. J Cell Biol 162(5):809–820. https://doi.org/10.1083/jcb.200304096
100. Ledford H (2010) Big science: The cancer genome challenge. Nature 464(7291):972–974. https://doi.org/10.1038/464972a
101. Gulley ML, Shea TC, Fedoriw Y (2010) Genetic tests to evaluate prognosis and predict therapeutic response in acute myeloid leukemia. J Mol Diagn 12(1):3–16. https://doi.org/10.2353/jmoldx.2010.090054

102. Okimoto RA, Bivona TG (2014) Recent advances in personalized lung cancer medicine. Per Med 11(3):309–321
103. Meaburn KJ, Misteli T (2008) Locus-specific and activity-independent gene repositioning during early tumorigenesis. J Cell Biol 180(1):39–50. https://doi.org/10.1083/jcb.200708204
104. Tian X, Wang Y, Zhao F, Liu J, Yin J, Chen D, Ma W, Ke X (2015) A new classification of interphase nuclei based on spatial organizations of chromosome 8 and 21 for t(8;21) (q22;q22) acute myeloid leukemia by three-dimensional fluorescence in situ hybridization. Leukemia Res 39(12):1414–1420. https://doi.org/10.1016/j.leukres.2015.09.013
105. Kozubek S, Lukasova E, Ryznar L, Kozubek M, Liskova A, Govorun RD, Krasavin EA, Horneck G (1997) Distribution of ABL and BCR genes in cell nuclei of normal and irradiated lymphocytes. Blood 89(12):4537–4545
106. Woudstra EC, Roesink JM, Rosemann M, Brunsting JF, Driessen C, Orta T, Konings AW, Peacock JH, Kampinga HH (1996) Chromatin structure and cellular radiosensitivity: a comparison of two human tumour cell lines. Int J Radiat Biol 70(6):693–703
107. Tomczak K, Czerwińska P, Wiznerowicz M (2015) The Cancer Genome Atlas (TCGA): an immeasurable source of knowledge. Contemporary Oncol 19(1A):A68–A77. https://doi.org/10.5114/wo.2014.47136

The Role of Chromosome Deletions in Human Cancers

9

Mei Chen, Yi Yang, Yu Liu, and Chong Chen

Abstract

Chromosome deletions are a hallmark of human cancers. These chromosome abnormalities have been observed for over than a century and frequently associated with poor prognosis. However, their functions and potential underlying mechanisms remain elusive until recently. Recent technique breakthroughs, including cancer genomics, high throughput library screening and genome editing, opened a new era in the mechanistic studying of chromosome deletions in human cancer. In this chapter, we will focus on the latest studies on the functions of chromosome deletions in human cancers, especially hematopoietic malignancies and try to persuade the readers that these chromosome alterations could play significant roles in the genesis and drug responses of human cancers.

Keywords

Chromosome deletion · Human cancer · Knudson's two-hit hypothesis · Haploinsufficient tumor suppressor · Genome editing

M. Chen · Y. Yang · Y. Liu · C. Chen (✉)
Department of Hematology, State Key Laboratory of Biotherapy and Cancer Center, West China Hospital, Sichuan University and National Collaborative Innovation Center, Chengdu, China

9.1 Introduction

Copy number variation (CNV) is one of the hallmarks of human cancers [1]. Deletions and amplifications of focal chromosome regions, chromosome arms or even whole chromosomes are frequent in both blood and solid cancers. Back to the end of the eighteenth century, German pathologist David Hansemann first observed asymmetric distribution of "chromatin loops" even though it was difficult to clearly see the chromosomes under the microscopes at his age [2, 3]. Following this seminal observation, another German pathologist Theodor Boveri proposed that "a particular, incorrect chromosome combination which is the cause of the abnormal growth characteristics passed on to daughter cells" [4].

Hansemann and Boveri's initial observations were further confirmed in the following more than 100 years. After the first chromosome abnormality in cancer, the Philadelphia chromosome, was discovered in 1960, sophisticated cytogenetics technologies have been developed to study the karyotyping of cancer, especially leukemia [5]. Given that CNVs are frequent and associated with poor prognosis, it is crucial to understand the functions of these chromosome alterations in tumorigenesis and drug response. It is generally believed that chromosome deletion regions contain tumor suppressors while chromosome amplification regions contain oncogenes [6, 7].

Y. Zhang (ed.), *Chromosome Translocation*, Advances in Experimental Medicine and Biology 1044,
https://doi.org/10.1007/978-981-13-0593-1_9

However, so far, majority of chromosome deletions don't contain obvious confirmed tumor suppressors. It has been argued that most chromosome abnormalities, including chromosome deletions, are the consequences of genome instability of cancer. In other words, cancers first have loss-of-function mutations on genes critical for genome integrity, such as TP53 and BRCA1/2, and then, as a consequence, these mutant cells acquire largely randomly chromosome deletions, together with many other genome abnormalities [1]. Though this hypothesis has been widely accepted, there are emerging evidences suggesting that at least it might not be the entire story. First, there are multiple cases of human cancer have chromosome large deletions or chromosome losses, while no mutation on any known genome integrity regulator genes [8]. More importantly, there are emerging evidences demonstrating the functions of chromosome deletions, as a whole, on tumor initiation and progress [9–12]. More and more tumor suppressors have been identified in these frequently deleted chromosome regions [13, 14]. Interestingly, co-deficiencies of these tumor suppressors in the same region promoted faster tumorigenesis than knockdown of any single tumor suppressor, suggesting the synergy of these tumor suppressors [8, 10]. Therefore, we propose a "synergy of multiple tumor suppressors" theory that there are multiple collaborating tumor suppressors in the common deleted regions in cancer, which make the chromosome large deletions more detrimental than single tumor suppressor mutations.

In this chapter we will focus on the studies of the functions and mechanisms of chromosome deletions on cancer and explain our "synergy of multiple tumor suppressors" theory.

9.2 Chromosome Deletions Are Frequent in Human Cancers and Associated with Poor Prognosis

9.2.1 The Boveri's Hypothesis on the Origin of Cancer

David Paul Hansemann was the first person to report unbalanced anaphases and telophases in freshly isolated epithelial cancer cells in 1890 [2]. He described in details the mitotic chromosomes of 13 cultured epithelial cancer cells and noted the aberrant multipolar mitoses and anaphases with asymmetrical distribution of "chromatin loops". However, Hansemann considered that these features were not unique to cancers. He thought that these chromosome alterations in tumor cells were the same process as in normal embryonic development [2, 3].

Soon after Hansemann's initial observation, Boveri made his own similar observations of hypo- and hyperchromacy and proposed his famous tumorigenesis hypothesis that "a tumor originates from a single cell in which there is a defined but incorrect combination of chromosomes" [4]. His work on sea urchin let him conclude that individual chromosome transmitted different inheritance factors. Therefore, Boveri is credited to the chromosome theory, together with Sutton [15]. Boveri applied his concept of chromosome to explain tumorigenesis and made many bold and accurate predictions, including the existence of tumour-suppressor genes ("teilungshemmende chromosomen") and oncogenes ("teilungsfoerdernde chromosomen"). Majority of Boveri's hypothesis and concepts have been approved and widely accepted by subsequent scientists.

Though chromosome alterations were frequently observed in cancers, Boveri's hypothesis on tumorigenesis was not well appreciated until the historic discovery of the Philadelphia chromosome in 1960 [5]. Nowell described the Philadelphia chromosome as the first consistent chromosome alteration in human cancers. Now the Philadelphia chromosome has become the golden standard to diagnose chronic myeloid leukemia and the resulting fusion protein BCR-ABL is the target of the first target therapy drug Gleevec [16–18]. The discovery of the Philadelphia chromosome and the target therapy against it opened new era for cancer research and clinical practice.

9.2.2 Chromosome Deletions Are a Hallmark of Human Cancers

Cancer cytogenetics is a new field ushered by the description of the Philadelphia chromosome for the diagnosis and prognosis of human cancer, especially hematopoietic malignancies [6]. Over the last half century, a series of technique advances have improved karyotyping with high resolution, accuracy and convenience. In the late 1960s, Torbjorn Caspersson developed Q-binding staining to reveal unique banding patterns of each chromosome [19]. This staining is generally applied to detect multiple types of chromosome abnormalities, including translocations, deletions and inversions. Later molecular cytogenetics was developed with fluorescent- or radioisotope-labeled molecular probes. Labeled sequence-specific probes were hybridized with chromosomes with the techniques as fluorescence in situ hybridization (FISH) [20, 21]. In 1990s, array comparative genomic hybridization (aCGH) was applied to analyze copy number variations of cancer cells compared to reference samples [22, 23]. Microarrays used for aCGH can contain limited customized probes or millions of probes for the whole genome. The increased number of probes will improve the resolution of CGH analysis and less than 100 kb focal copy number variations can be detected. With the introduction of high-throughput sequencing techniques, CNV-seq reaches the highest resolution to single nucleotides [24, 25]. With these advanced technologies, accumulating chromosome deletions in human cancers have been documented [26].

Right now, up to millions of human cancers have been analyzed for their chromosome alterations (Table 9.1). It has been estimated that averagely about 30% of the genome is affected by chromosome arm-level or focal deletions in a typical human cancer [24, 25]. It seems that chromosome deletions in human cancers involve all regions of the genome. It is interesting that there are significant "peaks" of deletions and amplifications, while these peaks vary among different types of cancer. Some of these chromosome deletions are common among human cancers or a specific type of cancer. For example, one third of human cancers contain chromosome 17 loss or 17p deletions [10]. Chromosome 1p and 16q loss are common in solid cancer [27]. Acute myeloid leukemia (AML) frequently have chromosome 5 loss or 5q deletions (-5/del(5q)) and chromosome 7 loss or 7q deletions (about 10% in de novo AML and 50% in relapsed or treatment-related AML, respectively) [28, 29], while chromosome 3p

Table 9.1 The common chromosome deletions and their frequencies in selected types of human cancers

Disease category	Chromosome abnormality	Frequency of occurrence	References
Acute Myeloid Leukemia (AML)	-7/del(7q)	~10%	Greenberg et al. [31]
	-5/del(5q)	~10%	Nimer et al. [32]
	Del(20q)	~5%	Haase et al. [33]
	Del(17p)	~3%–4%	Valerie Soenen et al. [34] and Yvon Sterkers et al. [35]
Therapy-related AML	-7/del(7q)/-5/del(5q)	~75%	Smith et al. [36]
Non-Hodgkin lymphomas	Del(17p)	~19%	Levine et al. [37]
B-chronic lymphocyte leukemia	Del(13q)	~30%	Caporaso et al. [38]
Multiple myeloma	-13/del(13q)	~40%	Chng et al. [39]
Lung cancer	Del(13q)	~32%	Jun Yokota et al. [40]
	Del(17p)	~25%	Jun Yokota et al. [40]
Ovarian cancer	Del(17q)	~39%	Hiroko Saito et al. [41]
	Del(8p)	~33%	Mitsuru Emi et al. [42]
Breast cancer	Del(17q)	~41%	Hiroko Saito et al. [41]
	Del(10q23)	~40–48%	Garcia et al. [43]
	Del(8p)	~9%	Mitsuru Emi et al. [42]
Hepatocellular carcinoma	Del(8p)	~47%	Mitsuru Emi et al. [42]
Colorectal cancer	Del(8p)	~46%	Mitsuru Emi et al. [42]

deletions are detected in almost all small cell lung cancers and 90% of non-small cell lung cancers [30]. The high frequent common chromosome deletions suggest that these phenomena might be important to these diseases and of clinical value.

9.2.3 Chromosome Deletions Are Associated with Poor Prognosis in Some Cancers

Chromosome deletions, and other chromosome abnormalities have been widely applied for cancer diagnosis, prognosis and guiding clinical treatments. Back to 100 years ago, Boveri has proposed to detect malignant cells with chromosome abnormalities [4]. The Philadelphia chromosome is the golden marker for chronic myeloid leukemia [5].

Following Chromosome 5q deletion syndrome (5q- syndrome) is a hematopoietic disorder called myelodysplastic syndrome characterized with acquired interstitial chromosome 5q33.1 deletion and macrocytic anemia. In1974, Van den Berghe et al. reported the first 5q- syndrome [44]. Though most of these patients have only moderate thrombocytosis, erythroblastopenia, and megakaryocyte hyperplasia with a good prognosis, 10% of them would transform to AML [45, 46]. Generally these patients have less than 5% blast count in their peripheral blood and lenalidomide is the standard therapy. Interestingly, -5/del(5q) are one of the most frequent chromosome abnormalities in AML, especially relapsed or treatment-related AML. -5/del(5q) is associated with very poor prognosis, with less than 10% 5-year survival rate [47]. Of note, the chromosome regions involved in 5q- syndrome (5q33.1) and -5/del(5q) AML (5q31) are close but exclusive [48]. Thus characterizing chromosome deletions in detail is critic for clinic diagnosis and prognosis.

-7/del(7q) is the most frequent chromosome abnormality in AML, found in more than 50% secondary and 10% de novo myeloid disorders [49, 50]. Two minimal deleted regions, 7q22 and 7q35–36, have been mapped in -7/del(7q) AML [51, 52]. Both of them are associated with poor prognosis. While -7/del(7q) can happen independently, they also frequently co-occur with many other chromosome alterations, especially -5/del(5q) and -17/del(17p). When these multiple chromosome abnormalities happen together, these AML are called complex karyotype AML and have the worst prognosis with a 5-year survival of less than 5% [47].

Chromosome 17p deletions, generally involving the whole short arm of chromosome 17 and containing the well-known tumor suppressor TP53, are frequent in almost all human cancers, including AML, CLL and non-Hodgkin's lymphoma [53, 54]. In all of these cases, del(17P) are associated with poor prognosis [49, 55].

9.3 Identifying Tumor Suppressors in Chromosome Deletions

9.3.1 Knudson Theory

Given the frequency and prognosis value of chromosome deletions in human cancer, it is critical to understand the mechanisms of these chromosome abnormalities in cancer initiation, progress, metastasis and drug response. According to Boveri's theory, chromosome deletions would be rich of tumor suppressors [4]. Great efforts have done to uncover these functionally important genes over the last 30 years.

Traditionally there were two major criteria to identify tumor suppressors in chromosome loss or deleted regions. First the candidate tumor suppressors should located in the commonly deleted regions among multiple patients, echoing Koch's postulates. Chromosome loss or deletions generally involve large chromosome regions of several hundreds of genes, or the whole arms and sometimes the entire chromosomes of up to thousands of genes. In these cases, identifying critical tumor suppressors in these chromosome loss and deletions would be challenging [6, 10, 24]. To narrow down the candidate genes involved in specific types of cancer, a lot of work has been done to identify minimal deleted regions or commonly deleted regions among these patients, taking the advantage of the variance of chromosome deletions and focal deletions in rare patients [56]. Recently, GISTIC (Genomic Identification of Significant Targets in Cancer), a powerful algorithm, is developed to identify tumor suppressors in chromosome deletion regions in cancer (and also oncogenic drivers in amplified regions) [57].

The second criterion is Knudson theory or the two-hit hypothesis. It was assumed that most of the mutations on tumor suppressors were loss-of-function mutations and recessive. Thus, both of the alleles of a putative tumor suppressor must be mutated. It is proposed that there is a first hit in a tumor suppressor, classically assumed to be a point mutation, and followed by a second hit, which is commonly thought as a chromosome deletion. This loss-of-heterozygosity hypothesis is called as the two-hit hypothesis, proposed by Alfred Knudson in 1971 [58]. Knudson theory has been the basis for identifying tumor suppressors during the last four decades [59].

The first example of Knudson theory is the retinoblastoma gene RB1 on chromosome 13q14. Knudson observed that retinoblastoma patients with bilateral retinoblastoma were first diagnosed at significantly earlier age than those patients with unilateral disease and sufferers of bilateral Rb1 were six times more likely to develop other cancer than those of a unilateral Rb1 [58]. Knudson explained that in the case of a bilateral Rb1 (familial form), one allele is already mutated in all somatic cells and only a second hit is needed to mutate the second working allele, a process of loss of Heterozygosity [60]. Thus, Knudson proposed his two-hit hypothesis through his studies on RB1.

Many negative regulators of cell cycle display similar mutation pattern as RB1. For example, cyclin-dependent kinase inhibitor 2A (CDKN2A) is a regulator of RB1 through inhibiting cyclin-dependent kinase 4 and 6, which in turn inhibiting RB1 [61]. Therefore CDKN2A blocks cells in from G1 phase to S phase. CDKN2A resides on chromosome 9p21, which is one of the most commonly deleted regions in human cancers, especially in melanoma, small cell lung cancer and lymphoma [62]. Similar to those with familial retinoblastoma, familiar melanoma patients are more likely to carry inherited mutations in one allele of CDKN2A gene, and the second allele of this loci is deleted through the loss-of-heterozygosity process.

TP53 is the most frequently mutated tumor suppressor in human cancers, which is also recognized as an example for Kudson theory. Interestingly, TP53 was first found to be overexpressed in many human cancers, which is in contrast to classic tumor suppressors. Therefore it was assumed to be an oncogene at the beginning instead of tumor suppressor. Later, it turned out that the overexpressed "TP53" is a gain-of-function mutant and TP53 fits to the classic two-hit tumor suppressor [63]. TP53 is located on chromosome 17p13. Generally one allele of TP53 carries missense or frameshift mutations, with hotspots on R175, R248 and R273, which have been confirmed as gain-of-function mutations, and the second allele is generally deleted together with the whole short arm of chromosome 17 [64]. Familial TP53 mutations count for about half of Li-Fraumeni syndrome, almost all of these patients would develop multiple types of cancers, including sarcoma, leukemia, breast cancer and brain cancers as results of loss of heterozygosity of TP53 [65].

Following these examples, great efforts have been applied to reveal putative tumor suppressors in chromosome deletions through mapping minimal deleted regions to narrow down the candidate genes and searching the point mutations or epigenetic silencing on the second allele as an evidence of loss of heterozygosity [59]. A long list of tumor suppressors, including PTEN on chromosome 10q23 [66], APC on chromosome 5q22 [54], NF1 on chromosome 17q11 [67], BRCA1 on chromosome 17q21 [68] and VHL on chromosome 3p25 [69], have been identified.

9.3.2 Haploinsufficient Tumor Suppressors

Despite the large success of Knudson theory, there are two obvious puzzles about chromosome deletions in human cancers. First there are no verified classic tumor suppressors in many chromosome deletions even after great efforts of searching. And second, chromosome deletions generally contain several hundreds genes while only one or very few of them have been validated as tumor suppressors [24, 25]. These contradictions suggest that classic tumor suppressors consistent with the two-hit hypothesis might not be the whole stories. Around 2000, a novel type of tumor suppressors, haploinsufficient tumor suppressor, has been proposed. Heterozygous loss of function of these genes, such as mutations or deletions on only one allele (and the second allele is still functioning), would contribute tumor genesis and progression [70, 71]. The new concept of haploinsufficiency dramatically expands the candidate genes for tumor suppressors, especially in chromosome deletion regions.

One of the first identified haploinsufficient tumor suppressor is the cyclin-dependent kinase inhibitor p27Kip1 [72]. p27kip1, a regulator of RB1-E2F pathway, is in chromosome 12p12, a region frequently deleted in pediatric acute lymphoblastic leukemia. All deletions involved chromosome 12p12 are heterozygous while neither missense nor truncated mutations were detected in the retained allele [73, 74]. And expression of p27Kip1 was detected in the nuclei of these effected cancer cells by immunostaining though at a reduced levels [72], suggesting a non-Knudson mechanism. With a genetically engineered mouse model, Fero et al. clearly demonstrated that p27Kip1 heterozygous loss resulted in spontaneous multiple organ tumors at a penetrance of 32% in mice. When exposed to X-ray irradiation, these mice developed dramatically more tumors than wildtype control mice, though fewer than those of p27Kip1 homozygous loss. More importantly, all of the tumors from p27Kip1+/− mice retained the wildtype allele and the expression of p27Kip1 were revealed by west blotting [72]. Thus p27Kip1 is a haploinsufficient tumor suppressor.

Haploinsufficent tumor suppressors may also residue in chromosome 7q, the most frequently deleted region in AML. Since its mapping by cytogenetics, great efforts of decades to identify classic tumor suppressors in this region have been in vain. By analyzing the big data of cancer genomics and in vivo function tests, we showed that the mixed lineage leukemia 3 gene, MLL3, was a haploinsufficient tumor suppressor in chromosome 7q36 [13]. MLL3 is a member of the MLL protein family with a SET domain capable of methylating lysine 4 on histone H3 and a core component of the COMPASS-like complex regulating transcription elongation [75]. MLL3 is one of the most frequently mutated chromatin modifiers in solid cancers. But all of these mutations are heterozygous [76, 77]. 7q is the most commonly deleted region in AML but so far no loss-of-function mutation of MLL3 (nor other genes) was found in 7q loss patients [49]. shRNAs knocking down Mll3, together with p53 and Nf1 loss, promoted full blown AML genesis, indicating Mlll3 as a tumor suppressor. Though these shRNAs could potently reduce the expression level of Mll3 in NIH3T3 cells at 90%, the inhibition of Mll3 expression by the same shRNAs in the resulting AML cells were only about 50%. Further CRISPR/Cas9-mediated genome editing of Mll3 leukemia cells also remained one intact wildtype allele. All of these evidences demonstrated that Mll3 is a haploinsufficient tumor suppressor in AML [13]. These results are striking given that MLL3 is an epigenetic regulator, which affects the expressions of many downstream genes but at a moderate level. The remaining questions would be how the moderate dosage change of an epigenetic gene would transform hematopoietic stem cells and whether restoring the expression of MLL3 (two-fold increase) in leukemia would be able to restrain the progression of the disease.

Interestingly, many of the putative classic tumor suppressors also show haploinsufficiency in preventing tumorigenesis. One example is PTEN, residing in chromosome 10q23 and encoding a lipid phosphatase that negatively regulates PI3K-AKT pathway [78]. It was estimated that up to 70% prostate cancer patients carried a heterozygous loss of PTEN, generally covered by a large deletion of one copy of chromosome 10 similar to MLL3 in chromosome 7q, while only less than 10% of the patients had homozygous deletions or mutations at diagnosis [79]. Consistent with the human clinic genetics, Pten Heterozygosity dramatically increased the rate of prostate cancer progression in TRAMP mice

[80]. Later, with a Pten hypermorphic mouse model whose expression level of Pten was 80% of that in wildtype control mice Alimonti et al. reported that even such subtle reduction of Pten dosage would promote the development of a wide spectrum of cancers [81]. Thus haploinsufficiency is a general principle for tumorigenesis.

Arguably all potential tumor suppressors in chromosome deletions without loss-of-function mutations on the second allele may be haploinsufficient tumor suppressors, which would strikingly deep our understanding of the molecular mechanisms of chromosome deletions in human cancers. It is also interesting to test whether these haploinsufficient tumor suppressors might be valuable therapeutic targets for the cancers with the corresponding defects.

9.4 The Role of Chromosome Deletions as a Whole in Carcinogenesis

9.4.1 Modeling Chromosome Deletions with Genetic Engineered Mouse Models

Identifying tumor suppressors in chromosome deletions is very important to study the functions of chromosome deletions in tumorigenesis. However, given the broad effects of chromosome deletions with generally several hundreds genes and structural abnormalities, none of any single tumor suppressor could recapitulate all of the phenomena of a chromosome deletion in cancer. Thus the full functions of chromosome deletions must be studied as a whole. Investigating the biological roles of chromosome deletions as a whole has been significantly delayed due to lack of available techniques to precisely model these chromosome configurators and confused by the results of spontaneous aneuploidies. At odds to being a hallmark of cancer, aneuploidy, including chromosome loss and large chromosome deletions, has been shown to be detrimental to normal cells, specifically yeast cells and mouse embryonic fibroblast cells, in some context [82]. It is argued that both gene-specific and general non-gene-specific effects of aneuploidy could interfere cell proliferation through "aneuploidy associated stresses". These experimental observations seem at odds with the frequent chromosome alterations associated with human cancers and Boveri's chromosome theory of carcinogenesis [83]. Therefore it is critical to provide direct evidences that chromosome deletions are able to drive tumorigenesis.

Recent technique advances including sophisticated genetically engineered mouse modeling, genome editing and high throughput library screening, made it possible to reveal the biological consequences of chromosome deletions in cancer [14, 84–87]. The first example is chromosome 17p deletion [10]. 17p loss is one of the most, if not the most, frequently genetical abnormalities found in various cancers and associates with tumor aggressiveness and poor prognosis [88]. Given the well-studied tumor suppressor TP53 on chromosome 17p13, it was generally assumed that chromosome 17p loss is to loss of Heterozygosity of the second allele of TP53, following the classic Knudson theory [63]. However, by analyzing the CNV and mutation data of more than 4000 human cancers, we found that one third of cases with TP53 alterations had heterozygous chromosome 17p loss but didn't have any detectable mutation of TP53 on the other allele [10]. Therefore it is very important to investigate whether chromosome 17p has more tumor suppression capacity beyond TP53 only. Taking the advantage of the high synteny between mouse chromosome 11B3 and human chromosome 17p13, which share the exact same over than 100 coding genes and noncoding microRNA genes even at the same order, we genetically engineered a conditional 11B3 knockout mouse model. Compared to p53 deleted tumors, heterozygous deletion of chromosome 11B3 can promote either Myc-driven lymphomagenesis or Nf1; Mll3-defective leukemogenesis with shorter tumor latency and overall survival. Moreover, the resulting 11B3-deleted tumor cells are more resistant to chemodrug like cyclophosphamide, vincristine and methotrexate. Interestingly, many lymphomas generated from heterozygous deletion of 11B3 carry spontaneously missense or frameshift mutations on the wildtype p53 allele, likely resulting from the procession of *Trp53* loss-of-heterozygosity. Other 11B3-deleted lymphomas keep wildtype Trp53 allele. Together, 11B3 tumors represent the chromosome 17p deletion configurations in human cancers [10]. These findings would not only shed light on understanding the molecular mechanisms under which chromosome 17p deletions impact on cancer biology, but also provide a platform to develop new therapeutic methods.

Chromosome 7q22 is another frequently deleted region in AML and so far no classic tumor suppressor has been validated in the context of AML [89]. To shed light on the sealed function of 7q22 deletions to Myelodysplastic Syndrome (MDS) pathogenesis, Wong et al. generated mice with a heterozygous germ line deletion of a 2 Mb interval of the murine chromosome band 5A3, which removing 13 genes correspondent to a commonly deleted segment of human 7q22 [12]. The resulting 5A3+/del mice exhibited typical characterizations of MDS. The 5A3+/del mouse model provided a novel platform for the studies of human 7q22 deletion MDS or AML.

These genetically engineered mouse models provide clear and direct evidences that chromosome deletions as a whole can be drivers of tumorigenesis and experimentally prove the 100-year-old Boveri's cancer theory. However, the big limitation of this strategy is that, though 99% of human and mouse genes are identical, the synteny between human and mouse chromosomes are poor [90, 91]. Therefore it is difficult to model chromosome large deletions of human cancers in mouse models.

9.4.2 Modeling Chromosome Deletions in Human Cell Models

Obviously human cells can be the best model to study chromosome alterations in human cancers. The efficiency of genome editing made it feasible [92]. It is widely known that chromosome 8p loss recurrently occurs in human breast cancer patients and it is tightly associated with poor patient survival. In order to elucidate the role of 8p loss in tumorigenic transformation, Cai et al. made a good use of TALEN-directed genomic engineering technology to generate human cellular models based on an non-malignant MCF10A mammary epithelial cell line, which mimicking 8p loss of heterozygosity and avoiding introducing other genomic abnormalities [9]. Though the entire loss of 8p chromosome showed limited tumor transformation capacity alone or cooperating with other driver genes like MYC, ERBB2 or loss of TP53, these cells displayed abnormal fatty acid and ceramide metabolism. The shift of fatty acid metabolism led to actin filament reorganization and further contributed to cell invasiveness. Besides, alterations in ceramide metabolism rendered cells increased autophagy capacity and better growth ability under hypoxia context. Primary human breast cancers with 8p loss deriving from clinical patients bear these metabolic changes as well. These discoveries suggest that models of chromosomal large deletions could be used to predict the responsiveness of cancer patients to anticancer therapies and could help to improve our understandings of human cancer [93].

Taking advantage of induce pluripotent stem cells (iPSCs), Papapetrou's laboratory investigated the biological consequences of chromosome 7q loss, the most frequent chromosome abnormalities in AML [11, 94]. First they generated iPS cells from chromosome 7q loss and intact cells from the same patients and showed that iPS cells with chromosome 7q deletions had defects to differentiate into hematopoietic cells and had increased apoptosis, similar to those observed in MDS patients with chromosome 7q deletions. Then using AAV-delivered CRISPR/Cas9, they generated chromosome 7q deletions in normal human iPS cells. These genome edited iPS cells also displayed reduced capacity to differentiate into CD45+ hematopoietic cells while increased percentage of CD34+ (a marker of hematopoietic stem and progenitor cells) population. These phenotypes are consistent with those in chromosome 7q deleted MDS patients [11]. It is of interest that spontaneous correction of chromosome 7q by a chromosome 7 trisomy largely rescued most of these abnormalities associated with the disease [94]. These studies indicate that chromosome 7q deletions as a whole are responsible for the pathology of MDS with chromosome 7q loss. The combination of iPS cells and genome editing opens a new era to study chromosome alterations in human cancers. In principle, this strategy could model all kinds of chromosome deletions in various types of human cancers [95]. A shortcoming is that in patients somatic chromosome deletions assumably occur in tissue-specific stem or progenitor cells while genome edited iPS cells are not physiologically related. Thus direct genome editing of cell-of-origin of human cancers might be more accurate to investigate the biological functions of chromosome abnormalities in the right context.

A new era in preclinical cancer research is emerging, in which human-based models are taking center stage and patient-derived cells are increasingly being used as primary discovery platforms. In this modern era of basic cancer research and precision oncology, iPSCs derived from patients with cancer can substantially expand the experimental repertoire applicable to human cells in ways that were hitherto restricted to model organisms. We envision that models for at least some cancers can be developed using iPSC technologies, and that these will occupy a unique place in this new era, bridging primary cells with immortalized cell lines by combining the physiological relevance of the former with the amenability to experimentation of the latter. Interdisciplinary collaborations between stem cell researchers, cancer researchers, physicians, translational scientists, bioengineers and drug developers will be paramount to harness the full potential of iPSCs as a new tool in this modern era of cancer research.

9.4.3 The Collaborative Effect of Multiple Tumor Suppressors in Chromosome Deletions

There are accumulating evidences indicating that chromosome deletions are powerful drivers for carcinogenesis and distinguishable to deficiency of single tumor suppressors. A plausible explanation is that there are multiple tumor suppressors in a chromosome deletion region and these tumor suppressors collaborate to inhibit tumor genesis and progress. To dissect these cooperative tumor suppressors, shRNA, CRISPR/Cas9 and ORF library screening have been successfully performed on several commonly deleted chromosome regions. Since chromosome 17p has tumor suppression capacity beyond TP53, it was proposed that there were other tumor suppressors besides TP53 in this region. To identify potential new tumor suppressors in chromosome 17p, Liu et al. generated a shRNA library against all of the coding genes except p53 and performed a high throughput in vivo screening. Multiple candidate tumor suppressors, including a cluster of Alox genes, were scored. After validating Eif5a and Alox15b as tumor suppressors in lymphoma, they further showed that simultaneously knocking down Eif5a and p53, or Alox15b and p53 led to shorter tumor-free survival of recipient mice compared to knocking down any single of these genes, indicating the collaboration between Eif5a and p53, and Alox15b and p53, respectively [10]. Kotini et al. applied ORF screening to identify key players in chromosome 7q with an iPS cell-blood cell differentiation assay. Multiple candidate tumor suppressors were hit and further work is needed to validate them in the context of AML genesis (Fig. 9.1) [11].

More high throughput library screenings have been performed in multiple cancer types. Zender et al. did in vivo shRNA library screening for genes recurrently deleted in human HCC cells in a mouse HCC model and identified 12 novel tumor suppressors [14]. Further they showed that these tumor suppressors from chromosome 8p could synergistically restrained HCC growth at least in mice [96]. A survey of genes in 82 recurrently focal deletions from 3131 tumors, Solimini demonstrated that these regions are rich of so called STOP genes than GO genes, which negatively and positively regulated cell growth and proliferation. They proposed that though majority of these STOP genes were hemizygously deleted and each of them had moderate effects on tumorigenesis, the cumulative haploinsufficien-

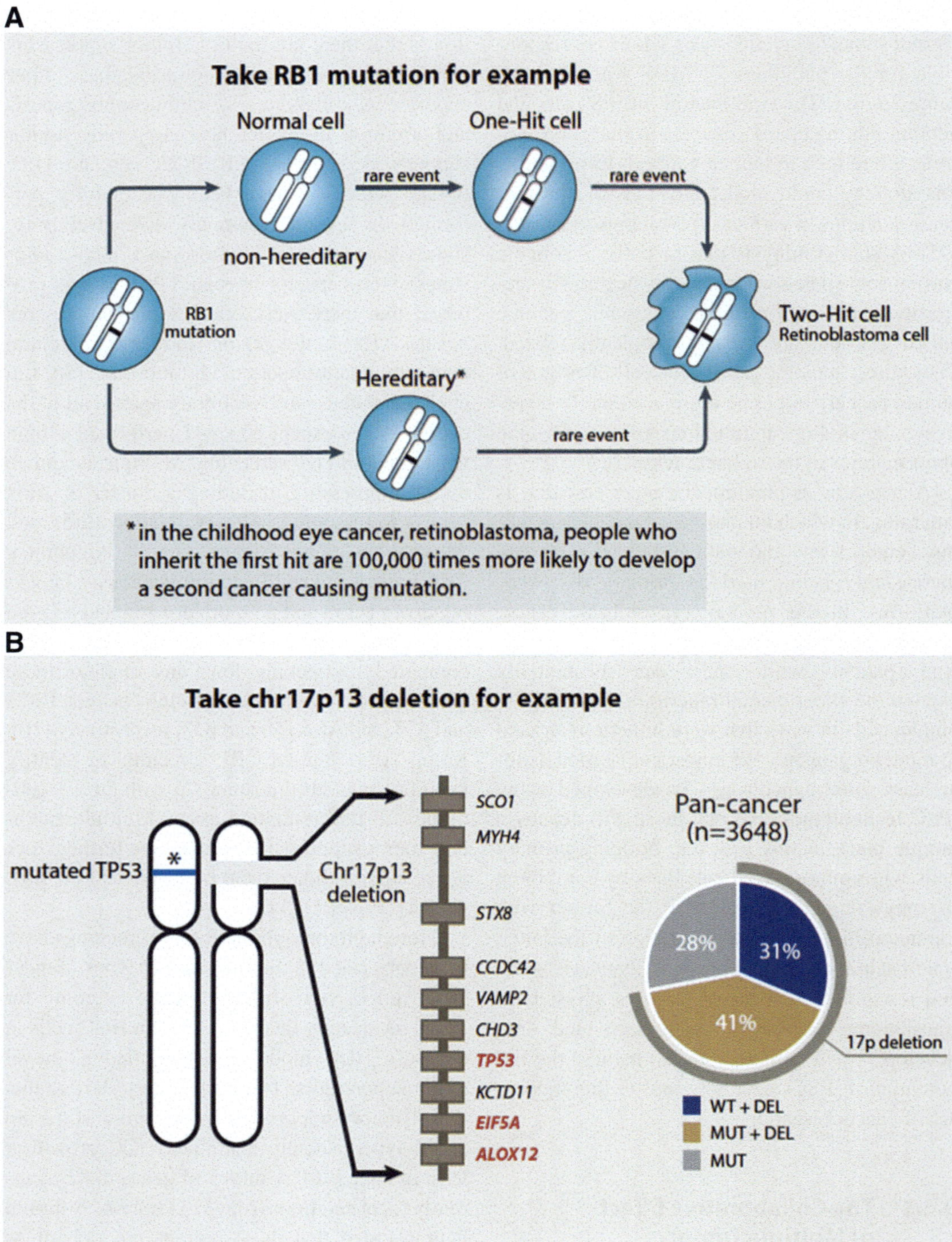

Fig. 9.1 (**a**) Knudson "Two-hit" theory of tumorigenesis. (**b**) "synergy of multiple tumor suppressors" theory on the role of chromosome large deletions in human cancers

cies led to tumorigenesis, which explained the driver role of chromosome deletions in human cancer [8, 97].

9.5 Perspective

It has been over 100 years since Hansemann's initial observations of chromosome abnormalities in cancer and Boveri's seminal hypothesis of chromosome alterations as drivers of cancer. Amounting data have documented them as a hallmark and association with pathology and prognosis of cancer. However, partially due to the technical challenges, we just start to understand the mechanisms of this critical phenomenon in cancer with both conceptual and technic breakthroughs. Solid evidences have provided that chromosome deletions are distinguishable and powerful drivers of cancers. These critical drivers display significant characteristics in terms of genetic configurations, biological consequences and more important, treatment vulnerabilities [98]. For example, passenger deletions of ENO1 in chromosome 1p36 give rise to sensitivity of the mutant GBM cells to ENO2 inhibition [99]. Chromosome deletions, together with other chromosome abnormalities, might also change the expressions of certain immune markers through unknown mechanisms, rendering the affected cancer cells resistance to immunotherapies [100]. Thus further efforts are in need to fully understand the biological functions, molecular mechanisms and vulnerabilities for the treatment of the diseases driven by these numerous and notorious chromosome abnormalities.

References

1. Hanahan D, Weinberg RA (2011) Hallmarks of cancer: the next generation. Cell 144:646–674. https://doi.org/10.1016/j.cell.2011.02.013
2. Hansemann D (1890) Asymmetrical cell division in epithelial cancers and its biological significance. Arch Pathol Anat etc Berl 119:299–326
3. Hansemann D (1891) Festschr. Rudolf Virchow, von seine Assistenten, Berl:1–12
4. Boveri T (1929) The origin of malignant tumors. Williams and Wilkins, Boston
5. Nowell PCH (1960) D.A. a minute chromosome in human chronic granulocytic leukemia. Science 132:1497
6. Shlien A, Malkin D (2009) Copy number variations and cancer. Genome Med 1:62. https://doi.org/10.1186/gm62
7. Sherr CJ (2004) Principles of tumor suppression. Cell 116:235–246
8. Solimini NL et al (2012) Recurrent hemizygous deletions in cancers may optimize proliferative potential. Science 337:104–109. https://doi.org/10.1126/science.1219580
9. Cai Y et al (2016) Loss of chromosome 8p governs tumor progression and drug response by altering lipid metabolism. Cancer Cell 29:751–766. https://doi.org/10.1016/j.ccell.2016.04.003
10. Liu Y et al (2016) Deletions linked to TP53 loss drive cancer through p53-independent mechanisms. Nature 531:471–475. https://doi.org/10.1038/nature17157
11. Kotini AG et al (2015) Functional analysis of a chromosomal deletion associated with myelodysplastic syndromes using isogenic human induced pluripotent stem cells. Nat Biotechnol 33:646–655. https://doi.org/10.1038/nbt.3178
12. Wong JC et al. (2015) Functional evidence implicating chromosome 7q22 haploinsufficiency in myelodysplastic syndrome pathogenesis. Elife 4. https://doi.org/10.7554/eLife.07839
13. Chen C et al (2014) MLL3 is a haploinsufficient 7q tumor suppressor in acute myeloid leukemia. Cancer Cell 25:652–665. https://doi.org/10.1016/j.ccr.2014.03.016
14. Zender L et al (2008) An oncogenomics-based in vivo RNAi screen identifies tumor suppressors in liver cancer. Cell 135:852–864. https://doi.org/10.1016/j.cell.2008.09.061
15. Mader SS (2007) Biology. 9th edn. McGraw Hill Higher Education, New York
16. Rowley JDL (1973) A new consistent chromosomal abnormality in chronic myelogenous leukaemia identified by quinacrine fluorescence and Giemsa staining. Nature 243:290–293
17. Druker BJ et al (2001) Activity of a specific inhibitor of the BCR-ABL tyrosine kinase in the blast crisis of chronic myeloid leukemia and acute lymphoblastic leukemia with the Philadelphia chromosome. N Engl J Med 344:1038–1042. https://doi.org/10.1056/NEJM200104053441402
18. Druker BJ et al (2001) Efficacy and safety of a specific inhibitor of the BCR-ABL tyrosine kinase in chronic myeloid leukemia. N Engl J Med 344:1031–1037. https://doi.org/10.1056/NEJM200104053441401
19. Caspersson T, Zech L, Johansson C, Modest EJ (1970) Identification of human chromosomes by DNA-binding fluorescent agents. Chromosoma 30:215–227
20. Speicher MR, Carter NP (2005) The new cytogenetics: blurring the boundaries with molecular biology.

Nat Rev Genet 6:782–792. https://doi.org/10.1038/nrg1692
21. Langer-Safer PR, Levine M, Ward DC (1982) Immunological method for mapping genes on Drosophila polytene chromosomes. Proc Natl Acad Sci U S A 79:4381–4385
22. Kallioniemi A et al (1992) Comparative genomic hybridization for molecular cytogenetic analysis of solid tumors. Science 258:818–821
23. Pinkel D, Albertson DG (2005) Array comparative genomic hybridization and its applications in cancer. Nat Genet 37(Suppl):S11–S17. https://doi.org/10.1038/ng1569
24. Beroukhim R et al (2010) The landscape of somatic copy-number alteration across human cancers. Nature 463:899–905. https://doi.org/10.1038/nature08822
25. Bignell GR et al (2010) Signatures of mutation and selection in the cancer genome. Nature 463:893–898. https://doi.org/10.1038/nature08768
26. Meyerson M, Gabriel S, Getz G (2010) Advances in understanding cancer genomes through second-generation sequencing. Nat Rev Genet 11:685–696. https://doi.org/10.1038/nrg2841
27. Albertson DG, Collins C, McCormick F, Gray JW (2003) Chromosome aberrations in solid tumors. Nat Genet 34:369–376. https://doi.org/10.1038/ng1215
28. Kenneth Kaushansky ML, Prchal J, Levi MM, Press O, Burns L, Caligiuri M (2015) Williams hematology, 9th edn. McGraw-Hill Education, New York
29. Neuman WL et al (1992) Chromosomal loss and deletion are the most common mechanisms for loss of heterozygosity from chromosomes 5 and 7 in malignant myeloid disorders. Blood 79:1501–1510
30. Zabarovsky ER, Lerman MI, Minna JD (2002) Tumor suppressor genes on chromosome 3p involved in the pathogenesis of lung and other cancers. Oncogene 21:6915–6935. https://doi.org/10.1038/sj.onc.1205835
31. Greenberg P et al (1997) International scoring system for evaluating prognosis in myelodysplastic syndromes. Blood 89:2079–2088
32. Nimer SD (2006) Clinical management of myelodysplastic syndromes with interstitial deletion of chromosome 5q. J Clin Oncol 24:2576–2582
33. Haase D et al (2007) New insights into the prognostic impact of the karyotype in MDS and correlation with subtypes: evidence from a core dataset of 2124 patients. Blood 110:4385–4395
34. Soenen V et al (1998) 17p deletion in acute myeloid leukemia and myelodysplastic syndrome. Analysis of breakpoints and deleted segments by fluorescence in situ. Blood 91:1008–1015
35. Sterkers Y et al (1998) Acute myeloid leukemia and myelodysplastic syndromes following essential thrombocythemia treated with hydroxyurea: high proportion of cases with 17p deletion. Blood 91:616–622
36. Smith SM et al (2003) Clinical-cytogenetic associations in 306 patients with therapy-related myelodysplasia and myeloid leukemia: the University of Chicago series. Blood 102:43–52
37. Levine EG et al (1990) Sequential karyotypes in non-Hodgkin lymphoma: their nature and significance. Genes Chromosom Cancer 1:270–280
38. Caporaso N et al (2007) Chronic lymphocytic leukaemia genetics overview. Brit J Haematol 139:630–634
39. Chng W, Glebov O, Bergsagel P, Kuehl W (2007) Genetic events in the pathogenesis of multiple myeloma. Best Pract Res Clin Haematol 20:571–596
40. Yokota J, Wada M, Shimosato Y, Terada M, Sugimura T (1987) Loss of heterozygosity on chromosomes 3, 13, and 17 in small-cell carcinoma and on chromosome 3 in adenocarcinoma of the lung. Proc Natl Acad Sci 84:9252–9256
41. Saito H et al (1993) Detailed deletion mapping of chromosome 17q in ovarian and breast cancers: 2-cM region on 17q21. 3 often and commonly deleted in tumors. Cancer Res 53:3382–3385
42. Emi M et al (1992) Frequent loss of heterozygosity for loci on chromosome 8p in hepatocellular carcinoma, colorectal cancer, and lung cancer. Cancer Res 52:5368–5372
43. Garcia JM et al (1999) Allelic loss of the PTEN region (10q23) in breast carcinomas of poor pathophenotype. Breast Cancer Res Treat 57:237–243
44. Van den Berghe H et al (1974) Distinct haematological disorder with deletion of long arm of no. 5 chromosome. Nature 251:437–438
45. List A et al (2006) Lenalidomide in the myelodysplastic syndrome with chromosome 5q deletion. N Engl J Med 355:1456–1465. https://doi.org/10.1056/NEJMoa061292
46. Boultwood J, Pellagatti A, McKenzie AN, Wainscoat JS (2010) Advances in the 5q- syndrome. Blood 116:5803–5811. https://doi.org/10.1182/blood-2010-04-273771
47. Dastugue N et al (1995) Prognostic significance of karyotype in de novo adult acute myeloid leukemia. The BGMT group. Leukemia 9:1491–1498
48. Qian Z et al (2010) Cytogenetic and genetic pathways in therapy-related acute myeloid leukemia. Chem Biol Interact 184:50–57. https://doi.org/10.1016/j.cbi.2009.11.025
49. Johnson E, Cotter FE (1997) Monosomy 7 and 7q--associated with myeloid malignancy. Blood Rev 11:46–55
50. Jerez A et al (2012) Loss of heterozygosity in 7q myeloid disorders: clinical associations and genomic pathogenesis. Blood 119:6109–6117. https://doi.org/10.1182/blood-2011-12-397620
51. Le Beau MM et al (1996) Cytogenetic and molecular delineation of a region of chromosome 7 commonly deleted in malignant myeloid diseases. Blood 88:1930–1935
52. Liang H et al (1998) Molecular anatomy of chromosome 7q deletions in myeloid neoplasms: evidence for multiple critical loci. Proc Natl Acad Sci U S A 95:3781–3785

53. Jary L et al (1997) The 17p-syndrome: a distinct myelodysplastic syndrome entity? Leuk Lymphoma 25:163–168. https://doi.org/10.3109/10428199709042506
54. Sankar M et al (1998) Identification of a commonly deleted region at 17p13.3 in leukemia and lymphoma associated with 17p abnormality. Leukemia 12:510–516
55. Miller LD et al (2005) An expression signature for p53 status in human breast cancer predicts mutation status, transcriptional effects, and patient survival. Proc Natl Acad Sci U S A 102:13550–13555. https://doi.org/10.1073/pnas.0506230102
56. Rowley JD, Le Beau MM (1989) Cytogenetic and molecular analysis of therapy-related leukemia. Ann N Y Acad Sci 567:130–140
57. Mermel CH et al (2011) GISTIC2.0 facilitates sensitive and confident localization of the targets of focal somatic copy-number alteration in human cancers. Genome Biol 12:R41. https://doi.org/10.1186/gb-2011-12-4-r41
58. Knudson AG Jr (1971) Mutation and cancer: statistical study of retinoblastoma. Proc Natl Acad Sci U S A 68:820–823
59. Knudson AG (2001) Two genetic hits (more or less) to cancer. Nat Rev Cancer 1:157–162. https://doi.org/10.1038/35101031
60. Machiela MJ et al (2016) Mosaic 13q14 deletions in peripheral leukocytes of non-hematologic cancer cases and healthy controls. J Hum Genet 61:411–418. https://doi.org/10.1038/jhg.2015.166
61. Witkiewicz AK, Knudsen KE, Dicker AP, Knudsen ES (2011) The meaning of p16(ink4a) expression in tumors: functional significance, clinical associations and future developments. Cell Cycle 10:2497–2503. https://doi.org/10.4161/cc.10.15.16776
62. Sulong S et al (2009) A comprehensive analysis of the CDKN2A gene in childhood acute lymphoblastic leukemia reveals genomic deletion, copy number neutral loss of heterozygosity, and association with specific cytogenetic subgroups. Blood 113:100–107. https://doi.org/10.1182/blood-2008-07-166801
63. Levine AJ, Oren M (2009) The first 30 years of p53: growing ever more complex. Nat Rev Cancer 9:749–758. https://doi.org/10.1038/nrc2723
64. Olivier M, Hollstein M, Hainaut P (2010) TP53 mutations in human cancers: origins, consequences, and clinical use. Cold Spring Harb Perspect Biol 2:a001008. https://doi.org/10.1101/cshperspect.a001008
65. Varley J, Germline M (2003) TP53 mutations and Li-Fraumeni syndrome. Hum Mutat 21:313–320. https://doi.org/10.1002/humu.10185
66. Steck PA et al (1997) Identification of a candidate tumour suppressor gene, MMAC1, at chromosome 10q23.3 that is mutated in multiple advanced cancers. Nat Genet 15:356–362. https://doi.org/10.1038/ng0497-356
67. Costa RM, Silva AJ (2002) Molecular and cellular mechanisms underlying the cognitive deficits associated with neurofibromatosis 1. J Child Neurol 17:622–626; discussion 627–629, 646–651. https://doi.org/10.1177/088307380201700813
68. Albertsen H et al (1994) Genetic mapping of the BRCA1 region on chromosome 17q21. Am J Hum Genet 54:516–525
69. Nordstrom-O'Brien M et al (2010) Genetic analysis of von Hippel-Lindau disease. Hum Mutat 31:521–537. https://doi.org/10.1002/humu.21219
70. Santarosa M, Ashworth A (2004) Haploinsufficiency for tumour suppressor genes: when you don't need to go all the way. Biochim Biophys Acta 1654:105–122. https://doi.org/10.1016/j.bbcan.2004.01.001
71. Quon KC, Berns A (2001) Haplo-insufficiency? Let me count the ways. Genes Dev 15:2917–2921. https://doi.org/10.1101/gad.949001
72. Fero ML, Randel E, Gurley KE, Roberts JM, Kemp CJ (1998) The murine gene p27Kip1 is haplo-insufficient for tumour suppression. Nature 396:177–180. https://doi.org/10.1038/24179
73. Komuro H et al (1999) p27KIP1 deletions in childhood acute lymphoblastic leukemia. Neoplasia 1:253–261
74. Kawamata N et al (1995) Molecular analysis of the cyclin-dependent kinase inhibitor gene p27/Kip1 in human malignancies. Cancer Res 55:2266–2269
75. Shilatifard A (2012) The COMPASS family of histone H3K4 methylases: mechanisms of regulation in development and disease pathogenesis. Annu Rev Biochem 81:65–95. https://doi.org/10.1146/annurev-biochem-051710-134100
76. Zang ZJ et al (2012) Exome sequencing of gastric adenocarcinoma identifies recurrent somatic mutations in cell adhesion and chromatin remodeling genes. Nat Genet 44:570–574. https://doi.org/10.1038/ng.2246
77. Kandoth C et al (2013) Mutational landscape and significance across 12 major cancer types. Nature 502:333–339. https://doi.org/10.1038/nature12634
78. Song MS, Salmena L, Pandolfi PP (2012) The functions and regulation of the PTEN tumour suppressor. Nat Rev Mol Cell Biol 13:283–296. https://doi.org/10.1038/nrm3330
79. Chen Z et al (2005) Crucial role of p53-dependent cellular senescence in suppression of Pten-deficient tumorigenesis. Nature 436:725–730. https://doi.org/10.1038/nature03918
80. Kwabi-Addo B et al (2001) Haploinsufficiency of the Pten tumor suppressor gene promotes prostate cancer progression. Proc Natl Acad Sci U S A 98:11563–11568. https://doi.org/10.1073/pnas.201167798
81. Alimonti A et al (2010) Subtle variations in Pten dose determine cancer susceptibility. Nat Genet 42:454–458. https://doi.org/10.1038/ng.556
82. Tang YC, Amon A (2013) Gene copy-number alterations: a cost-benefit analysis. Cell 152:394–405. https://doi.org/10.1016/j.cell.2012.11.043
83. Sheltzer JM et al (2017) Single-chromosome gains commonly function as tumor suppressors.

Cancer Cell 31:240–255. https://doi.org/10.1016/j.ccell.2016.12.004
84. Adams DJ et al (2004) Mutagenic insertion and chromosome engineering resource (MICER). Nat Genet 36:867–871. https://doi.org/10.1038/ng1388
85. Cong L et al (2013) Multiplex genome engineering using CRISPR/Cas systems. Science 339:819–823. https://doi.org/10.1126/science.1231143
86. Mali P et al (2013) RNA-guided human genome engineering via Cas9. Science 339:823–826. https://doi.org/10.1126/science.1232033
87. Shalem O et al (2014) Genome-scale CRISPR-Cas9 knockout screening in human cells. Science 343:84–87. https://doi.org/10.1126/science.1247005
88. Petitjean A, Achatz MI, Borresen-Dale AL, Hainaut P, Olivier M (2007) TP53 mutations in human cancers: functional selection and impact on cancer prognosis and outcomes. Oncogene 26:2157–2165. https://doi.org/10.1038/sj.onc.1210302
89. Johnson EJ et al (1996) Molecular definition of a narrow interval at 7q22.1 associated with myelodysplasia. Blood 87:3579–3586
90. Schwartz S et al (2003) Human-mouse alignments with BLASTZ. Genome Res 13:103–107. https://doi.org/10.1101/gr.809403
91. Wasserman WW, Palumbo M, Thompson W, Fickett JW, Lawrence CE (2000) Human-mouse genome comparisons to locate regulatory sites. Nat Genet 26:225–228. https://doi.org/10.1038/79965
92. Sanchez-Rivera FJ, Jacks T (2015) Applications of the CRISPR-Cas9 system in cancer biology. Nat Rev Cancer 15:387–395. https://doi.org/10.1038/nrc3950
93. Tschaharganeh DF, Bosbach B, Lowe SW (2016) Coordinated tumor suppression by chromosome 8p. Cancer Cell 29:617–619. https://doi.org/10.1016/j.ccell.2016.04.011
94. Kotini AG et al (2017) Stage-specific human induced pluripotent stem cells map the progression of myeloid transformation to transplantable leukemia. Cell Stem Cell 20:315–328 e317. https://doi.org/10.1016/j.stem.2017.01.009
95. Papapetrou EP (2016) Patient-derived induced pluripotent stem cells in cancer research and precision oncology. Nat Med 22:1392–1401. https://doi.org/10.1038/nm.4238
96. Xue W et al (2012) A cluster of cooperating tumor-suppressor gene candidates in chromosomal deletions. Proc Natl Acad Sci U S A 109:8212–8217. https://doi.org/10.1073/pnas.1206062109
97. Davoli T et al (2013) Cumulative haploinsufficiency and triplosensitivity drive aneuploidy patterns and shape the cancer genome. Cell 155:948–962. https://doi.org/10.1016/j.cell.2013.10.011
98. Nijhawan D et al (2012) Cancer vulnerabilities unveiled by genomic loss. Cell 150:842–854. https://doi.org/10.1016/j.cell.2012.07.023
99. Muller FL et al (2012) Passenger deletions generate therapeutic vulnerabilities in cancer. Nature 488:337–342. https://doi.org/10.1038/nature11331
100. Davoli T, Uno H, Wooten EC, Elledge SJ (2017) Tumor aneuploidy correlates with markers of immune evasion and with reduced response to immunotherapy. Science 355. https://doi.org/10.1126/science.aaf8399

Processing-Challenges Generated by Clusters of DNA Double-Strand Breaks Underpin Increased Effectiveness of High-LET Radiation and Chromothripsis

10

Emil Mladenov, Janapriya Saha, and George Iliakis

Abstract

Whereas most endogenous and exogenous DNA damaging agents typically generate lesions that are relatively isolated and can be repaired easily, ionizing radiation (IR) also induces clustered lesions causing DNA double strand breaks (DSBs). Moreover, forms of IR characterized by high linear energy transfer (LET) induce not only isolated DSBs but also DSB clusters – multiple DSBs in close proximity -that pose increased risks for the cell. DSB clusters can destabilize chromatin locally and compromise processing of individual DSBs within the cluster. Since the discovery of chromothripsis, a phenomenon whereby multiple DSBs locally generated by a catastrophic event causes genomic rearrangements that feed carcinogenesis, DSB clusters receive increased attention also in the field of cancer. While formation of DSB clusters after exposure to high LET is a direct and inherent consequence of the spatial distribution of the constituting energy deposition events, also called track structure, the sources of local genomic shattering underpinning chromothripsis are under investigation. Notably, many consequences of DSB clusters in the affected genome reflect processing by pathways that have evolved to repair DSBs, but which operate with widely different degrees of fidelity. The molecular underpinnings and the basis of the underlying repair pathway choices that ultimately lead to the observed consequences from DSB clusters remain unknown. We developed a tractable model of DSB clustering that allows direct analysis in cells of the consequences of certain configurations of DSB clusters. We outline the rationale for the development of this model and describe its key characteristics. We summarize results suggesting that DSB clusters compromise the first-line DSB-processing pathways of c-NHEJ and HRR, increasing as a consequence the contribution of alt-EJ, which has high propensity of generating chromosomal rearrangements. The results suggest a mechanism for the increased toxicity of high LET radiation and the extensive genomic rearrangements associated with chromothripsis.

Keywords

High LET · DSB clusters · Chromosomal translocations

E. Mladenov · J. Saha · G. Iliakis (✉)
Institute of Medical Radiation Biology, University of Duisburg-Essen Medical School, Essen, Germany
e-mail: Georg.Iliakis@uk-essen.de

Y. Zhang (ed.), *Chromosome Translocation*, Advances in Experimental Medicine and Biology 1044,
https://doi.org/10.1007/978-981-13-0593-1_10

10.1 Introduction

Ruptures in the genome in the form of DNA double strand breaks (DSBs) and the consequences of their erroneous processing are central not only to carcinogenesis but also to the toxicity of ionizing radiation (IR). This is because although IR is an oxidizing agent in a broad sense, it generates most of its adverse biological consequences in a unique way. Specifically, whereas many of the ionizing events associated with exposures to X-rays or γ-rays are spatially well-separated, a proportion of them occurs as clusters of ionizations that generate clusters of damage when they hit the DNA. One such form of clustered DNA damage is the DSB, which is generated when two single strand breaks are induced in opposite DNA strands, but only a few base pairs apart. It is this clustering of ionizations and the associated clustering of DNA single strand breaks that combine to form a DSB that makes IR a higher risk agent than many common oxidizing agents. As we discuss below, owing to the presence of damages in both DNA strands, repair of DSBs cannot rely on the complementary strand, as other DNA repair pathways do, and resorts therefore to unique solutions.

Whereas ionizations by electrons generated after absorption of the photons constituting X-rays or γ-rays are relatively randomly distributed in the irradiated space and typically cluster only at the ends of electron tracks (Fig. 10.1a, b), ionizations produced by heavier charged particles, such as protons and particularly by even heavier nuclei, tend to localize along the particle track and to be highly clustered –especially at the end of the particle's range (Fig. 10.1c, d). Notably, as the mass and charge of a particle increases, the clustering of ionizations along its track also increases. This clustering is described by the parameter of linear energy transfer (LET) that gives the energy imparted by the particle in the irradiated medium close to the particle track per unit of track-length (Fig. 10.1). Thus, after exposure to high LET radiation, a higher proportion of the constituting ionizing events will be clustered.

The effects of IR are uniformly described as a function of the administered radiation dose, which is defined as energy in Joules absorbed per Kg of mass and is given in Gray (1Gy = 1 J/Kg). One key discovery in the field of radiation biology is that the same dose of two radiation modalities – expected to produce approximately the same number of ionizations – causes dramatically different biological effects (Fig. 10.1e). For example deposition of energy to cells from α-particles causes orders of magnitude more killing than deposition of the same energy from X-rays (Fig. 10.1e), i.e. the biological effect at the same radiation dose, increases dramatically with increasing LET of radiation.

This unique and in many ways unprecedented increase in effect directly demonstrates that the increased clustering of ionizations (Fig. 10.1d) that is inherent in high LET IR increases biological efficacy. Considering that DNA is the main target for IR-induced killing, the inference that clustered DNA damage from ionization clusters underpins increased biological efficacy is obvious. Yet, half a century after discovery of these effects (Fig. 10.1e), the nature of DNA damage-clustering and the reasons for the dramatically increased severity of biological consequences remain poorly characterized. In the following chapters, we systematically analyze consequences of damage clustering in the DNA, introduce the concept of DSB complexity and outline how certain levels of DSB complexity affect DSB repair pathway choice and contribute to the observed biological effects.

10.2 Forms of DSBs and DSB Clusters

While DSBs are often regarded as a single lesion, observations such as those discussed above suggest that further classification is required for improved understanding of the spectrum and magnitude of IR-induced biological effects. To facilitate such analyses, we generated a classification of events causing DSBs that considers the underlying "event complexity": defined as number of events underpinning the ultimate breakage of the DNA molecule (Fig. 10.2) [102]. In this classification, Type 1 (T1) DSBs, the simplest form, are those typically generated by restriction

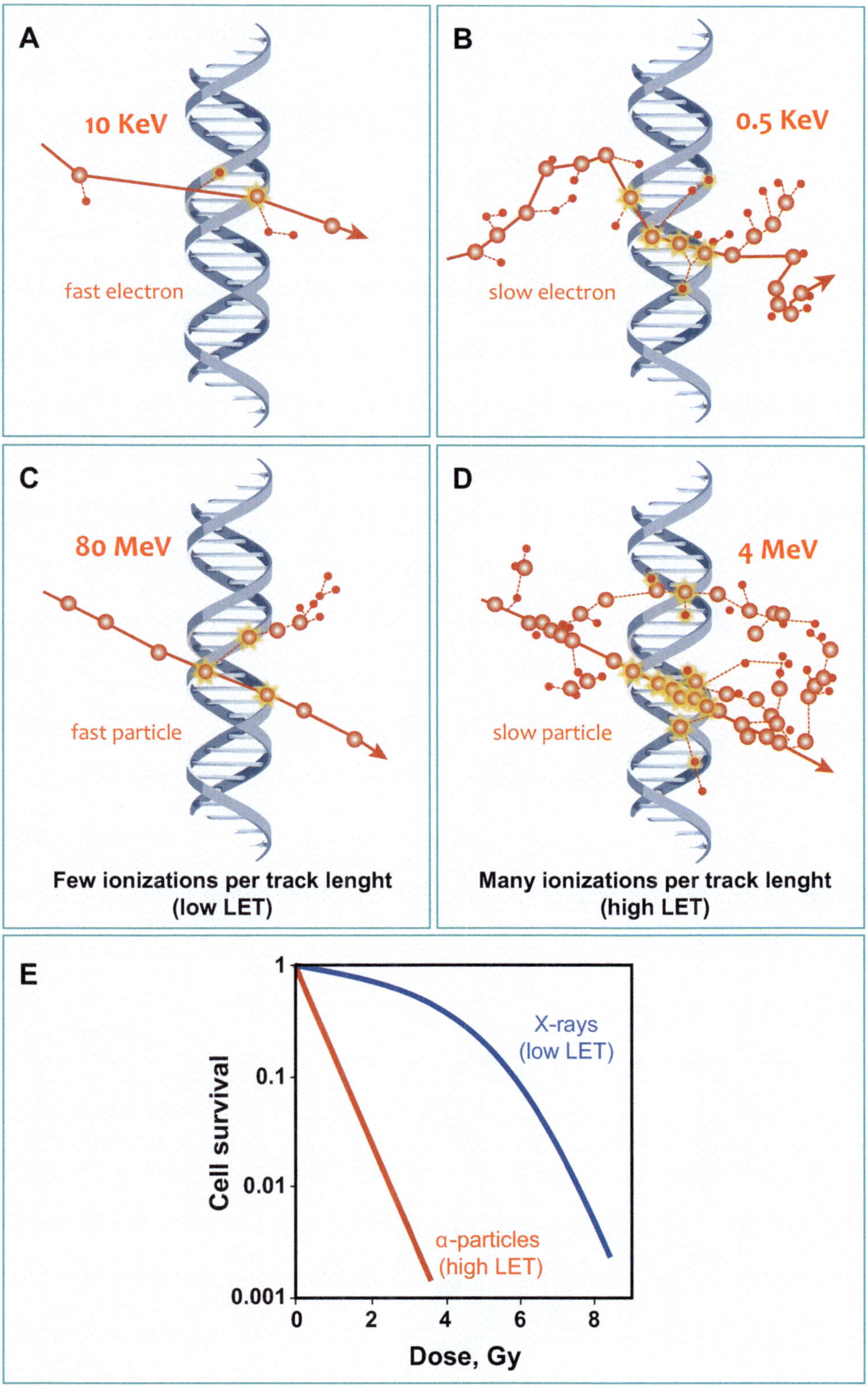

Fig. 10.1 Generation of ionization clusters by low and high LET irradiation (**a**) after irradiation with fast electrons, (**b**) after irradiation with slow electrons, (**c**) after irradiation with fast particles and (**d**) after irradiation with slow particles. (**e**) Survival curves after exposure of cells to low and high LET radiation

Type	DNA presentation	Properties
T1	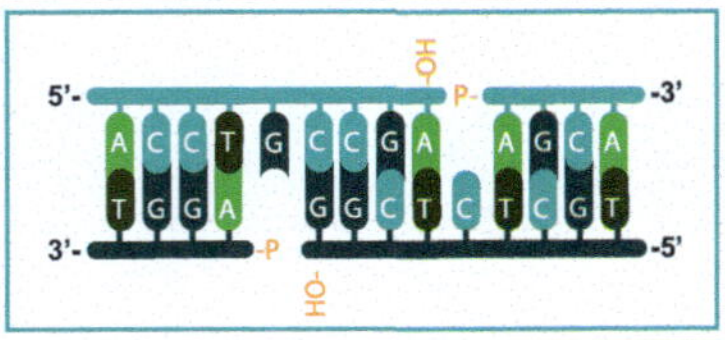	Enzymatically generated DSB. No lesions in DNA. Direct ligation possible.
T2	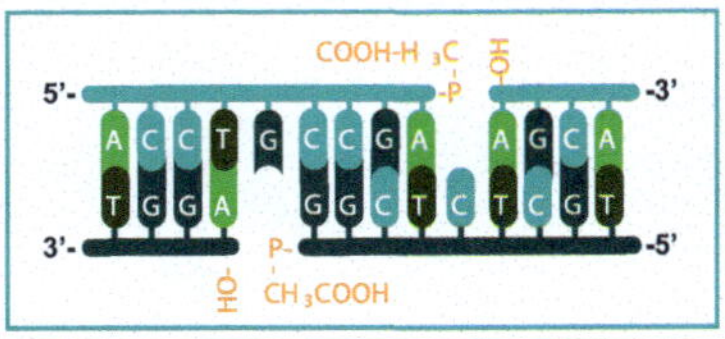	Chemical disruption of sugar-phosphate backbone. DNA end modifications. End processing before ligation required.
T3	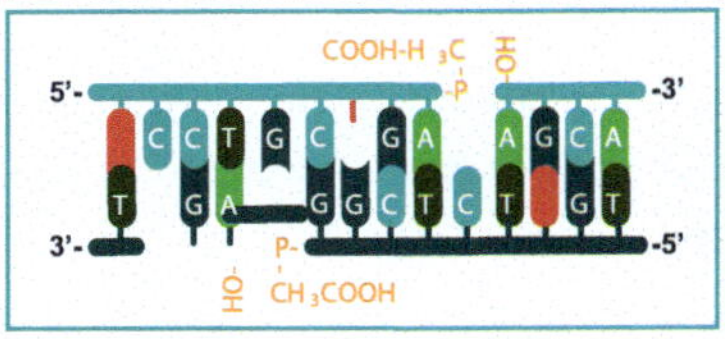	T2-DSB accompanied by SSB or/and base lesions ( ). Extensive processing required before ligation. Repair patway conflict possible.
T4	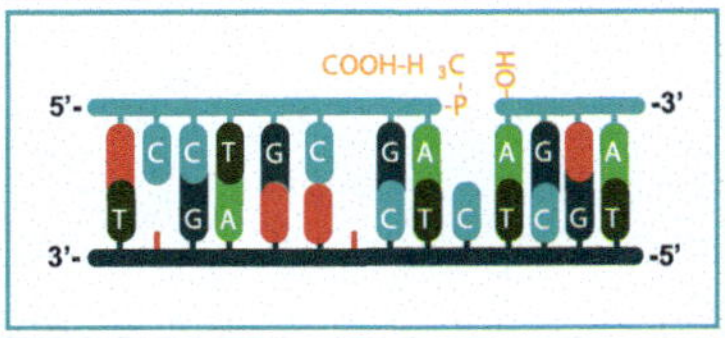	DSB formation only after enzymatic processing of base lesions on intact strand. Extensive processing required before ligation. Repair patway conflict possible.
T5	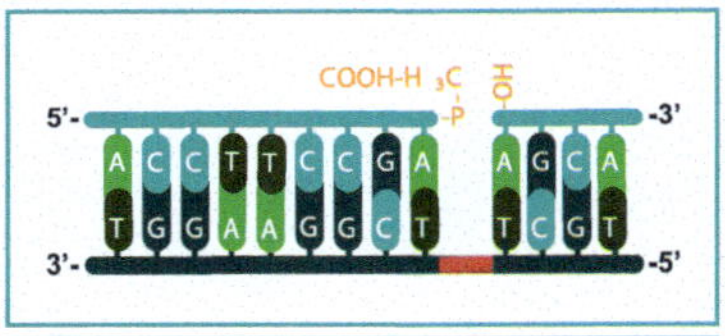	DSB formation only after chemical evolution of thermaly-labille sugar lesions on intact strand to generate strand break. Repair patway conflict possible.
T6	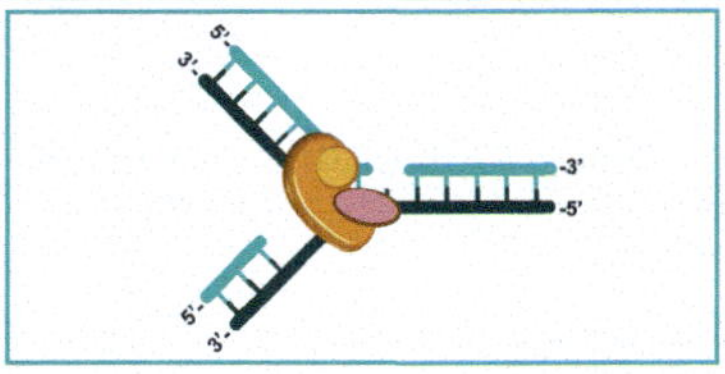	DSB formation at stalled or collapsed DNA replication forks. Formation of one ended DSB.
T7	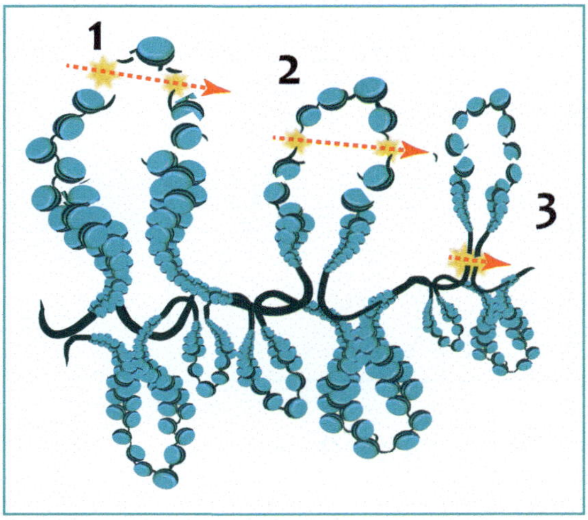	DSB - clusters 1. Single nucleosome loss by DSB pair 2. Multiple nucleosome loss by DSB pair 3. Chromatin loop loss by DSB pair Chromatin destabilization. Clusters with multiple DSB also possible in condensed chromatin.

Fig. 10.2 Nomenclature of DSB types (See text for details)

endonucleases (RE), including meganucleases and the CRISPR/Cas9 system. These families of proteins simultaneously disrupt the phosphodiester bonds on both DNA strands to generate directly-ligatable blunt or staggered ends without chemically modifying DNA constituents. RE are frequently used as model reagents to generate DSBs in cells and study their biological consequences. Advantage of this approach is that DSBs are generated at a defined/known genomic location, in a known sequence context. The outcome of DSB processing can be analyzed using appropriately constructed reporters, or approaches, such as those described below [102].

When DSBs are induced by IR instead of RE, the alterations generated in the DNA molecule go beyond the clean, sequence-defined digestion of the phosphodiester backbone. They are random and include damaged sugar and base moieties. It is known that chemically active DNA end modifications are frequently generated by the extraction of a hydrogen atom at 4′-C or 5′-C position of the sugar moiety and the addition of oxygen [91]. The DNA ends thus generated are frequently modified with nucleoside 5′-aldehyde on the 5′-terminus; at the 3′-terminus phosphoglycolate (PG), phosphoglycoaldehyde, formyl phosphate, and 3′-keto-2′-deoxynucleotide frequently occur [47, 53]. Such residues are recurrently unstable and are eventually converted to a hydroxyl (-OH) group with the exception of PG, which frequently generates a 3′ phosphoglycolate and a 5′-OH (Fig. 10.2). This form of ends precludes direct DNA ligation and necessitates end-processing for repair [127]. We, therefore term this form of more complex DSB Type 2 or, T2-DSB. The added level of chemical complexity of a T2-DSB at its ends, necessitates additional processing and thus more steps in the ensuing repair activities.

As mentioned above, DSBs generated by low LET IR, such as X-rays or γ-rays, form from individual ionization clusters produced at the ends of electron tracks [38, 125] rather than by the coincidence of independent ionizations on opposite DNA strands, hence, the linear induction of DSBs as a function of IR dose. In Type 3 (T3) DSBs, complexity increases further by the presence of additional lesions in the immediate vicinity of the DNA break – such as base damages or SSBs induced by ionizations in the same cluster. After exposure to high LET IR, ionization clusters are even more frequent and larger; as a result the induction of T3-DSBs will increase. Indeed, whereas only about 30% of DSBs are expected to contain lesions in addition to the two strand breaks after exposure to low energy electrons, this fraction increases up to 70% after exposure to α-particles. Also, the ratio of the number of SSBs to DSBs decreases from 22.8 for ^{60}Co γ-rays to 3.4 for 50 MeV ^{12}C-ions [3, 80–82]. Since increases in the yield of DSBs alone cannot explain the increased killing observed after exposure to high versus low LET radiation, it is frequently hypothesized that increased clustering of damage at the DSB, as in T3-form, is an important determinant of the resulting biological effect [37].

Type 4 (T4) and 5 (T5) DSBs (Fig. 10.2) comprise DNA damaging events from ionization clusters that do not break the DNA directly, but do so after enzymatic opening of the damaged strand to repair a lesion (T4-DSBs), or the chemical evolution of a non-strand-breaking sugar lesion to a DNA strand-breaking entity (T5-DSBs). In T5-DSBs, the chemical evolution of the sugar lesion is accelerated at temperatures above 10 °C. The importance of this form of DSBs in the biological consequences of IR has been reviewed earlier [13, 102].

Another form of DSB derives from DNA single strand breaks or base damages at the sites of replication forks in replication competent cells. The initiating lesions can be induced in addition to IR by radiomimetic drugs, topoisomerase or polymerase inhibitors etc. They can be converted to one-ended DSBs during DNA replication, which we classify as Type 6 (T6) DSBs (Fig. 10.2).

A further level of DSB complexity, that is increasingly considered highly relevant for high LET effects, are clusters of DSBs (T7-DSBs), where the individual DSBs can in principle belong to any of the above defined types. DSB clustering as a peculiarity of high LET IR and a cause of adverse radiation effects has been experimentally considered by several investigators. Bryant and colleagues [56] observed that DSB clustering in higher order chromatin loops affects DSB reparability. Fast repair is measured in loops containing a single DSB, but slow repair occurs in loops containing multiple DSBs. Small DNA fragments presumably originating from DSB clusters were detected using pulsed-field gel electrophoresis after high LET IR and implicated in the effects observed [65, 99]. Atomic force microscopy imaging also shows the induction of clustered DSBs even when irradiating "naked" DNA, and indeed such fragments inhibit DNA-PK activity in vitro [87]. Small (<70 bp) DNA fragments generated from clustered DSBs have also been implicated by Wang et al. [123] in the enhanced killing observed after exposure of cells to high LET IR and are proposed to act by virtue of their inability to accommodate bi-directional binding of Ku-protein.

Immunofluorescence, super-resolution microscopy and transmission electron microscopy (TEM) approaches for detecting DSBs after high LET exposure further support the formation of DSB clusters [67–69, 118]. Indeed, multiple DSBs are visualized as ionizing radiation-induced foci (IRIF) along the tracks of charged particles by γ-H2AX, pS1981-ATM or 53BP1 detection [2, 23, 35, 54, 55]. Foci forming after high LET IR are brighter and larger and may comprise multiple DSBs [48, 66, 100]. Also, high resolution analysis by TEM using gold-labeled antibodies against a phosphorylated form of Ku70 reveals that high-LET irradiation generates clusters with multiple DSBs that can reach, depending on local state of chromatin condensation, densities of ~500 DSBs per μm^3 track volume [67].

The generation of DSB clusters and their contribution to high LET IR-induced cell killing has also been the subject of extensive mathematical modeling. Holley and Chatterjee considered DSB clusters as a particularly consequential form of radiation damage and performed Monte Carlo simulations for the induction of such clusters in chromatin with increasing LET [45]. The calculations showed an increase in the production of DSB clusters with increasing LET according to rules that were compatible with the revolution period of the DNA about the histone core and the periodicity of nucleosomes in a solenoid model of chromatin. Ostashevsky et al. [84, 85] considered in his model that small DNA fragments are generated by IR but are unstable and can be lost from the chromatin context, thus compromising repair. A more specialized induction of DSB clusters within chromatin loops, similar to that considered by Bryant and Johnston, has been used to develop alternative mathematical models by Scholz et al. [26, 31, 32], as well as by Friedland et al. [27–30] and Cucinotta et al. [89, 90].

It is evident from this outline that DSB clustering and the associated production of small DNA fragments have been extensively considered for describing IR effects, particularly high LET IR effects. Notably, all approaches taken hitherto to understand the consequences of T7-DSBs are indirect and mainly based on mathematical modeling. Conversely, experiments documenting the formation of small DNA fragments are rarely accompanied by demonstrations of directly linked biological effects.

It is highly relevant to mention at this point that the nature of DSB induction by IR precludes mechanistic experiments on T7-DSBs using IR as inducing agent. This is because in an irradiated cell population, each of the irradiated cells sustains DSBs in a stochastic manner at different numbers (spread around an average) and severity, which are randomly distributed throughout the genome. As a result, it is impossible to analyze the processing of each type of DSB individually. This has been a major handicap in all attempts to implicate DSB clusters in the effects of high LET IR. We outline below a model developed in our laboratory for the generation of DSB-clusters by *I-SceI* endonuclease that overcomes some of these limitations and allows the generation of results that help our understanding of the underlying molecular processes underpinning the observed biological effects.

10.3 Physiological Processes That Require the Generation of DSB Clusters

There are physiological process that require or are associated with the formation of DSB clusters. Two essential processes, V(D)J recombination and class-switch recombination, which are required for the maturation of the immune system are mediated by the programmed and highly regulated induction of clustered DSBs [1, 25].

It has also been reported that the DSBs together with the molecular factors involved in their repair are critical for neural development, and that brain cells frequently display somatic genomic variations that might involve DSB intermediates [126]. Multiple recurrent DSB clusters have been characterized within gene structures of primary neural stem/progenitor cells (NSPCs) by unbiased, high-throughput approaches designed to identify genomic regions with such DSB clusters [126]. Along those lines, the physiological functions of programmed DNA breaks in signal-induced transcription reveals that indeed DSBs are present near transcription start sites (TSSs) in neural stem and progenitor cells. In addition, recurrent DSB clusters are located within long, transcriptionally active, late-replicating genes [93].

Notably, the formation of DSBs by Spo11 nuclease is a prerequisite for initiation of meiotic recombination, which plays an important role in genetic diversity [61, 95]. Pulse-field gel electrophoresis (PFGE) experiments indicate that meiotic recombination associated DSBs occur at numerous locations within the genome, and could be organized in clusters, separated by about 50–200 nucleotides [133] with no sequence preference [19]. Clustering of meiotic DSBs has been also shown by experimental analysis of recombination hot-spots mapping along chromosome 3 of *S. cerevisiae* [4].

10.4 DSB Cluster Formation and Carcinogenesis: The Phenomenon of Chromothripsis

A phenomenon of relevance to the above discussion that has been the focus of attention since its recent discovery is chromothripsis. Indeed, analysis of cancer genomes by next generation sequencing frequently reveals the presence of massive genomic rearrangements, acquired through a single catastrophic event affecting a small proportion of the genome [96, 114]. There is evidence that single chromothriptic events affecting one or few chromosomes occur in malignant carcinomas, and such events are considered critical for the evolution of cancerous phenotype [57, 71]. On the basis of its constitution, Type 7 DSBs are a form of local chromothripsis, and strikingly chromosome shattering is detectable after high-LET radiation exposures [40].

How such massive chromosomal pulverization, or chromothripsis events are generated remains unknown, but recent models consider premature chromosome condensation in cell-cycle-progression-lagging micronuclei a likely cause [17, 135]. Combination of live-cell imaging and single-cell genome sequencing suggests that micronuclei formation is a prerequisite for generation of the specific spectrum of genomic rearrangements, which recapitulate all known features of chromothripsis [16, 135]. Importantly, these events are found restricted to mis-segregated chromosomes and are revealed in just one cell division [16].

The above outline summarizes physiological and exogenous sources of DSB clustering and shows its relevance to normal cell physiology, the evolution of cancer, as well as the biological effects of high LET IR. Yet, the underlying causes for the risks posed by DSB clusters are not immediately obvious. Indeed, one could postulate that DSBs within clusters can be repaired with the same ease as individual DSBs.

Below, we describe a model system that helps to experimentally address consequences of defined constellations of DSB clusters in living cells and in the processing of constituent DSBs. However, before discussing results obtained with this model system, it will be useful to outline relevant properties of the repair pathways engaged in DSB processing and their contributions to genome stability.

10.5 DSB Processing Pathways

Three pathways are mainly implicated in the processing of DSBs in mammalian cells. Homologous recombination repair (HRR), classical non-homologous endjoining (c-NHEJ) and alternative end-joining (alt-EJ) [78, 102].

HRR [43, 101] can be divided into three main stages: presynaptic, synaptic and postsynaptic (Fig. 10.3). After sensing of the DSB by MRN complex (Mre11-Rad50-Nbs1) in the presynaptic stage, the DNA is resected to form an extended region of single stranded DNA (ssDNA) with 3′-overhangs. This step utilizes amongst other proteins MRN, CtIP, Exo1 and Dna2, as well as the BLM helicase [115], and generates an RPA coated filament that initiates HRR [18]. In the next step, RPA is replaced by the Rad51 protein to generate a presynaptic nucleoprotein filament. For efficient Rad51 filament formation, different classes of mediator proteins like the Rad51 paralogs (Rad51B, Rad51C, Rad51D, Xrcc2, Xrcc3), as well as Rad52 and Brca2 are utilized. The presynaptic Rad51 filament is essential for homology search and strand invasion [64, 72, 116]. During synapsis, the Rad51 nucleoprotein filament searches for homology in the sister chroma-

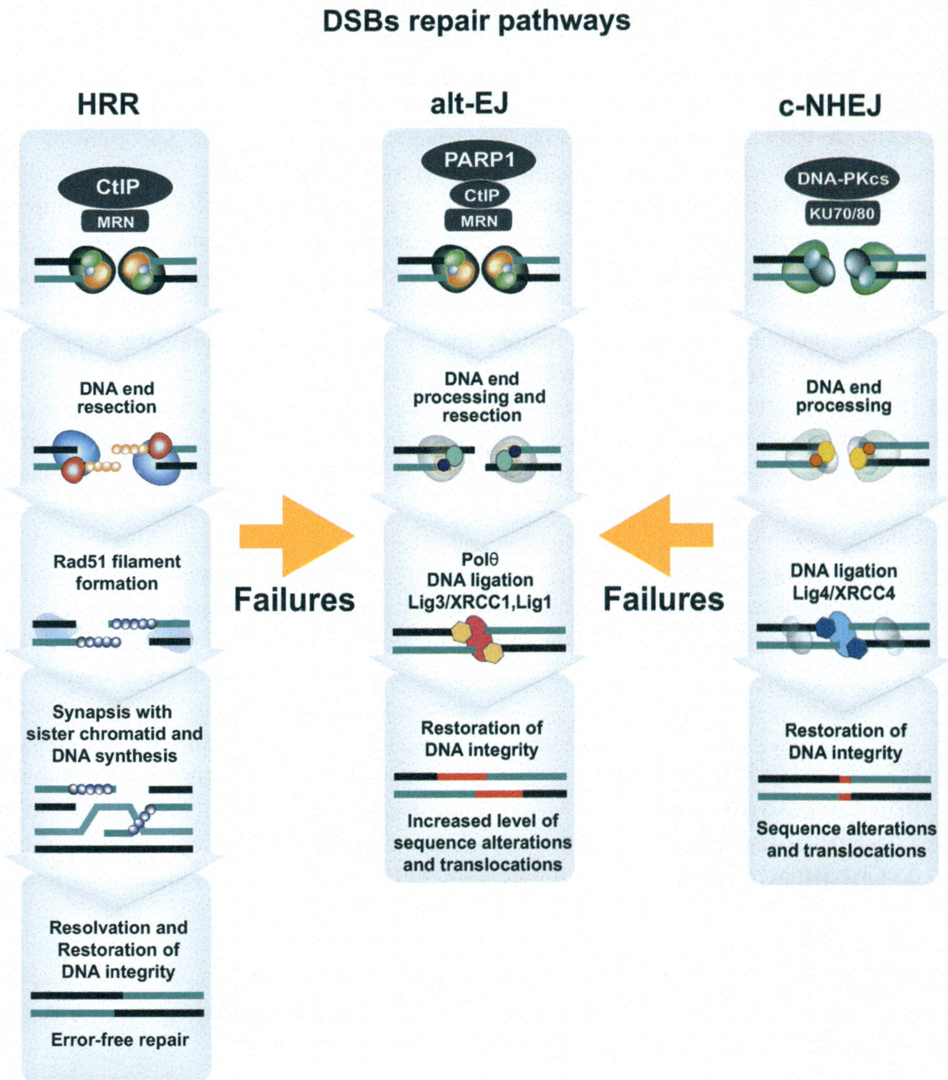

Fig. 10.3 Repair of DSBs by homologous recombination repair (HRR), clasical non homologous end joining (c-NHEJ) and alternative end joining alt-EJ)

tid and performs strand invasion to form a Holliday junctions and to initiate DNA synthesis with the help of Rad54 [8, 9, 46, 75, 120, 121]. In one version of the postsynaptic stage [12], the extended Holliday junction is resolved by synthesis-dependent strand annealing (SDSA). This enables the annealing of the newly synthesized strand with the resected strand of the second DNA end and the restoration of the broken DNA molecule by DNA synthesis and ligation. Alternatively, a double Holliday junction forms, which is subsequently resolved by resolvases [5, 11, 94].

The templated nature of DSB-repair by HRR not only ensures the structural restoration of the DNA molecule, but also enables the preservation of the DNA sequence at the break. As a result, HRR is an, in principle, error-free repair pathway on every count. The events initiating HRR imply that a wide spectrum of structural DNA-end substrate configurations at the DSB, like variations in the overhang length, DNA-end sequence and DNA-end chemistry (e.g. 3′ phosphoglycolate or 5′-OH present in T2 and T3-DSBs) can be accommodated. This is because many of the altered or missing bases will be removed during resection and those present at the non-resected 3′-ends may be removed by limited resection, or remain for processing after completion of HRR. HRR can thus function well for T1-T5 DSBs. It plays a major part in the processing of T6-DSBs that are generated during DNA replication. On the other hand, HRR is likely to be disrupted by DSB clusters that may cause the loss of DNA segments and thus compromise chromatin stability at the DSB.

C-NHEJ is widely considered as the prevalent DSB repair pathway in higher eukaryotes [50, 58, 78]. It mediates the fast ligation of broken DNA ends to ensure chromosome integrity [62] (Fig. 10.3). It is initiated by the binding of the Ku70/Ku80 heterodimer to DSB termini, which in-turn recruits and activates the large protein kinase, DNA-PKcs, to generate a binding scaffold for other NHEJ factors. DNA-PKcs also enables the regulation by phosphorylation of participating proteins [128]. The process culminates with the ligation of the two DNA ends by the Ligase 4/Xrcc4/Xlf/PAXX protein complex after displacement from the ends of DNA-PKcs through autophosphorylation [77, 119]. When required, various DNA end-processing functions, including the addition of a 5′-phosphate by Pnk and the removal of 3′-phosphoglycolates by Tdp1, Pnk or Artemis ensure the generation of ligatable DNA ends [91]. Filling of occasionally missing nucleotides is mediated by DNA polymerases λ and μ.

C-NHEJ enzymes tolerate a wide spectrum of structural DNA-end substrate configurations. These include variations in the overhang length, DNA-end sequence and DNA-end chemistry. It, thus, can also function as an important integrator of DNA ends with different chemistry, such as those of DSBs of types T1-T5. The pathway gains its speed by the functional coordination of the cooperating factors that operate like a highly efficient molecular machine, which reduces the probability that DNA ends will diffuse away and will engage in the formation of chromosomal translocations [39, 51, 52, 63]. Yet, c-NHEJ has no built-in means to ensure re-joining of the original DNA ends and translocations are in principle possible, nor does it possess mechanisms to ensure restoration of DNA sequence at the DSB site. However, it seems, plausible that c-NHEJ will be severely compromised by DSB clusters, which are likely to impair de-novo buildup of the associated molecular machinery.

Alt-EJ is an alternative form of DNA end-joining [7, 58, 76] thought to function as backup to c-NHEJ and HRR – hence its frequent designation by us as backup or B-NHEJ [50, 51] (Fig. 10.3). Although it functions on similar principles as c-NHEJ, alt-EJ is slower and less efficient and as a result, more error-prone on two counts. First, deletions and other modifications at the junction are larger than after processing with c-NHEJ. Second, and of particular relevancy here, the joining probability of unrelated ends is markedly increased. Thus, while the differences in the type of errors generated by c-NHEJ and alt-EJ are quantitative rather than qualitative, alt-EJ is considered an important source of chromosomal translocations [20, 39, 50–52, 63, 106, 113, 136].

Alt-EJ is rather diverse and may even have distinct sub-pathways [15, 20, 22, 24, 33, 41, 50, 51, 130]. Alt-EJ utilizes DNA Ligases I and III (Lig1 and Lig3) [34, 88, 106, 107], as well as Parp-1 [124] and possibly also Xrcc1 [21] and histone H1 [97]. Some aspects of alt-EJ benefit from resection at the DNA ends that exposes microhomologies facilitating repair and therefore, many factors described above for the resection step of HRR including CtIP and MRN complex are also implicated in alt-EJ [20, 76, 86, 104, 112] (Fig. 10.3).

Like c-NHEJ, alt-EJ is also active throughout the cell cycle [49, 70, 108, 131, 132]. However, unlike c-NHEJ, alt-EJ shows strong cell cycle dependent fluctuations with increased activity in G2 [131, 132], reduced in G1 and markedly ablated in resting cells [6, 49, 108, 111, 129]. A recently identified and highly relevant component of alt-EJ is Polθ [10, 73, 74]. Polθ mediates DNA end joining of two resected DNA ends with 3′- tails, harboring small sequence homology. The utilization of microhomologies at DSBs and the insertion of nucleotides at the joining sites are prominent signatures of Polθ activity. Polθ is involved in the formation of chromosome translocations and preserves genome integrity by limiting large deletions [130, 134].

By virtue of its molecular makeup, alt-EJ could accommodate all forms of DSBs described in Fig. 10.2. Indeed alt-EJ is frequently implicated in the processing of DSBs whenever other pathways fail to engage – for whatever reason. As we will see below, we consider that alt-EJ is particularly relevant for the processing of T7 –DSBs, where other pathways are likely to fail (Fig. 10.3).

10.6 DSB Repair Pathway Choice in Cells Exposed to High LET IR

Considering the cell cycle dependence and fidelity divergence of the DSB repair pathways described above and the striking biological effectives of high LET IR outlined in Sects. 10.1 and 10.2, the question whether DSB repair pathway choice manifests LET-dependence becomes central. Indeed, some characteristics of DSBs that differ between high and low LET IR may force the cell to choose the "wrong" pathway causing thus its demise. In this regard, it is striking that in contrast to the large increase in radiosensitivity observed in wild-type cells, the radiosensitivity of mutants defective in c-NHEJ is similar after exposure to high LET IR and X-rays [44, 60, 79, 83, 92, 105, 117, 122, 137]. This is illustrated by the results shown in Fig. 10.4a, b for the DNA-PKcs proficient M059 K cells and DNA-PKcs deficient M059 J cells after exposure to iron ions [109]. Direct conclusion from this observation is that lesions contributing to the survival of cells exposed to high LET IR are not processed by c-NHEJ.

Although cells defective in HRR show a clear enhancement in cell killing after exposure to high versus low LET IR [44], the effect is considerably smaller than that observed in wild-type cells. Thus, although HRR contributes to the processing of high LET IR lesions determining cell survival, the contribution is smaller than during processing of low LET lesions.

Strikingly, the results obtained at the cell survival level with HRR and c-NHEJ mutants after exposure to high LET IR and X-rays are not duplicated when repair of DSBs is analyzed using various techniques including PFGE (Fig. 10.4c, d). Here, active repair of DSBs is observed after exposure to high LET IR and X-rays, not only in wild-type cells, but also in c-NHEJ and HRR deficient mutants [92, 109, 110]. We conclude that only a subset of high-LET-IR-induced DSBs contributes to cell lethality and that this fraction has properties strongly compromising c-NHEJ and partly impairing HRR. In the following section, we will show that forms of T7-DSBs representing DSB clusters fulfill the biological-effect requirements of this subset of DNA damage.

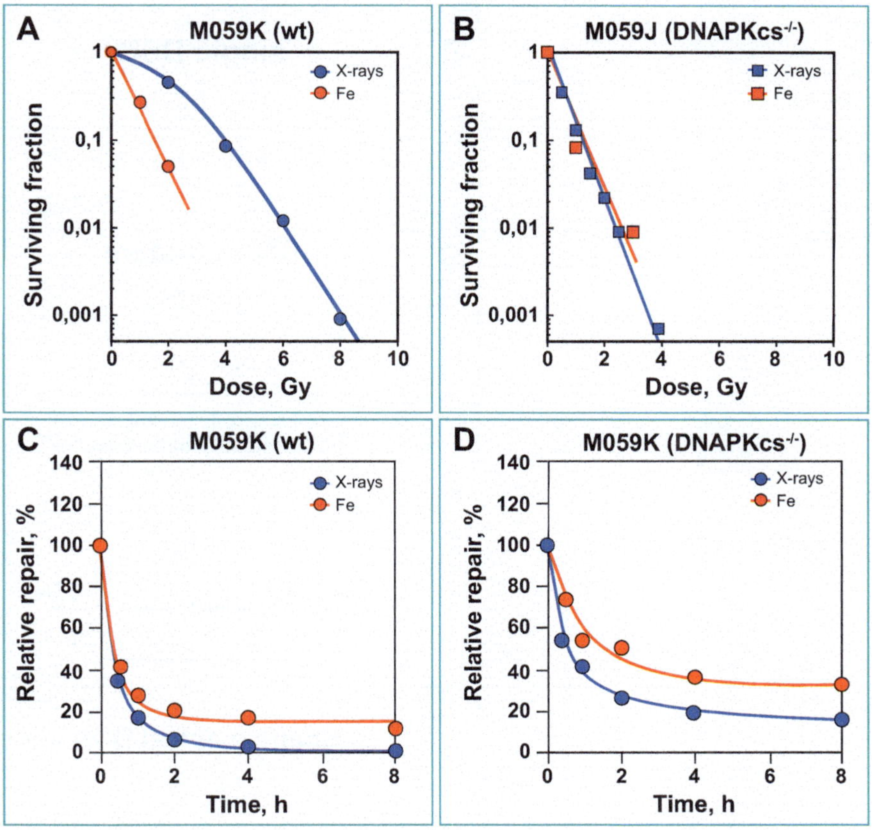

Fig. 10.4 (**a, b**) Representative survival experiments with DNAPKcs proficient (M059K) and DNAPKcs deficient (M059J) cells exposed to x-ray and iron ions. (**c, d**) PFGE experiments of M059K andM059J cells exposed to x-rays and iron ions

10.7 A Model of Defined DSB-Clusters at Fixed Genomic Locations

To study the biological consequences of DSB-clustering in a conclusive manner, we constructed cell lines harboring DNA sequences at which single DSBs and DSB clusters of known constitution can be enzymatically generated. We utilized the sequence of the *I-Sce*I meganuclease, which we engineered in plasmids at combinations and orientations as shown in Fig. 10.5. In addition to constructs harboring a single *I-Sce*I site, we generated constructs harboring pairs of *I-Sce*I sites locates 100 and 200 bp apart – the latter to mimic the approximate inter-nucleosomal distance. When these *I-Sce*I pairs are placed in direct orientation in the construct, compatible ends are generated in the apical ends. However, by selecting the reverse relative orientation, incompatible apical ends are generated upon digestion. Finally, we engineered clusters of four *I-Sce*I sites at the distances and orientations shown in Fig. 10.5.

With these constructs at hand, we generated in Chinese hamster cells (CHO) clones, harboring multiple copies of each construct in their genomes [103]. We wished to generate a system in which, similar to IR-exposure, multiple individual DSBs, or multiple, defined DSB clusters are induced after *I-Sce*I expression, in order to study their biological consequences. Each clone harbors a known number of integrations at fixed genomic locations. We have not yet determined the sites of integration of the indicated constructs in the selected clones.

To achieve multiple *I-Sce*I-construct integrations in the CHO genome, we utilized the

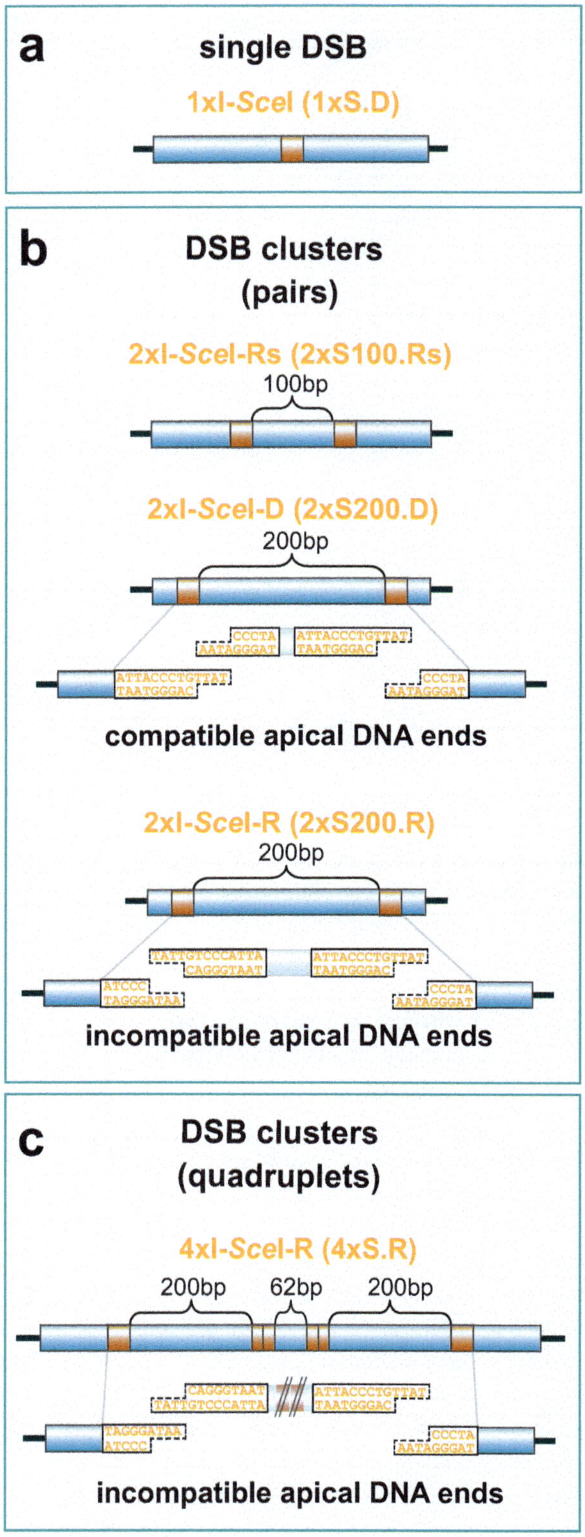

Fig. 10.5 (a–c) Schematic representation of constructs containing I-SceI recognition sequences to generate single DSBs, or DSB clusters comprising DSB pairs or quadruplets

Table 10.1 List of clonal cell lines used to characterize the biological effects of DSB clusters.

Name	Properties
CHO-1xS. D8	CHO cells with 8 integrations of single I-SceI site
CHO-2xS. Ds8	CHO cells with 8 integrations of I-SceI pairs at 10bp distance in direct orientation
CHO-2xS. D12	CHO cells with 12 integrations of I-SceI pairs at 200bp distance in direct orientation
CHO.2xS. R14	CHO cells with 14 integrations of I-SceI pairs at 200bp distance in reverse orientation
CHO.4xS. R12	CHO cells with 12 integrations of I-SceI quadruplets at 200, 62, 200bp distance in orientation resulting in incompatible apical ends

Sleeping-Beauty (SB) transposon system [103]. Clones were analyzed by Southern blotting to determine the number of integrations. From a pool of integration-characterized clones, we selected for experiments those shown in Table 10.1. In these clones, we studied the consequences of single-DSBs and DSB-clusters at different endpoints. SSBs and base damages, which far outnumber DSBs in cells exposed to IR, and contribute to cell lethality and translocation-formation to degrees that cannot be quantitatively separated from those of DSBs, are not confounding factors in this model system as they are completely absent.

Figure 10.6a shows results from colony formation assay when cells are plated immediately after transfection for transient *I-Sce*I expression to generate the corresponding DSB-clusters in the indicated clones. Single-DSBs and DSB-pairs separated by 100 bps cause relatively low cell killing (~30%), suggesting low efficacy in generating lethal events. Generation of DSB pairs with compatible apical ends separated by 200 bps causes more cell killing (~40%). Notably, a significant increase in cell killing (to 70%) is observed with a comparable number of DSB-pairs in the clone where incompatible apical ends are generated. Strikingly, DSB quadruplets with incompatible ends kill over 90% of cells. This result becomes even more impressive if one considers that transfection efficiency is about 90%. Thus, the vast majority of cells harboring *I-Sce*I-quadruplets and sustaining, therefore, DSB clusters of this complexity, succumb to this form of DNA damage.

Analysis of DSB recognition by the cellular DNA damage response system shows that single DSB and DSB clusters of the type shown here are detected as single events and form single γ-H2AX or MDC1 foci [103]. We assessed formation of chromosome translocations at metaphase, 12–24 h after *I-Sce*I expression. Figure 10.6b shows representative metaphases from these clones and demonstrates extensive formation of translocations at numbers and complexities that increase with increasing DSB-clustering. While low levels of chromosomal translocations are detected in the clones harboring 8 integrations of single *I-Sce*I sites or *I-Sce*I pairs separated by 100 bps, a statistically significant increase is noted in the clone harboring 12 *I-Sce*I pairs with compatible apical ends (Fig. 10.4c). Further increase is noted when 14 DSB pairs with incompatible ends are generated. Notably, the highest incidence of chromosomal translocations is found in the clone harboring 12 DSB-quadruplets (Fig. 10.4c). We conclude that DSB-clusters kill cells by destabilizing chromatin thus generating gross genomic rearrangements, frequently manifesting as chromosomal translocations.

Inhibition of c-NHEJ with NU7441 (a selective DNA-PK inhibitor) allows estimates of the contributions of c-NHEJ and alt-EJ to translocation formation. Figure 10.6d shows the relative change in translocation formation following incubation of each clone with NU7441. This number is obtained by dividing the incidence of translocations measured in cells treated with the inhibitor by the number of translocations measured in untreated controls. Treatment of WT cells with NU7441 causes an increase in translocation formation by over a factor of two. An increase in translocation formation by about 50% is observed with the clone harboring DSB pairs engineered 100 bps apart. Also, treatment of clones harboring pairs of DSBs in compatible or incompatible orientation causes an over two-fold

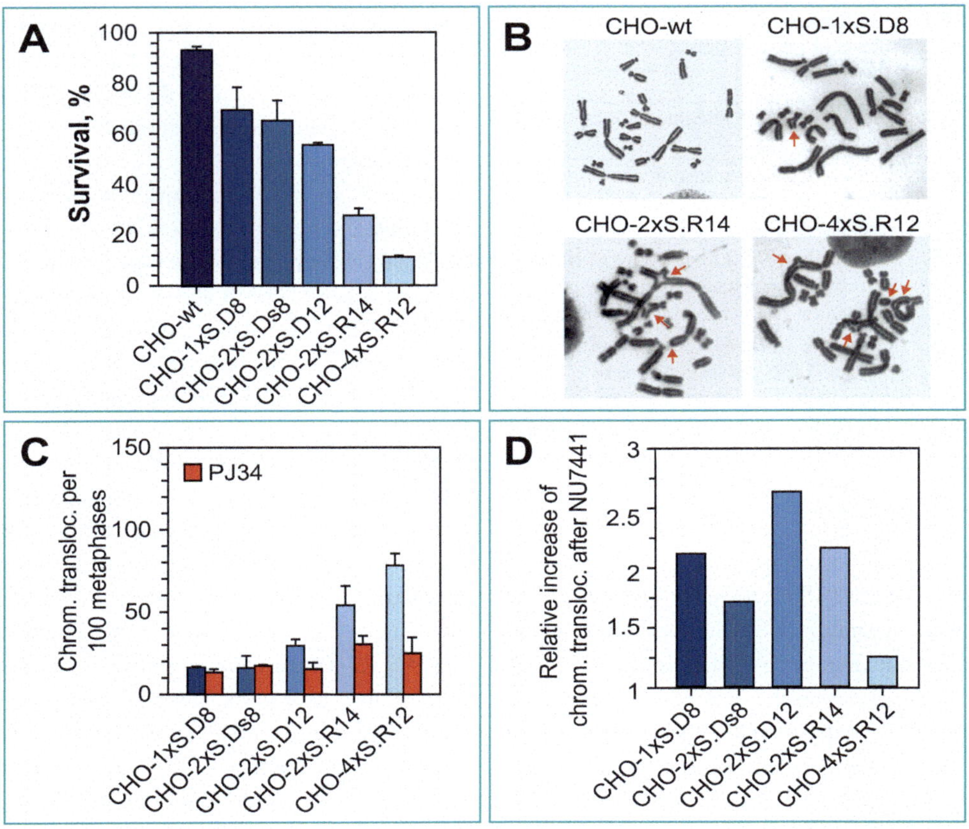

Fig. 10.6 Processing of single DSBs and DSBs clusters in CHO cells. (**a**) Survival experiments of CHO-cells harboring different numbers of I-SceI recognition sequences, designed to produce after I-SceI expression single DSBs or DSB clusters with increased complexity. (**b**) Representative images of chromosomal aberration formation in CHO-I-SceI clones as indicated. (**c**) Quantification analysis of chromosomal translocations in CHO-I-SceI clones in the presense or absense of PARP1 inhibitor (PJ-34). (**d**) Relative increased in chromosomal translocations after inhibition of c-NHEJ by the DNAPKcs inhibitor NU7441

increase in translocation formation. We conclude that c-NHEJ operates on these forms of DSBs and contributes to the suppression of translocations. Pharmacological inhibition of c-NHEJ impairs this processing and shunts DSBs to alt-EJ causing the translocations observed. Strikingly, treatment of cells harboring DSB-quadruplets with NU7441 has only a minor effect on translocation formation suggesting that with increasing DSB-clustering the engagement of c-NHEJ is inherently compromised and lesions are shunted by default to alt-EJ. As a consequence, pharmacological c-NHEJ inhibition by NU7441 generates only a minor effect.

This conclusion is further supported by the results obtained by inhibiting Parp1, a component of alt-EJ (Fig. 10.6c). Indeed treatment with the Parp1 inhibitor PJ34 has no effect, or has only a small effect on translocations forming from single-DSBs and DSB-pairs located 100 bps apart. A stronger PJ34 effect is noted for DSB pairs located 200 bps apart either when they present in direct or in inverse orientation. Notably, PJ34 has a strong inhibitory effect on translocations forming in cells harboring DSB-quadruplets (Fig. 10.6c). We, therefore, conclude that in the case of single-DSBs and DSB-pairs, c-NHEJ protects cells from chromosome translocation-formation and that its inhibition causes alt-EJ-mediated increase in translocations. In contrast, c-NHEJ is inherently compromised in the case of DSB-quadruples, which are constitutively processed by alt-EJ giving rise with high probability to chromosomal translocations.

10.8 Summary and Conclusions

The above outline provides strong evidence that clusters of DSBs pose increased risks for cells and may have adverse consequences that go beyond those of individual DSBs. This can be particularly relevant for cells exposed to high LET IR, but may also occur during chromothripsis and in all cases where physiological processes involving the generation of DSB clusters somehow fail. The precise mechanisms underpinning these effects are not well understood.

Results obtained using defined clusters of *I-Sce*I generated DSBs show that DSB clusters compromise c-NHEJ, allowing thus alt-EJ to promote chromosome translocation formation that is known to be lethal or carcinogenic. While both c-NHEJ and alt-EJ are candidates for the formation of chromosomal translocations, alt-EJ may contribute more prominently, although the actual contribution may depend on cell type, cell-cycle-phase and DSB-location [36, 50, 113].

Requirement for translocation-formation is that the ends of the participating DSBs drift apart and join with ends from neighboring DSBs also experiencing processing complications [14, 42, 59, 98]. It is evident that the chromatin destabilization caused by DSB clusters can perfectly facilitate such drifting. We envision similar mechanisms for the genomic alterations that underpin chromothripsis.

Our observations explain the enhanced adverse effects of forms of radiation such as radon and space radiation, and define DSB-clustering as a determinant of radiation-induced cell killing and possibly also carcinogenesis. The results suggest a mechanism for the increased toxicity of high LET radiations and the extensive genomic rearrangements associated with chromothripsis.

Acknowledgments Work supported by funds from the BMBF, the DFG, the University of Duisburg-Essen, Medical School and EU COST action CM1201.

References

1. Alt Frederick W, Zhang Y, Meng F-L, Guo C, Schwer B (2013) Mechanisms of programmed DNA lesions and genomic instability in the immune system. Cell 152:417–429. https://doi.org/10.1016/j.cell.2013.01.007
2. Asaithamby A, Chen DJ (2011) Mechanism of cluster DNA damage repair in response to high-atomic number and energy particles radiation. Mutat Res 711:87–99. https://doi.org/10.1016/j.mrfmmm.2010.11.002
3. Aziz K, Nowsheen S, Pantelias G, Iliakis G, Gorgoulis VG, Georgakilas AG (2012) Targeting DNA damage and repair: embracing the pharmacological era for successful cancer therapy. Pharmacol Ther 133:334–350
4. Baudat F, Nicolas A (1997) Clustering of meiotic double-strand breaks on yeast chromosome III. Proc Natl Acad Sci U S A 94:5213–5218
5. Bennett RJ, Dunderdale HJ, West SC (1993) Resolution of Holliday junctions by RuvC resolvase: cleavage specificity and DNA distortion. Cell 74:1021–1032
6. Bindra RS, Goglia AG, Jasin M, Powell SN (2013) Development of an assay to measure mutagenic non-homologous end-joining repair activity in mammalian cells. Nucleic Acids Res 41:e115. https://doi.org/10.1093/nar/gkt255
7. Boboila C, Alt FW, Schwer B (2012) Chapter one – classical and alternative end-joining pathways for repair of lymphocyte-specific and general DNA double-strand breaks. In: Frederick WA (ed) Advances in immunology, vol 116. Academic, New York, pp 1–49. https://doi.org/10.1016/b978-0-12-394300-2.00001-6
8. Bugreev D, Hanaoka F, Mazin AV (2007) Rad54 dissociates homologous recombination intermediates by branch migration. Nat Struct Mol Biol 14:746–753
9. Bugreev DV, Mazina OM, Mazin AV (2006) Rad54 protein promotes branch migration of Holliday junctions. Nature 442:590–593
10. Ceccaldi R et al (2015) Homologous-recombination-deficient tumours are dependent on pol[thgr]-mediated repair. Nature 518:258–262. https://doi.org/10.1038/nature14184. http://www.nature.com/nature/journal/v518/n7538/abs/nature14184.html#supplementary-information
11. Chan YW, West S (2015) GEN1 promotes Holliday junction resolution by a coordinated nick and counter-nick mechanism. Nucleic Acids Res 43:10882–10892. https://doi.org/10.1093/nar/gkv1207
12. Chapman JR, Taylor Martin RG, Boulton Simon J (2012) Playing the end game: DNA double-strand break repair pathway choice. Mol Cell 47:497–510. https://doi.org/10.1016/j.molcel.2012.07.029

13. Cheng Y, Li F, Mladenov E, Iliakis G (2015) The yield of DNA double strand breaks determined after exclusion of those forming from heat-labile lesions predicts tumor cell radiosensitivity to killing. Radiother Oncol 116:366–373. https://doi.org/10.1016/j.radonc.2015.08.004
14. Chiarle R et al (2011) Genome-wide translocation sequencing reveals mechanisms of chromosome breaks and rearrangements in B cells. Cell 147:107–119. https://doi.org/10.1016/j.cell.2011.07.049
15. Cho NW, Greenberg RA (2015) DNA repair: familiar ends with alternative endings. Nature 518:174–176. https://doi.org/10.1038/nature14200
16. Crasta K et al (2012a) DNA breaks and chromosome pulverization from errors in mitosis. Nature 482:53–58. http://www.nature.com/nature/journal/v482/n7383/abs/nature10802.html#supplementary-information
17. Crasta K et al (2012b) DNA breaks and chromosome pulverization from errors in mitosis. Nature 482:53–58. https://doi.org/10.1038/nature10802
18. Daley JM, Kwon Y, Niu H, Sung P (2013) Investigations of homologous recombination pathways and their regulation. Yale J Biol Med 86:453–461
19. de Massy B, Baudat F, Nicolas A (1994) Initiation of recombination *Saccharomyces cerevisiae* haploid meiosis. Proc Natl Acad Sci U S A 91:11929–11933
20. Decottignies A (2013) Alternative end-joining mechanisms: a historical perspective Front Genet 4:Article 48. https://doi.org/10.3389/fgene.2013.00048
21. Della-Maria J, Zhou Y, Tsai M-S, Kuhnlein J, Carney JP, Paull TT, Tomkinson AE (2011) Human Mre11/human Rad50/Nbs1 and DNA ligase IIIα/XRCC1 protein complexes act together in an alternative nonhomologous end joining pathway. J Biol Chem 286:33845–33853. https://doi.org/10.1074/jbc.M111.274159
22. Deriano L, Roth DB (2013) Modernizing the nonhomologous end-joining repertoire: alternative and classical NHEJ share the stage. Annu Rev Genet 47:451–473
23. Desai N, Davis E, O'Neill P, Durante M, Cucinotta FA, Wu H (2005) Immunofluorescence detection of clustered γ-H2AX foci induced by HZE-particle radiation. Radiat Res 164:518–522
24. Doetsch PW (1995) What's old is new: an alternative DNA excision repair pathway. Trends Biochem Sci 20:384–386
25. Dudley DD, Chaudhuri J, Bassing CH, Alt FW (2005) Mechanism and control of V(D)J recombination versus class switch recombination: similarities and differences. Adv Immunol 86:43–112
26. Elsässer T, Brons S, Psonka K, Scholz M, Gudowska-Nowak E, Taucher-Scholz G (2008) Biophysical modeling of fragment length distributions of DNA plasmids after X and heavy-ion irradiation analyzed by atomic force microscopy. Radiat Res 169:649–659
27. Friedland W, Dingfelder M, Jacob P, Paretzke HG (2005) Calculated DNA double-strand break and fragmentation yields after irradiation with he ions. Radiat Phys Chem 72:279–286. https://doi.org/10.1016/j.radphyschem.2004.05.053
28. Friedland W, Dingfelder M, Kundrát P, Jacob P (2011) Track structures, DNA targets and radiation effects in the biophysical Monte Carlo simulation code PARTRAC. Mutat Res/Fund Mol Mech Mutagen 711:28–40. https://doi.org/10.1016/j.mrfmmm.2011.01.003
29. Friedland W, Jacob P, Bernhardt P, Paretzke HG, Dingfelder M (2003) Simulation of DNA damage after proton irradiation. Radiat Res 159:401–410. https://doi.org/10.1667/0033-7587(2003)159[0401:soddap]2.0.co;2
30. Friedland W, Jacob P, Paretzke HG, Stork T (1998) Monte Carlo simulation of the production of short DNA fragments by low-linear energy transfer radiation using higher-order DNA models. Radiat Res 150:170–182
31. Friedrich T, Durante M, Scholz M (2012a) Modeling cell survival after photon irradiation based on double-strand break clustering in Megabase pair chromatin loops. Radiat Res 178:385–394. https://doi.org/10.1667/rr2964.1
32. Friedrich T, Scholz U, Elsässer T, Durante M, Scholz M (2012b) Calculation of the biological effects of ion beams based on the microscopic spatial damage distribution pattern. Int J Radiat Biol 88:103–107. https://doi.org/10.3109/09553002.2011.611213
33. Frit P, Barboule N, Yuan Y, Gomez D, Calsou P (2014) Alternative end-joining pathway(s): bricolage at DNA breaks. DNA Repair 17:81–97. https://doi.org/10.1016/j.dnarep.2014.02.007
34. Gao Y, Katyal S, Lee Y, Zhao J, Rehg JE, Russell HR, McKinnon PJ (2011) DNA ligase III is critical for mtDNA integrity but not Xrcc1-mediated nuclear DNA repair. Nature 471:240–244
35. Gerelchuluun A, Zhu J, Su F, Asaithamby A, Chen DJ, Tsuboi K (2014) Homologous recombination pathway may play a major role in high-LET radiation-induced DNA double-strand break repair. J Radiat Res 55:i83–i84. https://doi.org/10.1093/jrr/rrt181
36. Ghezraoui H et al (2014) Chromosomal translocations in human cells are generated by canonical nonhomologous end-joining. Mol Cell 55:829–842. https://doi.org/10.1016/j.molcel.2014.08.002
37. Goodhead DT (1994) Initial events in the cellular effects of ionizing radiations: clustered damage in DNA. Int J Radiat Biol 65:7–17
38. Goodhead DT, Nikjoo H (1989) Track structure analysis of ultrasoft X-rays compared to high- and low-LET radiations. Int J Radiat Biol 55:513–529
39. Gostissa M, Alt FW, Chiarle R (2011) Mechanisms that promote and suppress chromosomal translocations in lymphocytes. Annu Rev Immunol 29:319–350. https://doi.org/10.1146/annurev-immunol-031210-101329

40. Gudowska-Nowak E, Nasonova E, Ritter S, Scholz M (2004) Chromosome fragmentation after irradiation with C ions. Radiother Oncol 73(Suppl 2):S123–126
41. Haber JE (2008) Alternative endings. Proc Natl Acad Sci U S A 105:405–406. https://doi.org/10.1073/pnas.0711334105
42. Hakim O et al. (2012) DNA damage defines sites of recurrent chromosomal translocations in B lymphocytes. Nature 484:69–74. http://www.nature.com/nature/journal/v484/n7392/abs/nature10909.html#supplementary-information
43. Heyer W-D, Ehmsen KT, Liu J (2010) Regulation of homologous recombination in eukaryotes. Annu Rev Genet 44:113–139
44. Hill MA, Herdman MT, Stevens DL, Jones NJ, Thacker J, Goodhead DT (2004) Relative sensitivities of repair-deficient mammalian cells for Clonogenic survival after α-particle irradiation. Radiat Res 162:667–676
45. Holley WR, Chatterjee A (1996) Clusters of DNA damage induced by ionizing radiation: formation of short DNA fragments. 1. Theoretical modeling. Radiat Res 145:188–199
46. Holloway JK, Morelli MA, Borst PL, Cohen PE (2010) Mammalian BLM helicase is critical for integrating multiple pathways of meiotic recombination. J Cell Biol 188:779–789. https://doi.org/10.1083/jcb.200909048
47. Hutchinson F (1985) Chemical changes induced in DNA by ionizing radiation. Prog Nucleic Acid Res Mol Biol 32:115–154
48. Ibañez IL et al (2009) Induction and rejoining of DNA double strand breaks assessed by H2AX phosphorylation in melanoma cells irradiated with proton and Lithium beams. Int J Radiat Oncol Biol Phys 74:1226–1235
49. Iliakis G (2009) Backup pathways of NHEJ in cells of higher eukaryotes: cell cycle dependence. Radiother Oncol 92:310–315
50. Iliakis G, Murmann T, Soni A (2015) Alternative end-joining repair pathways are the ultimate backup for abrogated classical non-homologous end-joining and homologous recombination repair: implications for the formation of chromosome translocations. Mutat Res/Genet Toxicol Environ Mutagen 793:166–175
51. Iliakis G et al (2004) Mechanisms of DNA double strand break repair and chromosome aberration formation. Cytogenet Genome Res 104:14–20
52. Iliakis G, Wu W, Wang M, Terzoudi GI, Pantelias GE (2007) Backup pathways of nonhomologous end joining may have a dominant role in the formation of chromosome aberrations. In: Obe G, Vijayalaxmi (eds) Chromosomal alterations. Springer, Berlin/Heidelberg/New York, pp 67–85
53. Isildar M, Schuchmann MN, Schulte-Frohlinde D, von Sonntag C (1981) gamma-Radiolysis of DNA in oxygenated aqueous solutions: alterations at the sugar moiety. Int J Radiat Biol Relat Stud Phys Chem Med 40:347–354
54. Jakob B, Splinter J, Durante M, Taucher-Scholz G (2009) Live cell microscopy analysis of radiation-induced DNA double-strand break motion. Proc Natl Acad Sci U S A 106:3172–3177
55. Jakob B, Taucher-Scholz G (2017) Live cell imaging to study real-time ATM-mediated recruitment of DNA repair complexes to sites of ionizing radiation-induced DNA damage. Methods Mol Biol 1599:287–302. https://doi.org/10.1007/978-1-4939-6955-5_21
56. Johnston PJ, MacPhail SH, Banáth JP, Olive PL (1998) Higher-order chromatin structure-dependent repair of DNA double-strand breaks: factors affecting elution of DNA from nucleoids. Radiat Res 149:533–542. https://doi.org/10.2307/3579899
57. Jones MJ, Jallepalli PV (2012) Chromothripsis: chromosomes in crisis. Dev Cell 23:908–917. https://doi.org/10.1016/j.devcel.2012.10.010
58. Kinner A, Wu W, Staudt C, Iliakis G (2008) γ-H2AX in recognition and signaling of DNA double-strand breaks in the context of chromatin. Nucleic Acids Res 36:5678–5694
59. Klein Isaac A et al (2011) Translocation-capture sequencing reveals the extent and nature of chromosomal rearrangements in B lymphocytes. Cell 147:95–106. https://doi.org/10.1016/j.cell.2011.07.048
60. Kysela BP, Arrand JE, Michael BD (1993) Relative contributions of levels of initial damage and repair of double-strand breaks to the ionizing radiation-sensitive phenotype of the Chinese hamster cell mutant, XR-V15B. Part II. Neutrons. Int J Radiat Biol 64:531–538
61. Lam I, Keeney S (2014) Mechanism and regulation of meiotic recombination initiation. Cold Spring Harb Perspect Biol 7:a016634. https://doi.org/10.1101/cshperspect.a016634
62. Lieber MR (2010a) The mechanism of double-strand DNA break repair by the nonhomologous DNA end-joining pathway. Annu Rev Biochem 79:1.1–1.31
63. Lieber MR (2010b) NHEJ and its backup pathways in chromosomal translocations. Nat Struct Mol Biol 17:393–395
64. Liu Y, Tarsounas M, O'Regan P, West SC (2007) Role of RAD51C and XRCC3 in genetic recombination and DNA repair. J Biol Chem 282:1973–1979. https://doi.org/10.1074/jbc.M609066200
65. Löbrich M, Cooper PK, Rydberg B (1996) Non-random distribution of DNA double-strand breaks induced by particle irradiation. Int J Radiat Biol 70:493–503
66. Lopez Perez R et al (2016) Superresolution light microscopy shows nanostructure of carbon ion radiation-induced DNA double-strand break repair foci. FASEB J 30:2767. https://doi.org/10.1096/fj.201500106R
67. Lorat Y, Brunner CU, Schanz S, Jakob B, Taucher-Scholz G, Rübe CE (2015) Nanoscale analysis of clustered DNA damage after high-LET irradiation

by quantitative electron microscopy – the heavy burden to repair. DNA Repair 28:93–106. https://doi.org/10.1016/j.dnarep.2015.01.007
68. Lorat Y, Schanz S, Schuler N, Wennemuth G, Rübe C, Rübe CE (2012) Beyond repair foci: DNA double-strand break repair in Euchromatic and heterochromatic compartments analyzed by transmission Electron microscopy. PLoS One 7:e38165. https://doi.org/10.1371/journal.pone.0038165
69. Lorat Y, Timm S, Jakob B, Taucher-Scholz G, Rube CE (2016) Clustered double-strand breaks in heterochromatin perturb DNA repair after high linear energy transfer irradiation Radiother Oncol 121:154–161 doi:https://doi.org/10.1002/bies.201600115 10.1016/j.radonc.2016.08.028
70. Magin S, Papaioannou M, Saha J, Staudt C, Iliakis GE (2015) Inhibition of homologous recombination and promotion of mutagenic repair of DNA double-strand breaks underpins arabinoside-nucleoside analog-radiosensitization. Mol Cancer Ther 14:1424–1433. https://doi.org/10.1158/1535-7163.mct-14-0682
71. Maher CA, Wilson RK (2012) Chromothripsis and human disease: piecing together the shattering process. Cell 148:29–32. https://doi.org/10.1016/j.cell.2012.01.006
72. Masson J-Y et al (2001) Identification and purification of two distinct complexes containing the five RAD51 paralogs. Genes Dev 15:3296–3307
73. Mateos-Gomez PA, Gong F, Nair N, Miller KM, Lazzerini-Denchi E, Sfeir A (2015) Mammalian polymerase theta promotes alternative NHEJ and suppresses recombination. Nature 518:254–257. https://doi.org/10.1038/nature14157
74. Mateos-Gomez PA et al. (2017) The helicase domain of Poltheta counteracts RPA to promote alt-NHEJ. Nat Struct Mol Biol. https://doi.org/10.1038/nsmb.3494
75. Mazin AV, Mazina OM, Bugreev DV, Rossi MJ (2010) Rad54, the motor of homologous recombination. DNA Repair 9:286–302
76. McVey M, Lee SE (2008) MMEJ repair of double-strand breaks (director's cut): deleted sequences and alternative endings. Trends Genet 24:529–538
77. Meek K, Dang V, Lees-Miller SP (2008) DNA-PK: the means to justify the ends? Adv Immunol 99:33–58. https://doi.org/10.1016/S0065-2776(08)00602-0
78. Mladenov E, Iliakis G (2011) Induction and repair of DNA double strand breaks: the increasing spectrum of non-homologous end joining pathways. Mutat Res 711:61–72. https://doi.org/10.1016/j.mrfmmm.2011.02.005
79. Nagasawa H, Little JB, Inkret WC, Carpenter S, Raju MR, Chen DJ, Strniste GF (1991) Response of X-ray-sensitive CHO mutant cells (xrs-6c) to radiation. II. Relationship between cell survival and the induction of chromosomal damage with low doses of alpha particles. Radiat Res 126:280–288
80. Nikjoo H, O'Neill P, Wilson WE, Goodhead DT (2001) Computational approach for determining the Spectrum of DNA damage induced by ionizing radiation. Radiat Res 156:577–583
81. O'Neill JP (2000) DNA damage, DNA repair, cell proliferation, and DNA replication: how do gene mutations result? [In Process Citation] Proc Natl Acad Sci USA 97:11137–11139
82. O'Neill P, Wardman P (2009) Radiation chemistry comes before radiation biology. Int J Radiat Biol 85:9–25
83. Okayasu R, Okada M, Okabe A, Noguchi M, Takakura K, Takahashi S (2006) Repair of DNA damage induced by accelerated heavy ions in mammalian cells proficient and deficient in the non-homologous end-joining pathway. Radiat Res 165:59–67
84. Ostashevsky JY (1989) A model relating cell survival to DNA fragment loss and unrepaired double-strand breaks. Radiat Res 118:437–466
85. Ostashevsky JY (2000) Higher-order structure of interphase chromosomes and radiation-induced chromosomal exchange aberrations. Int J Radiat Biol 76:1179–1187
86. Ottaviani D, LeCain M, Sheer D (2014) The role of microhomology in genomic structural variation. Trends Genet 30:85–94. https://doi.org/10.1016/j.tig.2014.01.001
87. Pang D et al (2011) Radiation-generated short DNA fragments may perturb non-homologous end-joining and induce genomic instability. J Radiat Res 52:309–319
88. Paul K et al (2013) DNA ligases I and III cooperate in alternative non-homologous end-joining in vertebrates. PLoS One 8:e59505
89. Ponomarev AL, Belli M, Hahnfeldt PJ, Hlatky L, Sachs RK, Cucinotta FA (2006) A robust procedure for removing background damage in assays of radiation-induced DNA fragment distributions. Radiat Res 166:908–916
90. Ponomarev AL, Cucinotta FA (2006) Chromatin loops are responsible for higher counts of small DNA fragments induced by high-LET radiation, while chromosomal domains do not affect the fragment sizes Int J Radiat Biol 82:293–305. https://doi.org/10.1080/09553000600637716
91. Povirk LF (2012) Processing of damaged DNA ends for double-strand break repair in mammalian cells ISRN Mol Biol 2012:Article: ID 345805
92. Prise KM, Davies S, Michael BD (1987) The relationship between radiation-induced DNA double-strand breaks and cell kill in hamster V79 fibroblasts irradiated with 250kVp X-rays, 2.3MeV neutrons or ^{238}Pu α-particles. Int J Radiat Biol 52:893–902
93. Puc J, Aggarwal AK, Rosenfeld MG (2017) Physiological functions of programmed DNA breaks in signal-induced transcription. Nat Rev Mol Cell Biol 18:471–476. https://doi.org/10.1038/nrm.2017.43
94. Rass U et al (2010) Mechanism of Holliday junction resolution by the human GEN1 protein. Genes Dev 24:1559–1569. https://doi.org/10.1101/gad.585310

95. Robert T, Vrielynck N, Mezard C, de Massy B, Grelon M (2016) A new light on the meiotic DSB catalytic complex. Semin Cell Dev Biol 54:165–176. https://doi.org/10.1016/j.semcdb.2016.02.025
96. Rode A, Maass KK, Willmund KV, Lichter P, Ernst A (2016) Chromothripsis in cancer cells: an update. Int J Cancer 138:2322–2333. https://doi.org/10.1002/ijc.29888
97. Rosidi B, Wang M, Wu W, Sharma A, Wang H, Iliakis G (2008) Histone H1 functions as a stimulatory factor in backup pathways of NHEJ. Nucleic Acids Res 36:1610–1623
98. Roukos V, Voss TC, Schmidt CK, Lee S, Wangsa D, Misteli T (2013) Spatial dynamics of chromosome translocations in living cells. Science 341:660–664. https://doi.org/10.1126/science.1237150
99. Rydberg B (1996) Clusters of DNA damage induced by ionizing radiation: formation of short DNA fragments. 11. Experimental detection. Radiat Res 145:200–209
100. Saha J, Wilson P, Thieberger P, Lowenstein D, Wang M, Cucinotta FA (2014) Biological characterization of low-energy ions with high-energy deposition on human cells. Radiat Res 182:282–291. https://doi.org/10.1667/RR13747.1
101. San Filippo J, Sung P, Klein H (2008) Mechanism of eukaryotic homologous recombination. Annu Rev Biochem 77:229–257
102. Schipler A, Iliakis G (2013) DNA double-strand-break complexity levels and their possible contributions to the probability for error-prone processing and repair pathway choice. Nucleic Acids Res 41:7589–7605. https://doi.org/10.1093/nar/gkt556
103. Schipler A, Mladenova V, Soni A, Nikolov V, Saha J, Mladenov E, Iliakis G (2016) Chromosome thripsis by DNA double strand break clusters causes enhanced cell lethality, chromosomal translocations and 53BP1-recruitment. Nucleic Acids Res 44:7673–7690. https://doi.org/10.1093/nar/gkw487
104. Sfeir A, Symington LS (2015) Microhomology-mediated end joining: a back-up survival mechanism or dedicated pathway? Trends Biochem Sci 40:701–714. https://doi.org/10.1016/j.tibs.2015.08.006
105. Shadley JD, Whitlock JL, Rotmensch J, Atcher RW, Tang J, Schwartz JL (1991) The effects of radon daughter alpha-particle irradiation in K1 and xrs-5 CHO cell lines. Mutat Res 248:73–83
106. Simsek D et al (2011a) DNA ligase III promotes alternative nonhomologous end-joining during chromosomal translocation formation. PLoS Genet 7:e1002080
107. Simsek D et al (2011b) Crucial role for DNA ligase III in mitochondria but not in Xrcc1-dependent repair. Nature 471:245–248
108. Singh SK, Bednar T, Zhang L, Wu W, Mladenov E, Iliakis G (2012a) Inhibition of B-NHEJ in plateau-phase cells is not a direct consequence of suppressed growth factor signaling. Int J Radiat Oncol Biol Phys 84:e237–e243. https://doi.org/10.1016/j.ijrobp.2012.03.060
109. Singh SK, Bencsik-Theilen A, Mladenov E, Jakob B, Taucher-Scholz G, Iliakis G (2013) Reduced contribution of thermally labile sugar lesions to DNA double strand break formation after exposure to heavy ions. Radiat Oncol 8:77
110. Singh SK, Wu W, Stuschke M, Bockisch A, Iliakis G (2012b) Reduced contribution of thermally-labile sugar lesions to DNA double-strand break formation after exposure to neutrons. Radiat Res 178:581–590. https://doi.org/10.1667/RR2996.1
111. Singh SK, Wu W, Zhang L, Klammer H, Wang M, Iliakis G (2011) Widespread dependence of backup NHEJ on growth state: ramifications for the use of DNA-PK inhibitors. Int J Radiat Oncol Biol Phys 79:540–548
112. Sinha S, Villarreal D, Shim EY, Lee SE (2016) Risky business: microhomology-mediated end joining. Mutat Res/Fundam Mol Mech Mutagen. https://doi.org/10.1016/j.mrfmmm.2015.12.005
113. Soni A, Siemann M, Pantelias GE, Iliakis G (2015) Marked cell cycle-dependent contribution of alternative end joining to formation of chromosome translocations by stochastically induced DNA double strand breaks in human cells. Mutat Res 793:2–8
114. Stephens PJ et al (2011) Massive genomic rearrangement acquired in a single catastrophic event during cancer development. Cell 144:27–40. https://doi.org/10.1016/j.cell.2010.11.055
115. Symington LS, Gautier J (2011) Double-strand break end resection and repair pathway choice. Annu Rev Genet 45:247–271. https://doi.org/10.1146/annurev-genet-110410-132435
116. Tarsounas M, Davies D, West SC (2003) BRCA2-dependent and independent formation of RAD51 nuclear foci. Oncogene 22:1115–1123
117. Thacker J, Stretch A (1985) Responses of 4 X-ray-sensitive CHO cell mutants to different radiations and to irradiation conditions promoting cellular recovery. Mutat Res 146:99–108
118. Tobias F, Durante M, Taucher-Scholz G, Jakob B (2010) Spatiotemporal analysis of DNA repair using charged particle radiation. Mutat Res 704:54–60. https://doi.org/10.1016/j.mrrev.2009.11.004
119. Uematsu N et al (2007) Autophosphorylation of DNA-PKCS regulates its dynamics at DNA double-strand breaks. J Cell Biol 177:219–229. https://doi.org/10.1083/jcb.200608077
120. Uringa E-J, Youds JL, Lisaingo K, Lansdorp PM, Boulton SJ (2011) RTEL1: an essential helicase for telomere maintenance and the regulation of homologous recombination. Nucleic Acids Res 39:1647–1655
121. Vannier J-B, Sarek G, Boulton SJ (2014) RTEL1: functions of a disease-associated helicase. Trends Cell Biol 24:416–425. https://doi.org/10.1016/j.tcb.2014.01.004
122. Wang H, Wang X, Zhang P, Wang Y (2008) The Ku-dependent non-homologous end-joining but not other repair pathway is inhibited by high lin-

ear energy transfer ionizing radiation. DNA Repair 7:725–733
123. Wang H et al (2010) Characteristics of DNA-binding proteins determine the biological sensitivity to high-linear energy transfer radiation. Nucleic Acids Res 38:3245–3251. https://doi.org/10.1093/nar/gkq069
124. Wang M, Wu W, Wu W, Rosidi B, Zhang L, Wang H, Iliakis G (2006) PARP-1 and Ku compete for repair of DNA double strand breaks by distinct NHEJ pathways. Nucleic Acids Res 34:6170–6182
125. Ward JF (1988) DNA damage produced by ionizing radiation in mammalian cells: identities, mechanisms of formation, and reparability. Prog Nucleic Acid Res 35:95–125
126. Wei P-C, Chang Amelia N, Kao J, Du Z, Meyers Robin M, Alt Frederick W, Schwer B (2016) Long neural Genes Harbor recurrent DNA break clusters in neural stem/progenitor cells. Cell 164:644–655. https://doi.org/10.1016/j.cell.2015.12.039
127. Weinfeld M, Mani RS, Abdou I, Aceytuno RD, Glover JNM (2011) Tidying up loose ends: the role of polynucleotide kinase/phosphatase in DNA strand break repair. Trends Biochem Sci 36:262–271
128. Weterings E, Chen DJ (2008) The endless tale of non-homologous end-joining. Cell Res 18:114–124
129. Windhofer F, Wu W, Wang M, Singh SK, Saha J, Rosidi B, Iliakis G (2007) Marked dependence on growth state of backup pathways of NHEJ. Int J Radiat Oncol Biol Phys 68:1462–1470
130. Wood RD, Doublie S (2016) DNA polymerase theta (POLQ), double-strand break repair, and cancer. DNA Repair (Amst) 44:22–32. https://doi.org/10.1016/j.dnarep.2016.05.003
131. Wu W, Wang M, Mussfeldt T, Iliakis G (2008a) Enhanced use of backup pathways of NHEJ in G_2 in Chinese hamster mutant cells with defects in the classical pathway of NHEJ. Radiat Res 170:512–520
132. Wu W, Wang M, Wu W, Singh SK, Mussfeldt T, Iliakis G (2008b) Repair of radiation induced DNA double strand breaks by backup NHEJ is enhanced in G2. DNA Repair 7:329–338
133. Xu F, Petes TD (1996) Fine-structure mapping of meiosis-specific double-strand DNA breaks at a recombination hotspot associated with an insertion of telomeric sequences upstream of the HIS4 locus in yeast. Genetics 143:1115–1125
134. Yousefzadeh MJ et al (2014) Mechanism of suppression of chromosomal instability by DNA polymerase POLQ. PLoS Genet 10:e1004654. https://doi.org/10.1371/journal.pgen.1004654
135. Zhang CZ et al (2015) Chromothripsis from DNA damage in micronuclei. Nature 522:179–184. https://doi.org/10.1038/nature14493
136. Zhang Y, Jasin M (2011) An essential role for CtIP in chromosomal translocation formation through an alternative end-joining pathway. Nat Struct Mol Biol 18:80–84
137. Zyuzikov NA, Prise KM, Zdzienicka MZ, Newman HC, Michael BD, Trott KR (2001) The relationship between the RBE of alpha particles and the radiosensitivity of different mutations of Chinese hamster cells. Radiat Environ Biophys 40:243–248

The manufacturer's authorised representative in the EU is Springer
Nature Customer Service Centre GmbH, Europaplatz 3, 69115 Heidelberg,
Germany. If you have any concerns regarding our products, please
contact ProductSafety@springernature.com

Printed and bound by CPI Group (UK) Ltd, Croydon, CR0 4YY
25/09/2026
02222899-0007